Industrial Arts
in General Education

This giant Apollo spacecraft and lunar module bound for a moon landing symbolizes American industry and technology. Industrial arts education seeks to develop insights, understandings, and appreciations of the industrial-technological environment in which we live. *(Courtesy, National Aeronautics and Space Administration)*

INDUSTRIAL ARTS IN GENERAL EDUCATION

Gordon O. Wilber
State University of New York, College at Oswego

Norman C. Pendered
East Carolina University

FOURTH EDITION

INTEXT EDUCATIONAL PUBLISHERS
New York

Third Printing

Copyright © 1967 by International Textbook Company

Copyright © 1973 by Intext Press, Inc.

Library of Congress Number: 72-84134

ISBN 0-7002-2425-4

All rights reserved. No part of this book may be reprinted, reproduced, or utilized in any form or by any electronic, mechanical, or other means, now known or hereafter invented, including photocopying and recording, or in any information storage and retrieval system, without permission in writing from the Publisher.

Contents

Foreword   *xi*
Preface   *xiii*

Chapter 1 **THE PURPOSES OF GENERAL EDUCATION**   1

    The Basic Precepts of Democracy, What Is General Education? Purposes of General Education, Summary, Discussion Topics and Assignments, Selected References.

Chapter 2 **THE RELATIONSHIP OF INDUSTRIAL ARTS TO GENERAL EDUCATION**   14

    Industrial Arts as General Education, Industrial Arts Defined, Nature of American Culture, Transmitting a Way of Life, Improving the Culture, Meeting the Needs of Individuals, Summary, Discussion Topics and Assignments, Selected References.

Chapter 3 **INDUSTRIAL ARTS IN OUR SCHOOLS**   36

    A Nationwide Picture of Industrial Arts, Industrial Arts—A Curriculum Area of Education, Industrial Arts at the College Level, The Subject Areas of Industrial Arts, Types of Laboratory Organization, Summary, Discussion Topics and Assignments, Selected References.

Chapter 4 **OBJECTIVES OF INDUSTRIAL ARTS**   64

    How to Derive Objectives, Use of Objectives, Importance of

Objectives, Analysis of the Relationships of Industrial Arts to General Education, Implications for Industrial Arts to Transmit a Way of Life, Implications for Industrial Arts to Improve the Emergent Culture, Implications for Industrial Arts to Meet the Needs of Individuals, Important Objectives of Industrial Arts, Summary, Discussion Topics and Assignments, Selected References.

Chapter 5 ANALYSIS OF OBJECTIVES 80

Behavior Changes as Outcomes, Typical Behavior Changes, What Objective Is of Most Worth? Summary, Discussion Topics and Assignments, Selected References.

Chapter 6 LEARNING ACTIVITIES IN INDUSTRIAL ARTS 101

Basis for Selecting Learning Activities, Learning Activities for Industrial Arts, Summary, Discussion Topics and Assignments, Selected References.

Chapter 7 ORGANIZING LEARNING ACTIVITIES 132

Analyze Objectives in Terms of Student Behavioral Outcomes, Organization of Learning Activities, The Unit Method of Teaching, General Phases of Unit Teaching, How to Plan a Teaching Unit, Unit Teaching in the Field, Lesson Planning, Course of Study Construction, Summary, Discussion Topics and Assignments, Selected References.

Chapter 8 THE INDUSTRIAL ARTS PROJECT 159

What Is a Project? Some Historical Aspects, Primary Function of the Project, Steps in the Project Method, Criteria for Project Selection, Methods for Using Projects, Types of Projects, Sources for Project Ideas, Planning the Industrial Arts Project, Advantages of Student Plan Sheets, Characteristics of Student Plan Sheets, Methods for Promoting the Use of Student Plan Sheets, Misuses of the Project and the Project Method, Project Evaluation, Summary, Discussion Topics and Assignments, Selected References.

Chapter 9 WRITTEN INSTRUCTIONAL MATERIALS 192

Origin of Use of Written Instructional Matter in Industrial

Contents

Education, Purposes of Written Instructional Materials, Limitations of Written Instructional Material, Types of Written Instruction Sheets, Programed Instruction, Features of Programed Instruction, Types of Programed Instruction, Program Format, Steps in Developing Programed Instruction Sheets, Example of a Programed Instruction Sheet, Recent Research Findings, Summary, Discussion Topics and Assignments, Selected References.

Chapter 10 THE DEMONSTRATION 224

Uses of the Shop Demonstration, Types of Demonstrations, Advantages of the Demonstration, Making the Demonstration Effective, Preparing for the Demonstration, Presenting the Demonstration, Terminating the Demonstration, Follow-Up to the Demonstration, The Demonstration Area, New Developments, Using Television in the Demonstration, Teaching Skills with Programed Instruction, Related Research Findings, Summary, Discussion Topics and Assignments, Selected References.

Chapter 11 LABORATORY MANAGEMENT—STARTING THE CLASS 256

Starting the Class, Rotating Pupils through Areas, Summary, Discussion Topics and Assignments, Selected References.

Chapter 12 LABORATORY MANAGEMENT—PERSONNEL ORGANIZATION 271

Purposes of the Personnel Organization, Organizing a Personnel Plan, Features of a Personnel Organization, Typical Examples of Personnel Organizations, Devices for Depicting Personnel Organizations, Summary, Discussion Topics and Assignments, Selected References.

Chapter 13 LABORATORY MANAGEMENT—RECORDS AND RECORD KEEPING 288

Scope of the Problem, The Importance of Records, General Types of Records, Administrative Records, Instructional Records, Financial Records, Designing a Records System, Keeping Records, Summary, Discussion Topics and Assignments, Selected References.

Chapter 14 **INSTRUCTIONAL MEDIA—Part I**   311

Instructional Media in Industrial Arts, Types of Instructional Media, Summary, Discussion Topics and Assignments, Selected References.

Chapter 15 **INSTRUCTIONAL MEDIA—Part II**   340

Types of Instructional Media (continued), Photography for the Industrial Arts Teacher, Summary, Discussion Topics and Assignments, Selected References.

Chapter 16 **COMMUNITY RESOURCES**   377

Roadblocks to Using Community Resources, Importance of Using Community Resources, Types of Community Resources, Locating Community Resources, Opportunities for Community Service, Summary, Discussion Topics and Assignments, Selected References.

Chapter 17 **EVALUATION IN INDUSTRIAL ARTS**   395

Traditional Practice, Evaluation in Terms of Objectives, Technique of Evaluation, An Example of Behavioral Change Evaluation, Examples of Evaluative Test Items, Standardized Testing, Evaluation and Grading, Discuss Progress with Students, Summary, Discussion Topics and Assignments, Selected References.

Chapter 18 **INDUSTRIAL ARTS AND PUBLIC RELATIONS**   418

The Importance of Public Relations, Developing a Public Relations Program, Keep the Gap Closed, Building the Industrial Arts Image, Summary, Discussion Topics and Assignments, Selected References.

Chapter 19 **LABORATORY PLANNING AND LAYOUT**   443

Educational Specifications, Preliminary Planning for the

Contents ix

Laboratory, Factors To Be Considered in Planning, General Layout Pointers, Planning for the Future, Typical Laboratory Layouts, Summary, Discussion Topics and Assignments, Selected References.

Chapter 20 **THE INSTRUCTIONAL AND PLANNING CENTERS** 469

The Instructional Center, The Planning Center, Location of the Centers, Equipping the Instructional Center, Equipping the Planning Center, Combined Instructional and Planning Center, Suggested Floor Plans, Summary, Discussion Topics and Assignments, Selected References.

Chapter 21 **EQUIPPING THE LABORATORY** 481

Factors Affecting Equipment Selection, Machine Tool Selection, Writing Specifications, Selection of Hand Tools, Tool Storage, Summary, Discussion Topics and Assignments, Selected References.

Chapter 22 **COMPLETING AND EVALUATING THE LAYOUT** 499

Color in the Environment, Selection of Shop Colors, Standard Safety Color Code, Safety Features in School Shop Layouts, Evaluation of Physical Facilities, Summary, Discussion Topics and Assignments, Selected References.

Appendix 1 **SAMPLE UNIT FROM NDEA INSTITUTE** 515

Appendix 2 **A PROGRAMED INSTRUCTION SHEET ON THE HISTORY OF MASS PRODUCTION** 523

Index *527*

Foreword

This excellent text, the original edition of which was a pioneer in the behavioral approach to industrial arts education, continues with the same underlying theme. Not only are behavioral outcomes listed but learning activities are included to accomplish these objectives.

The potential teacher will find this volume an excellent exposition of the why and the how of industrial arts education. The experienced teacher will find it provocative in reassessing his program.

Almost encyclopedic in its content, this edition has been rearranged in a logical organizational pattern with the addition of much new material. In fact, the revision process has resulted in the addition of a third more material than appears in the last edition.

Industrial arts is evolving to meet career education objectives and a closer liaison with occupational education through legislative and financial support. This volume wisely recognizes this trend and treats the development realistically, as well as covering in fine fashion the contribution of industrial arts to general education.

The updating of the content with coverage of newer instructional methods and materials, as well as the book's organization for teachability through the inclusion of discussion topics, assignments, and comprehensive references, makes it a valuable contribution to the literature and a welcome teacher-education text. Few such volumes are so comprehensive and well balanced between the idealogical and the practical.

Lawrence W. Prakken
Editor and Publisher,
School Shop Magazine

Preface

This book has been prepared for teachers—for prospective teachers, for young teachers, and for experienced teachers who have remained young in heart. In brief, it is intended for all who engage in the great adventure of teaching industrial arts—and especially for those who wish to keep abreast with progressive thinking in the field.

It is anticipated that this edition will continue to be of service as a textbook for teacher education classes. The content has been carefully selected for use in specific coursework dealing with philosophy and objectives, principles and methods of teaching, organizing and managing classes, laboratory planning, curriculum construction, and evaluation. Its broad coverage also makes it suitable for purposes of general orientation to the field. The volume should be of special interest and help to the school administrator who desires to keep up-to-date on the role of industrial arts in education, including its philosophy, objectives, trends, and practices—both contemporary and innovative.

The basic framework of the text remains unchanged from preceding editions. This unique approach is based on the derivation of industrial arts objectives from those of general education, the expression of these aims in terms of student behavioral changes, the selection and use of appropriate learning activities, and evaluation practices consistent with the original objectives. Increased emphasis in this edition is placed on the integration of new research findings into the textual content. Each chapter has been reorganized, revised, and expanded to provide more complete coverage of new and important topics. With few exceptions, all new photographs support and enrich the textual materials.

Hopefully, this Fourth Edition will continue to serve the needs of the profession.

Industrial Arts
in General Education

Chapter 1

The Purposes of General Education

This chapter and the one following represent a philosophy upon which can be built a dynamic, constantly evolving system of education that will progressively meet the needs of coming generations. Such a philosophy calls for actual participation and learning by experimentation. This experimentation should be considered as being directed toward the building of value systems. As such, it will normally include planning which will allow for a consideration of the comparative worth of various "ends-in-view." Thus evaluation, or the choice among values, also seems to be involved. Further, this statement of a basic philosophy implies an individual constantly reconstructing and enriching his life through activity and intelligent analysis. It further indicates a school organized and equipped to make this type of learning possible.

In the ensuing sections of this chapter it is proposed (1) to define the term *general education*, (2) to discuss the purposes of general education, and (3) to identify the central purpose of American education. Before considering these matters, however, it seems desirable to consider briefly the propositions on which our democratic society rests.

THE BASIC PRECEPTS OF DEMOCRACY

There are three basic precepts of democracy which should be clearly understood by every boy and girl in America. These are (1) the unique worth of the individual, (2) acceptance of responsibility for one's own development and that of his fellow man, and (3) dependence upon the intelligence of the common man.

1. The Unique Worth of the Individual

A cornerstone of our democracy is the recognition of the unique worth of each individual, regardless of race, color, or creed. All students should be led to see that in this country any person should be free to develop his individuality to its fullest potential. At the same time, each student should understand that freedom for individual development carries with it the responsibility to refrain from interfering in any manner with the development of others. Thus the necessity for fostering personal traits, such as tolerance, cooperation, and social sensitivity becomes clear.

2. Responsibility for One's Own Development and That of His Fellow Man

The democratic way of life is dependent upon individuals who are concerned with "Life, Liberty, and the Pursuit of Happiness" not only for themselves, but for others as well. Each individual must assume the responsibility not only for his own development, but also, to some degree, for that of his fellow man.

In 1963 the Educational Policies Commission of the National Educational Association noted the great need for the development of social responsibility in all pupils. The Commission stated:

> The continued health of the American society—perhaps its very survival—demands a high and rising awareness of social responsibility on the part of the people. For the socially responsible person, the welfare of his fellow man is of deep personal concern.[1]

To teach social responsibility in the schools requires that each student participate in a variety of "shared experiences." The extent and richness of such sharing is, to a degree, a measure of an individual's contribution to the program of a democratic way of life.

3. Dependence upon the Intelligence of Common Man

It should be further recognized that only in a democracy is there a recognition of, and dependence upon, the intelligence of the common man. If democracy means that each individual has the right and the responsibility to express his ideas concerning questions of government and social policy—national, state, and local—then it is evident that such

---

[1] *Social Responsibility in a Free Society,* Prepared by the Educational Policies Commission, National Educational Association of the United States and the Association of School Administrators (Washington, D.C.: National Education Association, 1963), p. 35.

a society can be no better than the best thinking of the combined people can make it. It is not a question here of a select few—an aristocracy of minds—ruling the people. Rather, the people rule themselves. Better and better education, therefore, becomes the only hope for continuing a strong and virile democracy.

WHAT IS GENERAL EDUCATION?

What is meant by the term *general education?* Instead of defining the term directly, most writers on the subject prefer the indirect method of stating a series of aims or purposes. The following definition seems an exception to common practice:

> General Education consists of the body of knowledge commonly designated as liberal arts. It aims at preparing one to assume responsibility for meeting one's own and society's needs and for becoming a more alert, cultivated, and rational individual. The term "general education" implies a commonality of experiences; however, adaptation to individual needs would be expected.[2]

General education is the sum total of all the common skills, knowledge, understandings, attitudes, and appreciations which every person should possess for effective citizenship in our democratic society. It represents a minimum achievement level or foundation out of which may grow education or training for special interests or purposes. General education seeks to provide broad understandings of our culture through experiences in the arts and humanities, social sciences, natural sciences, and mathematics.

PURPOSES OF GENERAL EDUCATION

In 1918 the Commission on the Reorganization of Secondary Education issued one of the most important statements ever made on the purposes of general education. The Commission formulated the now-historic *Seven Cardinal Principles of Secondary Education.*[3] These are (1) health, (2) command of fundamental processes, (3) worthy home membership, (4) vocation, (5) civic education, (6) worthy use of leisure time, and (7) ethical character.

---

[2] *Teaching in North Carolina: Certification, Employment, Procedures, Salary Policies* (Raleigh, N.C.: State Superintendent of Public Instruction, 1965).
[3] *Cardinal Principles of Secondary Education* Commission on the Reorganization of Secondary Education of the National Education Association of Secondary Education of the National Education Association, Bulletin No. 35 (Washington, D.C.: Government Printing Office, 1918), pp. 7–16.

This classic statement of educational purposes has withstood the test of time and is still regarded highly. A nationwide survey made in 1966 by the NEA Research Division revealed "an overwhelming verdict in favor of the seven cardinal principles as formulated in 1918."[4] More than 85 percent of the teachers surveyed indicated that the cardinal principles still represent a satisfactory list of major objectives for American education.

In 1938 the Educational Policies Commission classified the purposes of education under four major objectives:[5] (1) self-realization (personal growth and development of the individual), (2) human relationships (home, family, and community relationships), (3) economic efficiency (occupations, personal economics, and consumer judgment), and (4) civic responsibility (civic and social duties).

From time to time statements of educational purpose are made by local and state groups. Representative of these are the "Ten Goals of Quality Education" as issued by the Committee on Quality Education of the Pennsylvania State Board of Education (1968). These goals are (1) self-understanding, (2) understanding others, (3) basic skills, (4) interest in school and learning, (5) good citizenship, (6) good health habits, (7) creativity, (8) vocational development, (9) understanding human accomplishments, and (10) preparation for a changing world.

Some of the statements concerning the purpose of American education have been broad and inclusive, while others have been more specific. Careful consideration reveals, however, that stripped of verbiage and special applications, the various statements may be summed up as implying three basic purposes of general education: (1) to transmit a way of life, (2) to improve and reconstruct that way of life, and (3) to meet the educational needs of individuals.

Transmitting a Way of Life

From the very dawn of civilization (and even before), man has been concerned with transmitting or passing on to the rising generation a particular way of life. First, the simplest elements of survival were passed to the young; later, as tribes came into being, customs, cults, and mores were included. What was passed to the young represented the sum total of the culture of that society at that time. Each society must do this if its cultural and technological heritage is to be preserved. If the

---

[4] "A New Look at the Seven Cardinal Principles of Education," *National Educational Association Journal*, vol. 56, no. 1 (January 1967), p. 53.

[5] *The Purposes of Education in American Democracy*, Prepared by the Educational Policies Commission, National Education Association and the Association of School Administrators (Washington, D.C.: National Education Association, 1938), pp. 47–108.

way of life is passed from one generation to the next, the new generation can build, grow, and advance from where the old left off. Thus, it is unnecessary for each succeeding generation to, say, reinvent the wheel or make fire; instead new generations can reach for the stars by standing on the shoulders of those who have preceded them.

As society has become increasingly complex, so has the amount and nature of the culture to be transmitted. Part of our cultural heritage to be preserved by transmission to the young includes, for example, our knowledge for conquering disease and hunger, concepts of humane treatment of fellow human beings, concepts of regard for our ecology, and so on. The importance of transmitting a way of life may be realized when one considers that if mankind should fail in this important duty for a period of only two or three generations, the entire culture would revert to savagery. All of our knowledge, customs, and technical accomplishments would be lost.

Methods for transmitting the culture have varied through the ages. When civilization was young and life was comparatively simple, the task of inducting youth into the adult life could be effectively handled through the medium of the home and tribal council. Boys were taught to hunt, fish, and fight by their fathers and other adult members of the tribe or clan, while the girls learned from their mothers to cook, till the soil, and care for children. As soon as a body of beliefs concerning the supernatural was developed, certain members of the society were designated as priests or medicine men; and it was their duty to pass on such doctrines to the youths, as well as to interpret them for the adults.

As life became more complex, especially after the development of writing, there appeared a body of information which was not common to all members of society. It was no longer possible for the home effectively to pass on much knowledge and many of the skills which it seemed essential to perpetuate. Parents began to send their children to those who had mastered the desired accomplishments. At first these persons were usually the priests and scribes, but soon a new class known as teachers appeared. From these beginnings grew schools and the present method of transmitting the culture.

Important for present consideration is the fact that, once a body of subject matter is adopted by the schools, it tends to persist—even after it no longer represents accepted values in the way of life of a given place or time. For example, the most important factor in the way of life during and following the "Dark Ages" in Europe is described by the term *other-worldliness*. Civilization, as a whole, was primarily concerned with making the best possible preparation for the world to come. Little consideration was given to improving conditions or extending knowledge of a worldly nature, for such activity would be considered incompatible with preparation for the life to come. Schools of this period therefore consisted largely of training for a life of humility and consecration, such as was found in the monasteries of the various orders. The teaching of reading and writing was justified only because it was necessary for reading the sacred books and transcribing them for others.

With the coming of the Renaissance, however, there developed a new interest in the classical writings of early philosophers and scholars. The teachings of Aristotle, Plato, and others were accepted without question and became the basis for the new education. The fact that there were many conflicts between the writings of these scholars and the Bible did not seem to occur to students of that period. Neither did the fact that an entirely different way of life was represented by the teaching of these principles. Basic conflicts were ignored, and the schools of that period taught the concepts of both other-worldliness and the classics side by side.

Later, other purposes influenced the culture of advancing civilization. The age of chivalry, with its emphasis on physical prowess and courtly grace, was followed by a period of scientific discovery. Each period in its turn has had profound influence on the way of life and culture of the European world. The schools, however, failed to change with the changing aims of society. It is true that they accepted some new developments, but they also clung to the old beliefs and subject matter.

If, as appears evident, the schools should be concerned with transmitting a way of life, then one should be able to look at the schools and see mirrored there the more important aspects of modern culture. Instead, one finds a conglomeration of many ways of life that have been handed down all the way from early ages to the present. The result has been to produce a state of confusion to a point where it is difficult to discern what is meant by "our way of life." *The crying need of today is for reexamination of our present culture and of the school program, toward the end of bringing the two into harmony.*

Such a reexamination will show that, among other features, present civilization is characterized by the fact that it is *democratic*. The implication is clear, therefore, that the youth of the nation should be thoroughly acquainted with the meaning of the term and the duties and responsibilities of every person living under such a social organization. It should be pointed out that "democracy" means more than a political form of government—that it is indeed a complete "way of life" involving practically all activities of the people.

Providing for Improving the Culture

If society did nothing more than transmit its culture there would be no progress or improvement. Education has the further objective, therefore, to provide for extending and improving the way of life.

The logical question which follows concerns the nature of progress. Through what process does civilization advance? A study of the past seems to indicate that societal advancement can best be achieved through the development and practice of critical thinking. Wherever there is progress, whether it be in the field of technological inventions, social advances, literature, arts, or any other field, someone has taken a step beyond present knowledge. He has formulated a hypothesis and acted upon it. In other words, he has projected his thoughts one step into the "dark" through the process of critical thinking.

For example, when Eli Whitney noted the vast amount of time and labor required to remove seeds from cotton, he recognized a problem which needed solving. A careful study of the cotton boll led to a guess or hypothesis as to how the seed could be removed more readily. Acting upon this "guess," he produced (after several trials) the cotton gin. It would be generally admitted that this result of critical thinking represents progress.

There are important implications for education in this interpretation of the nature of progress. If society advances best through the practice of critical thinking, then a major aim of all education should be to promote this ability in every student.

In 1961 the Educational Policies Commission noted that our schools must be concerned with *all* of the objectives of education. At the same time they recognized that there was not enough time in the day for the schools, the teachers, or the students to engage in all of the activities necessary to achieve fully each and every one of these goals. Thereupon the Commission identified the central purpose of American education as the development of the rational powers of an individual. They stated:

> The purpose which runs through and strengthens all other educational purposes—the common thread of education—is the development of the ability to think. This is the central purpose to which the school must be oriented if it is to accomplish either its traditional tasks or those newly accentuated by recent changes in the world.[6]

The responsibility for developing a student's ability to think cannot be left to any one teacher or to any one field of subject matter. Neither

---

[6] *The Central Purpose of American Education*, Prepared by the Educational Policies Commission, National Education Association of the United States and the Association of School Administrators (Washington, D.C.: National Education Association, 1961), p. 12.

can this objective be accomplished through setting up a class exclusively for the purpose of stimulating critical thinking. Rather the development of such thinking must be the responsibility of all teachers in all subjects.

The study of an abstract subject alone will not ensure success in the development of the ability to think. While this ability may be developed in such subjects, it may be developed equally well in aesthetic or practical subjects, such as music, art, and industrial arts. Thus it is not the subject content itself which develops the ability to think. Rather, it is developed by methods which encourage the application of learning from one situation to another as well as by the reorganization of things learned.

The ability to think critically can be developed only through practice in solving problems. It, therefore, becomes the duty of every teacher to face his students with challenging situations and lead them, not to a predetermined solution, but to their own answers in terms of all available data. It is assumed by some teachers that thinking cannot be expected until the student has acquired a broad foundation of information in the field in which thinking is to be done. While it is true that facts are the basis for critical thinking, it is evident that no child is too young to use the reasoning process.

True, the more information a student may have at hand, other things being equal, the more logical the student's thinking will be, and the more likely he will be to arrive at a correct conclusion. The fact remains, however, that *one learns to think only through practice, and even a child in kindergarten may reason critically about topics on his own level of development.* If every teacher would assume direct responsibility for teaching his students how to analyze problem situations, how to formulate hypotheses, and then how to set about to prove them, the whole process of education and of social progress would be immeasurably advanced.

Education as the Meeting of Needs

If it is assumed that the needs of students are important in determining the purposes of education, then the question of how these needs are to be ascertained is at once raised. In the past, numerous solutions to this problem have been tried. One method suggested that the *supposed* needs of youth be determined by examining adult society to identify the activities which youths would eventually be called upon to pursue. The development of this point of view is reflected in what has sometimes been called the "scientific movement" in education. Much of the maladjustment so apparent in the traditional school undoubtedly has been due to the fact that this procedure gives little or no attention to the nature of the child, to his maturity level, to his interests, or to his abilities.

Another school of thought, reverting either consciously or unconsciously to the philosophy of Rousseau, expresses the theory that the *felt needs and interests* of the child are adequate guides to his developmental requirements. It is held that to go contrary to these is to endanger the personality of the individual. The implementation of this philosophy has led to such developments as the "child-centered school" and the "goal-less school." Modern thinking and experimentation seems to repudiate the theories underlying such practices.

The present tendency in determining the needs of students seems to be concerned with the findings of basic sciences. Thus one notes that the findings of *biology* have important bearings on a definition of needs. To the biologist, the individual is an *organism* which is much more than the sum of all its parts.[7] It is an organism which continually acts and reacts with its environment. As a result of this interaction, certain changes take place in the individual which constitute *growth*. It is conceded that growth is dependent on two sets of forces, one of these being inherited tendencies and the other the interaction with the environment. Research has tended to show that many traits once thought inherited are, in fact, subject to change through outside influence.[8] This concept constitutes at once a challenge and an opportunity for the schools to supply the type of surroundings best suited to meet individual needs. In other words, the human organism is now known to be *multipotential*. The effect of this change of viewpoint upon education cannot be overestimated.

The biologist also sees the individual as an *active* organism. He is no longer considered a mere passive collection of protoplasm waiting for an impression to be made from the outside. He is dynamic. He wants to do something and get results. This would seem to indicate that the traditional system, where teachers judged "goodness in terms of quietness" and where environment, instead of being enriched, was drained of everything that might distract attention from assigned lessons, and was running in direct opposition to biological nature.

Recent developments in the field of psychology also have important bearings on the nature of individual needs. The older theories of faculty, behavioristic, and atomistic concepts, which tended to support the necessity for drill and memorization, have largely been replaced by

[7]John L. Childs, *Education and the Philosophy of Experimentalism* (New York, N.Y.: The Century Company, 1931), Chapter IV.
[8]F. K. Berrien, *Practical Psychology* (New York, N.Y.: The Macmillan Company, 1944), Chapter III.

newer and seemingly more adequate hypotheses. For example, the breakdown in mechanistic psychology was the result, mainly, of observations which tend to show the organismic nature of the individual. It is noted that, given certain stimuli, a simple and direct reaction cannot be counted upon. These stimuli, affect not only the brain and the reacting muscles, but the whole organism.

One may no longer consider single isolated manifestations of human behavior as a basis for study. Rather, one must study the total organism in relation to all forces acting upon it. Thus emotions become important not in their peculiar manifestations but as they are indicative of disequilibrium within the individual. A further implication of the wholeness of learning situations makes it appear that an individual not only learns the lesson which the teacher is presenting but may be developing certain very definite attitudes—likes and dislikes—and emotional reactions in addition. The matter of concomitant learning, therefore, takes on unusual importance.

In like manner psychology seems to show that learning takes place much more quickly and easily when it is the result of experiences carried on in relationship to pupil goals and purposes. For example, a boy who had difficulty in solving even the simplest problems in arithmetic learned how to measure with a rule when he needed this skill to complete a gun rack which he was making in the industrial arts laboratory. This is but one of many implications of psychology which have a significance in determining the individual's requirements.

The needs of students have been studied at length by the Educational Policies Commission. Their analysis was based on an examination of society and a study of the daily lives of boys and girls of various ages. As a result the Commission has identified the common and essential needs which all youth has in a democratic society as follows.

    1. All youth need to develop salable skills and those understandings and attitudes that make the worker an intelligent and productive participant in economic life. To this end, most youths need supervised work experience as well as education in the skills and knowledge of their occupations.

    2. All youth need to develop and maintain good health and physical fitness.

    3. All youth need to understand the rights and duties of the citizen of a democratic society, and to be diligent and competent in the performance of their obligations as members of the community and citizens of the state and nation.

    4. All youth need to understand the significance of the family for the individual and society and the conditions conducive to successful family life.

5. All youth need to know how to purchase and use goods and services intelligently, understanding both the values received by the consumer and the economic consequences of their acts.

6. All youth need to understand the methods of science, the influence of science on human life, and the main scientific facts concerning the nature of the world and of men.

7. All youth need opportunities to develop their capacities to appreciate beauty, in literature, art, music, and nature.

8. All youth need to be able to use their leisure time well and to budget it wisely, balancing activities that yield satisfactions to the individual with those that are socially useful.

9. All youth need to develop respect for other persons, to grow in their insight into ethical values and principles, and to be able to live and work cooperatively with others.

10. All youth need to grow in their ability to think rationally, to express their thoughts clearly, and to read and listen with understanding.[9]

As one studies the various aspects of the democratic way of life, however, it becomes more and more evident that both the biological nature of the individual and the various aspects of the learning process are conditioned in many important aspects by the *nature of society*. It thus becomes necessary to consider the influence of such trends as the seeming breakdown of home life, the increasing number of employed mothers, and the economic status of the family as it is reflected in the needs of children. Another important aspect of this social picture relates to the change from agrarian to urban society, the population explosion, and the rapid increase in industrialization and technological change. These and other like forces are potent influences in determining the needs of students to which the school must administer. A more detailed discussion of some of the more important aspects of today's society is presented under the heading, *The Nature of American Culture*, in Chapter 2.

SUMMARY

The American way of life is best characterized as democratic. Our democracy rests upon three basic precepts: (1) a recognition of the unique worth of the individual; (2) an individual responsibility not only for personal development, but for that of all men as well; and (3) a recognition of, and dependence upon, the intelligence of the common man.

[9]"Imperative Needs of Youth," *The Bulletin of the National Association of Secondary School Principals*, vol. 31, no. 145 (Washington, D.C.: National Education Association, March 1947), p. 2.

General education is the sum total of all the common skills, knowledges, understandings, attitudes, and appreciations which every person should possess for effective citizenship in our democratic society.

General education has three principal objectives: (1) to transmit a way of life; (2) to improve that way of life, the most feasible method being by training for effective critical thinking; and (3) to meet the needs of individuals in the basic aspects of living.

DISCUSSION TOPICS AND ASSIGNMENTS

1. Compare the "Seven Cardinal Principles of Secondary Education" with the aims of general education as outlined in this chapter. Show the extent to which they can be reconciled.

2. Compare the cardinal principles with the "Ten Imperative Needs of Youth." Note similarities and differences.

3. If a high school student asked you, "What are the purposes of education?", how would you answer him in language he could understand?

4. How would you rearrange the list of imperative needs of youth in order of importance? Would you make any additions or deletions to the list?"

5. How can our schools better achieve the central purpose of education as identified by the Educational Policies Commission? Be specific.

6. Analyze the organization of a school with which you are familiar from the standpoint of the extent to which it is based on democratic principles. Indicate ways in which it could be made more democratic.

7. What specific preparation can a teacher make which will develop his ability to determine and meet the needs of his pupils?

8. Cite several examples in the school curriculum (K through 12) where tradition has largely been responsible for retaining certain practices and/or subject matter, but these are no longer particularly related to today's way of life.

SELECTED REFERENCES

"A New Look at the Seven Cardinal Principles of Education," *National Education Association Journal*, vol. 56, no. 1 (January 1967), pp. 53–54.

"Imperative Needs of Youth," *The Bulletin of the National Association of Secondary School Principals*, vol. 31, no. 145. Washington, D.C.: National Education Association, March, 1947.

Policies for Education in American Democracy. Educational Policies Commission, National Education Association of the United States and the American

Association of School Administrators. Washington, D.C.: National Education Association, 1946.
Social Responsibility in a Free Society. Educational Policies Commission, National Education Association of the United States and the American Association of School Administrators. Washington, D.C.: National Education Association, 1963.
Teaching in North Carolina: Certification, Employment, Procedures, Salary Policies. Raleigh, N.C.: State Superintendent of Public Instruction, 1965.
The Central Purpose of American Education. Educational Policies Commission, National Education Association of the United States and the American Association of School Administrators. Washington, D.C.: National Education Association, 1961.

Chapter 2

The Relationship of Industrial Arts to General Education

The type of school work under discussion in this text has passed through three phases since its introduction into American schools during the latter part of the nineteenth century: manual training, manual arts, and now industrial arts. While each phase has distinct characteristics, no sharp line of demarcation can be drawn between them from a chronological point of view.

At first, manual training was justified on the basis of the training of "hand and eye," a theory that agreed well with the principles of faculty psychology then prevailing.

Later, manual arts emphasized design or the application of art to the shop project. The word *arts* opened the way for a gradual expansion of shop activities into areas other than woodworking and metalworking.

INDUSTRIAL ARTS AS GENERAL EDUCATION

Shortly after the turn of the century, leaders like Russell, Richards, and Bonser began to see in industrial arts a medium for enriching the offerings and extending the values of the regular school program. In 1904 Charles R. Richards called for a fundamental change in content as well as a new name, *industrial arts*. A few years later James E. Russell and Frederick G. Bonser formulated their Social-Industrial Theory of Industrial Arts which laid the foundations for the present-day concept of industrial arts. Although their work was in elementary education, their concepts of industrial arts in the schools applied equally well to the secondary level. From that time on it has become commonplace to regard industrial arts as a part of general education.

Today industrial arts is accepted by all as an integral part of general education. Its unique contributions have been noted by many general

The Relationship of Industrial Arts to General Education

educators, especially since the establishment of the junior high school. The literature of the field is replete with recognition of the values of industrial arts by those in the field of general education. The following examples are typical:

> Industrial arts is a part of General Education and no amount of good General Education can compensate for the lack of Industrial Arts . . . Industrial Arts is here to stay and must be developed as an integral and a coordinate part of a sound system of public education.[1]

> The practical side of school life needs to be increasingly recognized for the contribution it is able to make to general education—it is an opportunity for children to develop their inherent manual and artistic skills, without in any way leading to a specific vocation.[2]

> . . . industrial arts poses the only means of demonstrating what man has traditionally done through the ages, that is, to take nature's materials and their power potential in the raw state and convert them into useful goods and services for mankind.
> No other body of knowledge within the public-school curriculum can make this claim.[3]

In his book, *Education in the Junior High School Years*, Conant stated:

> All girls should receive instruction in home economics and all boys instruction in industrial arts.[4]

Somewhat earlier the Report of the Harvard Committee on *General Education in a Free Society* noted that

> The direct contact with materials, the manipulation of simple tools, the capacity to create by hand from a concept in the mind—all these are indispensable aspects of the general education of everyone.[5]

Such claims have not been made, of course, without foundation. A rationale for the inclusion of industrial arts in general education follows.

Industrial arts is an essential part of general education. It is conceived as an answer to the problem of educating boys and girls to live in a world which may be accurately characterized as industrial and

---

[1] Earl C. Funderburk, "A Superintendent Looks at Industrial Arts," *The Industrial Arts Teacher*, vol. 22, no. 5 (May–June 1964), p. 16.

[2] "Education in a Technological Society," published by the United Nations Scientific and Cultural Organization (Paris, France: UNESCO 1960), p. 36.

[3] W. L. Johnston, "Interpreting Industrial Practice in the Modern I-A Lab," *School Shop*. vol. XXIX, no. 4 (December 1969), p. 36.

[4] James B. Conant, *Education in the Junior High School Years* (Princeton, N.J.: Educational Testing Service, 1960), p. 16.

[5] Report of the Harvard Committee, *General Education in a Free Society* (Cambridge, Mass.: Harvard University Press, 1945), p. 175.

technological. From a nation which was largely agrarian and in which industries were simple and widely decentralized, the United States has moved rapidly to a position of world leadership in industrial development. Children and adults are now living in a civilization that has surrounded itself with electromechanical marvels which must be understood and used. At the same time, industry through increasing centralization has been removed from the everyday experience of the average individual. This complexity has made difficult a comprehension of the organization, products, processes, and occupations in industry. Hence it becomes a function of the schools to give every student an appreciation and understanding of our industrial civilization as a vital segment of American life.

INDUSTRIAL ARTS DEFINED

Industrial arts has been defined by many writers and in various ways. Bonser and Mossman say that industrial arts is "a study of the changes made by man in the forms of materials to increase their values and of the problems of life related to these changes."[6]

According to the American Industrial Arts Association, "Industrial arts is that phase of general education which offers individuals an insight into our industrial society through laboratory-classroom experiences."[7]

Another professional organization defines industrial arts as "the study of our technology, including industrial tools, materials, processes, products, occupations, and related problems. It involves activities in shops, laboratories, drafting rooms, and elementary school classrooms."[8]

For purposes of this book, industrial arts will be defined as *those phases of general education that deal with industry—its evolution, organization, materials, occupations, processes, and products—and with the problems resulting from the industrial and technological nature of society*. This definition does not differ materially from those given previously, but it does tend to stress the place and function of industrial arts as a school subject and its relationship to general education.

---

[6]Frederick G. Bonser and Lois Coffey Mossman, *Industrial Arts for Elementary Schools* (New York, N. Y.: The Macmillan Company, 1923), p. 5.

[7]*A Career in Teaching Industrial Arts* (Washington, D.C.: American Industrial Arts Association, 1965), p. 2.

[8]*Industrial Arts in Education*, A Statement by the Industrial Arts Policy and Planning Committee of the American Vocational Association, Inc., 4th ed. (Washington, D.C.: American Vocational Association, Inc., 1965), p. 3.

NATURE OF AMERICAN CULTURE

It was shown in the preceding chapter that the three primary objectives of education are (1) to transmit a way of life to the rising generation, (2) to improve our culture, and (3) to meet the needs of individuals. Before considering the relationship of industrial arts to these goals of general education, it seems desirable to examine some of the essential features of the culture in question. Certain of the features inherent in the American way of life are traditional in character, such as, for example, our belief in education. Other features are new.

Looking beyond the multiplicity of detail for something fundamental, one discovers that basic to most other considerations is the fact that the American way of life is democratic. Conceived in its broader implications (as outlined in Chapter 1), this fact explains, coordinates, and makes significant most of the other features which are typical of American civilization.

A careful study of this democratic organization reveals that it may be accurately characterized as an *industrial* democracy. Scenes like that shown in Fig. 2–1 are indicative of the importance of industry in the American way of life.

Our industrial might can be traced to the fact that America has always been a nation of tool users. As Thomas Carlyle aptly said:

> Man is a Tool-using Animal—he can use Tools, can devise Tools: with these the granite mountain melts into light dust before him; he kneads glowing iron, as if it were soft paste; seas are his smooth highway, winds and fire his unwearying steeds. Nowhere do you find him without Tools. without Tools he is nothing, with Tools he is all.[9]

Since the Civil War period, and especially since the turn of the century, there has been a continuous growth in industrial development and technological advancement. The "knowledge explosion" has resulted in the doubling of man's knowledge about every decade. Increased demands for goods, fostered by the availability of working capital within the country, and the effects of new inventions have brought about a society more highly industrialized than any the world has yet seen.

A good indicator of American industrial development is the number of patents issued by the United States Patent Office. In 1860 this office issued 5,000 patents. During the past decade (1960–69) the yearly average number of patent applications exceeded 85,000. In other words, for

---
[9]Thomas Carlyle, *Sartor Resartus* (London: Chapman and Hall, Ltd., originally published 1831), p. 32.

Fig. 2-1. Industry and technology are important influences in the American way of life. (Courtesy, Aluminum Company of America)

each application filed in 1860, 17 applications were filed each year for the period 1960–69. A new record was set in 1969 when over 96,000 applications were received and more than 61,000 patents were issued.

The increased demand for consumer goods is illustrated by the following examples:

The telephone was invented in 1876 and by 1969 more than 115 million phones were in use in this country.

In 1970 over 90 percent of the homes in this country which were wired for electricity had radios, black-white television, electric irons, toasters, vacuum cleaners, and clothes washers. Ninety-nine percent of the homes had radios, black-white television, and electric irons; nearly 40 percent had colored television.

Automobiles increased from 8,000 cars in 1900 to 105 million in 1970. Nearly 80 percent of American families own an automobile; 30 percent now own two or more.

Each new invention has had its effect upon the growth of industry. Huge machines have been turned to mass production; the demand for more power has been met by the building of larger and more efficient

The Relationship of Industrial Arts to General Education

generating plants utilizing hydro, steam, and atomic energy. For example, in 1940 the total horsepower for all prime movers was approximately 2.7 billion; in 1960 this figure rose to 11 billion and it reached 18.7 billion in 1969.

By far the greater part of this rapid growth in the mechanization of industry was abetted and made possible through the development of huge corporations. Vast amounts of capital under the control of a few groups led to the establishment of such great corporate industries as General Electric, General Motors, Standard Oil, and the United States Steel Company. In the past two decades industrial growth and expansion have occurred at an ever-increasing rate. To purchase new plants and equipment, industry expended some $9 billion in 1945 and over $50 billion in 1965.

Increases in industrial production have been made possible by a standarization of machines and processes and by a more complete and efficient division of labor. Under the leadership of such industrial geniuses as Ford, Westinghouse, and Carnegie, mass production has long been the accepted method in America. When automation and electronic computers are coupled with mass production, the results are nothing less than phenomenal. For example:

One man seated at a master console can control all operations of a copper-concentrating plant from crushing the copper ore to the storage of the final concentrate before it is shipped to a smelter.[10]

A computer is used to monitor steel during production at a new furnace. Sensitive instruments keep track of the chemical composition of each batch. If the desired characteristics as spelled out in the computer's memory are not maintained, the electronic brain calculates what additives are needed and issues the instructions.[11]

With a three-man team, a "robot" rolling mill in New Jersey turns out 217 miles of copper rod per day. This is more than twice the old rate at one-tenth the old labor cost.[12]

There is a 300-ton machine in a Detroit auto plant that is a city block long and a block wide. It is operated by one man and can turn out a finished engine block every 45 seconds. This is twice the old production rate and at one fifth the old labor cost.[13]

---

[10] Yale Brozen, "Automation: A Job Creator Not a Job Destroyer," *U.S. News and World Report* (March 8, 1965), p. 100.

[11] "The Black Box Plant Is Coming," *Duns Review*, Special Supplement–Part II (March 1965), p. 117.

[12] Robert E. Cubbedge, *Who Needs People?* (New York, N.Y.: Robert B. Luce, Inc., 1963), p. 22.

[13] *Ibid.*

20 The Relationship of Industrial Arts to General Education

One man in an engine plant in Cleveland operates a transfer machine which performs over 500 different operations. Before automation the job required 35 to 70 men.[14]

In a Detroit engine plant 76 engines are tested automatically at one time on a special test stand (see Fig. 2–2). The engines "seek out" a test station where they are connected to oil, water, and fuel lines automatically. Each engine is test run for 20 minutes, then it is disconnected, ejected, and moves down the line—all automatically controlled!

Fig. 2–2. Natural gas is used to test engines automatically on this Nankervis-designed stand. (Courtesy, Chrysler Corporation)

The machining of pistons is one of the most automated processes in the automotive industry. From the time of the first machining operation until final processing with a coating of tin, the piston is untouched by human hands.

Edwin F. Sheeley, inventor of automated equipment, states:

[14]Walter Buckingham, *Automation, Its Impact on Business and People* (New York, N.Y.: Harper & Row, 1961), p. 21.

> I think it not rash to predict that 20 years from today there will be no human beings, other than supervisory personnel engaged in the manufacture of the necessities of life.[15]

DeVore says it this way:

> the days of mass employment generated by mass industry are over and ... the phenomena of technology has but one purpose—the disemployment of human beings.[16]

From the preceding the reader may infer that man may soon displace himself with a machine. Nothing could be further from the truth! Technology enables man to control his environment and to improve his standard of living. Machines may be able to do more things faster than man, but they cannot think like a man. The human element will always be needed in conjunction with man-made machines. Besides, not all manufactured goods can be produced economically by computer-controlled, automated equipment. There are innumerable things necessary to man's existence which do not lend themselves to such sophisticated production. Even today it is not uncommon for production techniques to range all the way from automated production to hand processes in the same factory. No, there is still a place in the world of industry and technology for people.

If mass production has been the keystone in our industrial growth, certainly it has itself fathered a dilemma called mass consumption. Consumption must keep pace with production lest the wheels of industry slow down or even stop. Even a slight retardation in consumption brings with it a proportional threat of economic danger.

The problem of mass consumption is of such magnitude that industry invests huge sums in national, and even international, advertising. Radio and television commercials, advertisements of all kinds, and easy-payment plans seek to keep the public in a buying mood. To encourage mass consumption, industry has resorted in some instances to (1) planned obsolescence and (2) style changes made solely to court customer approval of newness.

Even this brief analysis would be incomplete without mentioning some of the related social aspects which have accompanied our industrial expansion. The rise of organized labor in this country has been a significant factor in the industrial scene. Labor's conflicts with those who own and control the facilities of production have brought into

---

[15]Cubbedge, *op. cit.*, p. 1.
[16]Paul W. DeVore, "Perspective," *School Shop*, vol. XXVII, no. 5 (January 1969), p. 17.

common use such terms as unions, open and closed shops, sitdown strikes, walkouts and lockouts, boycotts, picketing, injunctions, and bargaining and mediation boards.

However, the concern of labor, capital, and the government have established minimum-wage levels, unemployment insurance, and retirement plans for the American worker. The end result of all this has been to provide Americans with one of the highest living and working standards in the world. American workers enjoy a shortening work week, pay for holidays, vacations, and overtime, health and unemployment insurance, paid leave for retraining purposes, a lowering retirement age, and numerous other fringe benefits. Our industrial civilization has given the ordinary American of today a more abundant and more luxurious life than a king of some bygone era. The impact of industrial automation is only now beginning to be felt, but it promises to lead our industrial economy and our way of life to even greater heights in the future.

TRANSMITTING A WAY OF LIFE

Until a few generations ago every youth had an opportunity to watch industrial processes at close range, and to participate in industrial activities around the home. Figure 2-3 depicts a well-known scene of bygone days when the average youth was well acquainted with the work of the village blacksmith. He knew from first-hand observation how hot metal was shaped into horseshoes, how a wagon tire was shrunk onto the wheel, and how metals were united by welding. He visited the village grist mill and saw how whole grain was processed into meal, flour, and other food products. He saw at first hand how the cheesemaker changed milk into cheese; how the carpenter built a house or barn; how the stonemason "laid" a wall; and how the shoemaker repaired or made shoes from leather tanned in the local tannery.

In his book, *Transformation of the School*, Cremin describes it thus:

> Behind the older agrarian society lay the time-honored education of the agrarian household and neighborhood where every youngster shared in meaningful work and where the entire industrial process stood revealed to any observant child.[17]

Obviously, such participation and observation are no longer possible because of the magnitude and the centralization of industry. In some

---

[17] Lawrence A. Cremin, *Transformation of the School* (New York, N.Y.: Alfred A. Knopf, 1961), p. 117.

The Relationship of Industrial Arts to General Education 23

Fig. 2–3. In times past youth could easily observe industrial activities. (Courtesy, Turner Manufacturing Company, Chicago)

large industrial plants children under twelve years of age are denied admission even for purposes of supervised field trips or industrial visits. Thus, the necessity for furnishing similar experiences in the school becomes apparent. Figure 2–4 depicts one difficulty which children today encounter in trying to observe the world of work compared to their peers of yesteryear.

Industry and technology are keystones in today's way of life and our democratic society has been so characterized in this chapter. American culture is filled to overflowing with electromechanical wonders. To live intelligently in such a complex and highly industrialized economy, one must understand and appreciate it; this applies to all members of our society, young and old alike. Burl N. Osburn once wrote:

> One cannot truly appreciate any portion of the cultural heritage without that quality of understanding that comes only after experience has brought meaning to it.[18]

[18]Burl N. Osburn, "Basis for a Program of Industrial Arts" (unpublished paper, Millersville, Pa., c. 1960).

Fig. 2-4. This television monitor allows centralized control of access to company facilities. (Courtesy, HRB-Singer, Inc.)

The agency in our society which is best able to provide these experiences for both youth and adult is our public schools. If, as has been pointed out, an important purpose of education relates to the transmission of the social culture, then the vital place which industry holds in the American way of life should certainly call for major emphasis upon those phases of the program that deal with its exemplification in the schools.

Rapid and unprecedented changes in the nature of our society have naturally brought about strains and evidences of lack of adjustment. Youth is asked to adjust to conditions and occupations the very nature of which is strange and foreign. Thousands of new materials are offered for sale to a public which has no training in choosing among them. Serious social problems arising from the divergent interests of capital and labor, such as strikes, picketing, and boycotts, have also tended to complicate the problems of education. The results of rapid industrialization indicate a need for a new emphasis in the schools. The industrial evolution has changed American society, and in so doing has charged our schools with a vital responsibility—to provide orientation for life in an industrial-technological era.

Of all the offerings of general education in the public schools, industrial arts is unquestionably one of the most appropriate means for developing understandings and appreciations of our industrial environment. Figure 2–5 illustrates how students become acquainted with the industrial nature of American society.

Fig. 2–5. Students in industrial arts learn about our industrial and technological society. (Courtesy, Clausing Corporation)

It must be remembered that industrial arts does not provide specific job or trade training, but provides insights into the processes, tools, and materials of American industry. Industrial arts is designed to provide orientation to modern industrial society and its related problems through informative study and problem-solving experiences.

Industrial arts is for all in our society, because all must learn to adapt to a world of industry and technology. All members of our society must deal with, purchase, maintain, or consume the products of our industrial world. Because of its contributions to the general education of every boy and girl, industrial arts is not considered a single subject or course, but is recognized as an area of the curriculum beginning in the kindergarten and continuing through all grades and culminating in adult education. The educational values of industrial arts cumulate progressively.

Truly, industrial arts can make a unique contribution to one of the prime goals of general education, namely that of transmitting the American way of life.

IMPROVING THE CULTURE

Closely related to the process of transmitting a way of life is the necessity for continuously improving and enriching it. This is more than the mere passing on of what is already known. It calls for dissatisfaction with the present and a desire for better things—such as better living conditions, better tools, machines, and utensils; better recreation; better social relationships. Since American democracy is primarily industrial, progress in this area is essential if our present position of world leadership is to be maintained.

Advancement in industry goes on as the result of many discoveries, both large and small. These discoveries may be made in the great laboratories and research departments of large corporations, or they may be the result of experimental thinking by an unknown, obscure workman. Basically, however, the same process is involved: (1) a problem is recognized, (2) a guess or hypothesis is made concerning the solution, (3) data are gathered, and (4) the hypothesis is tested. If the results support the hypothesis, a discovery has been made and civilization has been advanced.

Analysis of this process indicates that it is merely an application of critical thinking. It *appears, therefore, that the ability of a society to advance and make progress depends largely on the extent to which its members are able and willing to do critical thinking.*

Critical thinking is known by several different names including the ability to think, reflective thinking, problem solving, and the scientific method. As pointed out in Chapter 1, the Educational Policies Commission recognized the ability to think as the central purpose of American education. *The development of the ability to think can well be regarded as the central purpose of industrial arts education.* This ability is an overlapping factor which contributes to the attainment of each of the objectives of industrial arts as well as to each of the objectives of general education.

If reflective thinking is so important, one may well inquire: *How does a teacher develop the ability to think in his students?* Unfortunately, educational research has not yet provided a formula for developing intellectual power. But research findings and the empirical experiences of teachers do provide some guidelines.

Learning to think is a power that develops gradually in a learner. It does not arise suddenly of its own accord or without effort on the part of a student. It seems to develop in proportion to the successful use which a student makes of it. There are no research findings that indicate this power can be developed in any other way.

As cited in Chapter 1, there is no special body of knowledge that will in itself facilitate the growth of the rational powers. Their development depends on teaching methods which encourage the application of knowledge from one context to another. Development depends on the reorganization of things learned with their subsequent reapplication to new situations.

Although subject content itself does not convey intellectual power, it is a necessary ingredient in the developmental process. Subject matter is the raw material of thought and the ability to think cannot be developed without it.

The pupil must have a desire to learn how to think. Motivation of this sort is reinforced by successful experience in reflective thinking. The old adage applies. Nothing breeds success like success.

It should be noted that the first step in the thinking process is the recognition of a problem. It is essential, then, that education continuously face the student with problem situations that require a solution and encourage individual critical thinking. Only under such conditions can maximum progress be made in teaching youth to think and to solve problems.

The implication of this fact for industrial arts teachers is clear. The pupil must be given every opportunity to solve his own problems relating to the planning and making of industrial arts projects. Figure 2-6 shows two students with an industrial arts project which required much planning and problem solving.

It is probably more important, from an educational standpoint, that a pupil be able to plan his project correctly than that he be able to carry it out skillfully. This implies that the teacher must give at least as much attention to helping a student plan his work as to directing his manipulative experiences. Effective instruction in industrial arts will make certain that planning always precedes construction. It is understood, of course, that rethinking and replanning during the execution stage are quite legitimate as new evidence is gathered and as new insights are developed. While a pupil's plan should represent careful thinking and be well executed in terms of his ability level, it need not always be complete in every minute detail before the experience begins. This is especially important in working with younger pupils. The

Fig. 2–6. An example of problem solving in industrial arts. Two Pennsylvania high school seniors constructed this replica of a 1903 Cadillac. (Courtesy, Don DiMarco, Deer Lakes High School, Pa. and *Pennsylvania Industrial Arts News.*)

point is that even after careful thinking and planning have taken place, a plan must not be regarded as unchangeable.

The teacher must supervise carefully the selection of problems. It is important that the pupil select a problem the solution of which will challenge his ability but will not be beyond his capacity. As mentioned previously, facility in critical thinking develops from continued success rather than from the facing of situations for which no satisfactory solution can be found. Thus the teacher must encourage a student in his early efforts to grapple with problems that cause him to think and he must enjoy success in his efforts.

Problem solving is not new in industrial arts because industrial education, together with agriculture, gave education the educational project or, as it is sometimes called, the project method of teaching.[19]

---

[19]John F. Friese, *Course Making in Industrial Education* (Peoria, Ill.: Charles A. Bennett Company, Inc.), p. 19.

The Relationship of Industrial Arts to General Education 29

Directly related to planning and problem solving are the opportunities presented by industrial arts education for thinking in terms of concrete materials. The boy shown in Fig. 2–7 is engaged in this type of activity. Many pupils who find difficulty in developing abstract ideas are able to solve problems readily in terms of tangible materials. For example, many who have been unable to master fractions, as presented in the academic classroom, find little difficulty in measuring accurately the stock for making an industrial arts project.

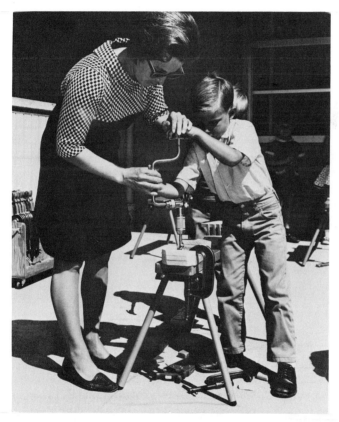

Fig. 2–7. Pupils think more readily in terms of concrete materials. (Courtesy, California State Department of Education)

Since industrial arts offers the opportunity for practice in problem solving and also provides for thinking in terms of concrete materials, it has much to offer to the general objective of teaching pupils to think and thereby tends to improve the emergent culture. Industrial arts can make a distinct contribution to the second major goal of general education, namely, improving our way of life. The strong contribution of industrial arts to this goal is through fostering the growth and development of critical thinking in students, especially as related to the materials, tools, and processes of industry.

MEETING THE NEEDS OF INDIVIDUALS

The importance of meeting individual needs was discussed in Chapter I. Industrial arts activities can make several unique contributions to the achievement of this aim of general education. There are two chief reasons for this. First, the informal nature of the organization of an industrial arts class fosters close personal relationships between the teacher and the students so that the teacher comes to know his pupils intimately and to learn about their problems. This is a prime requisite in meeting the needs of individuals. Secondly, numerous opportunities are afforded in the informal nature of laboratory work to develop those relationships between teacher and students and among students that promote the meeting of individual needs.

There are two groups of needs which can be especially well met by the industrial arts program. These involve (1) personal needs and (2) occupational needs.

1. Personal Needs

Pupils have a need for a feeling of belonging to and being accepted by a group and also they have a need for a feeling of success. The skillful teacher will try to organize his class so that each pupil will have a recognized contribution to make to the group, and so that each pupil will realize that the success of the organization depends not upon the efforts of any one individual but upon the teamwork of all working together. It is much easier for a pupil in an industrial arts laboratory, where students may work together informally, to feel that he is a member of the group and is being accepted by other members of the group than is the case in a formal classroom situation. More important is the fact that each pupil must accept certain responsibilities and that others in the class must recognize and appreciate his contribution.

So far as the need for a feeling of success is concerned, there is probably no other school offering which comes so near to making it possible for all students to attain a fair measure of success as industrial arts. Ideally, each member of the class will be solving a different problem; and if planning and guidance have been wise, each will be working on something within his ability to complete successfully. A pupil's objective is not to keep up with the rest of the class or to make as good a project as the student at the next work station, but is rather to accomplish successfully the task that he has set for himself with the help and advice of the teacher. It should, therefore, be possible for every pupil to feel that he is succeeding and making progress (see Fig. 2–8). *The possibility of promoting this feeling of success for all pupils is of such importance that industrial arts might be justified on this basis alone.*

Fig. 2–8. Pride in accomplishment and a feeling of success are important. (Courtesy, Bragg Stockton, Dallas Independent School District, Dallas, Texas)

An industrial arts program also plays an important part in meeting other needs of youth. Especially pertinent is the necessity for activity which has been amply demonstrated by biological research and child study. As a release from the relative inactivity of the classroom, industrial arts activities serve as both a physical and mental stimulus to a student of any age. The very nature of the work requires active participation in a program which releases the tensions and strains of concentrated attention required by certain learning situations, even in the more active classrooms of the modern school. Apparently there are important therapeutic values to industrial arts activities from the mental hygiene point of view. The experience of the Armed Services and the Veterans Administration in the use of "Manual Arts" activities as a means of relieving emotional strains and promoting mental stability amply demonstrates the effectiveness of such work. Industrial arts activities in the public schools doubtless have an equally important role in promoting the mental health of the pupil.

Opportunity is also provided for the pupil to learn through experimentation in constructive activities. Most students are inherently curious; they appear to have a need to learn about things by handling them and taking them apart. Undirected, this tendency frequently manifests itself in destructive behavior. Under competent guidance this same tendency may be channeled into constructive activity in the industrial arts laboratory.

2. Occupational Needs

Industrial arts activities contribute to the meeting of certain needs of youth in the economic-vocational field. It has been noted that youth have a basic need to feel that they are growing toward a position of economic independence and a place in the vocational scheme of things. An industrial arts program provides tryout opportunities where some of the important occupational fields may be sampled. While little attempt is made in most industrial arts classes to develop salable skills, the student is nevertheless impressed by the fact that he is using the tools and processes of industry and that he is being given an opportunity to study and select from the trades and occupations of adult society.

In the public schools today there is growing emphasis on career education for all students. This movement is aided and abetted by vast sums of money from federal and state agencies. Many innovative programs are being tried out. At both the junior high school and the middle school levels, pupils are being confronted with occupational information, "hands on" experiences, and individual counseling the purpose of which is to orient them to the world of work. Such career exploration for grades five through nine can serve as a steppingstone for some students to programs of vocational-technical education at high school and post-high school levels. It seems likely that the vocational-guidance objectives of industrial arts will receive even greater emphasis in the years immediately ahead.

Industrial arts can contribute to the fulfillment of the third goal of general education, namely, meeting the needs of individuals in the basic aspects of living. In this respect its major contribution is in the areas of meeting the personal and occupational needs of youth.

SUMMARY

Since an important objective of general education appears to be transmitting of a culture or a way of life, it is essential that those charged with the responsibility for this transmission should have a clear conception of the nature of the culture they desire to pass on. An analysis of the American way of life indicates that its main characteristic is that it is democratic. Further study shows it to be highly industrialized and technological. It appears, therefore, that industrial arts should be included in the school curriculum to orient youth to living in this highly industrialized society.

The further objective of improving and extending the culture seems to indicate a need for training in critical thinking and problem solving. While all school subjects should contribute to this end, industrial arts activities seem especially adapted for teaching by the problem-solving technique.

The industrial arts program also may make substantial contributions to the meeting of the basic requirements of individuals in the fields of group status needs, personal needs, and occupational needs.

It appears, therefore, that a close relationship exists between the objectives of general education and the nature of industrial arts. Industrial arts is in an enviable position to make a unique contribution to each of the three basic purposes of general education.

DISCUSSION TOPICS AND ASSIGNMENTS

1. Make a list of what you consider to be the ten most important features of the American way of life.

2. The United States has been characterized as an industrial democracy. List at least ten items or objective data to substantiate this statement.

3. Discuss the importance and social results of the change in the United States from a predominately agrarian culture to an industrial and urban culture.

4. Give specific examples to show how an industrial arts program can contribute to the development of the ability to do critical thinking.

5. Psychologists hold that most human behavior is motivated by a desire for self-esteem. Show how an industrial arts program can contribute to satisfying this basic need.
6. Find three new examples of computer-controlled automated mass production in our industrial world.
7. What unique contributions can industrial arts make to general education?
8. In your own words, justify the inclusion of industrial arts as an integral part of general education.
9. Criticize or defend this statement: Girls should study industrial arts in the public schools.

SELECTED REFERENCES

Arnstein, George E. *Automation—The New Industrial Revolution.* Washington, D.C.: American Industrial Arts Association, Bulletin No. 4, 1964.

Breidenstine, A. G. "Industrial Arts as General-Liberal Education," *Industrial Arts Teacher,* vol. 22, no. 5 (May-June 1963), pp. 10–13.

Brown, Walter C. and Richard M. Boone. "Designing Industrial Education for the 1980's," *School Shop,* vol. XXIX, no. 6 (February 1970), pp. 54–55, 62–63.

Chichura, Diane B. "A Woman's Place Is In the Shop!" *School Shop,* vol. XXXI, no. 4 (December 1971), pp. 32–33.

Conant, James B. *Education in the Junior High School Years.* Princeton, N. J.: Educational Testing Service, 1960.

Cremin, Lawrence A. *Transformation of the School.* New York, N.Y.: Alfred A. Knopf, 1961.

DeVore, Paul W. *Technology—An Intellectual Discipline.* Washington, D.C.: American Industrial Arts Association, Bulletin No. 5, 1964.

―――. "Perspective," *School Shop,* vol. XXVII, no. 5 (January 1969), p. 17.

Friese, John F. *The Role of Industrial Arts in Education.* (published privately by the author) 1964.

―――. *Course Making in Industrial Education.* Peoria, Ill.: Charles A. Bennett Company, Inc., 1966.

General Education in a Free Society. Report of the Harvard Committee. Cambridge, Mass.: Harvard University Press, 1945.

Hackett, Donald F. "Study of American Industry Is Essential to Liberalizing General Education," *Industrial Arts and Vocational Education,* vol. 53, no. 4 (April 1964), pp. 25–28, 70.

Hornbake, R. L. "What is the Place of Industrial Arts in the American Culture?" *Improving Industrial Arts Teaching.* Conference Report, June 1960. Washington, D.C.: U.S. Department of Health, Education, and Welfare. OE-33022. Circular No. 656, pp. 2–10.

Johnston, W. L. "Interpreting Industrial Practice in the Modern I-A Lab," *School Shop,* vol. XXIX, no. 4 (December 1969), pp. 36–37.

Kranzberg, Melvin. *Technology and Culture: Dimensions and Exploration.*

Washington, D.C.: American Industrial Arts Association, Bulletin No. 6, 1964.

Spence, William P. "The Vital Role of Industrial Arts," *School Shop*, vol. XXV, no. 9 (May 1966), pp. 40–41.

UNESCO, "Education in a Technological Society," published by the United Nations Educational, Scientific and Cultural Organization. Paris, France: UNESCO, 1960, p. 36.

Venable, Tom C. "Industrial Arts and the Central Purpose of American Education," *Journal of Industrial Arts Education*, vol. 24, no. 1 (October 1964), pp. 23–26.

Chapter 3

Industrial Arts in Our Schools

In seeking to describe industrial arts education in today's public schools, this chapter will present the following: (1) a nationwide picture of industrial arts as revealed by a recent national survey, (2) industrial arts as a curriculum area extending from kindergarten through programs of adult education, and (3) the areas of industrial arts and their organization.

A NATIONWIDE PICTURE OF INDUSTRIAL ARTS

In 1966 the Office of Education published the results of a national survey on the status of industrial arts education.[1] This investigation was the first of its kind ever attempted in the industrial arts field. The study was based on a stratified sample of selected schools in the United States on the basis of school size and type. The ten enrollment sizes of schools ranged from 1 to 99 through 2,500 and over. The four types of schools were: (1) junior-senior high schools (including 5-year and 6-year high schools); (2) junior high schools (including 2-year and 3-year schools); (3) traditional high schools (4-year schools preceded by 8 years of elementary schools); and (4) senior high schools (3-year and 4-year schools, preceded by a junior high school).

Questionnaire returns were recieved from the public schools (grades 7 through 12) from every state. Two questionnaires were employed in the study. One form was sent to 2,259 principals with a return of 95.1 percent while the second form was sent to 3,040 selected industrial arts teachers with a return of 93.7 percent.

[1] Marshall L. Schmitt and Albert L. Pelley, *Industrial Arts Education—A Survey of Programs, Teachers, Students, and Curriculum,* U.S. Department of Health, Education, and Welfare, OE-33038, Circular Number 791 (Washington, D.C.: Government Printing Office, 1966).

In the space available here it is impossible to report all of the findings of this survey, however, selected summary statements are reported under the following headings.[2]

1. Sizes and Types of Schools in Which Industrial Arts Is Found. Three-fourths of the American public secondary schools have industrial arts programs. Virtually all schools with enrollments of 2,500 students or more offer industrial arts. In general, as the size of the school decreases, the percentage of schools offering industrial arts decreases also. Nevertheless, more than 40 percent of schools with enrollments of fewer than 100 students have industrial arts programs.

The type of school seems to be related to the presence of industrial arts programs, but this relationship is not so influential as is the size of the school. More than two-thirds of the junior-senior high schools have industrial arts programs. Over 90 percent of the traditional high schools have industrial arts as do 83 percent of the junior high schools. The high percentage of industrial arts noted in the traditional high schools and in the junior high schools is probably due to the widespread acceptance of the exploratory functions which industrial arts performs at the beginning levels of the secondary school.

Of a total of 5,823 schools which currently do not have industrial arts facilities, 1,058 are planning to have them within three years following this study.

2. The Instructional Program. Most industrial arts courses are 30 or more weeks in session or a full year in length. Classes generally meet five days a week for one period a day. From 25 to 30 percent of the instructional time in industrial arts is devoted to theoretical or related instruction; conversely, 70 to 75 percent is devoted to laboratory activities.

The subject-matter content in most industrial arts programs is concentrated in three areas: woodworking, drafting, and metalworking. Analysis of the instructional content shows that drawing is an important part of industrial arts instruction. In fact, the high percentage of drawing or planning activities in most industrial arts courses reveals the interrelationship of problem solving and its practical application in each course. It is the combination of these two aspects which makes an industrial arts activity a real creative act, and the two processes reinforce one another.

3. Industrial Arts in the Elementary School. The study shows that 5.6 percent of the schools reported that industrial arts was required in all of their elementary schools while an equal number (6 percent) re-

[2] *Ibid.*

ported that it was required in some elementary schools. In addition, another 8 percent of the schools reporting indicated that although industrial arts was not required in any elementary school, some schools did have it.

4. The Enrollment. Total enrollment in industrial arts courses (boys and girls) in grades 7 through 12 is nearly 4 million students. Enrollments are heaviest in the junior high school and in schools whose enrollment size is over 1,000 students. The highest percentage of students is found in ninth grade where 24 percent of the enrollment is reported. More girls are enrolled in crafts courses than in any other kind of industrial arts course.

During the period 1954 to 1963 there has been a gradual increase in the compulsory requirements for both boys and girls to take industrial arts. Although these gains were not large, they are significant because the national concern at that time focused on meeting the needs of the academic student. Evidently, industrial arts instruction is recognized as meeting some of the basic educational needs of all students, academic and nonacademic alike.

5. The Industrial Arts Teacher. The study reveals that there are 40,428 industrial arts teachers in the United States. Little variation exists as to the number of teachers by type of school since the size of the school is the deciding factor.

Nearly 95 percent of the industrial arts teachers in the public secondary schools hold a professional teaching certificate. About 60 percent have a bachelor's degree and 35 percent have a master's degree.

The average number of years of teaching experience for industrial arts teachers is 9.5 years, with the larger schools attracting the more experienced teachers. The average number of years of industrial experience or its equivalent is 5.6 years.

Almost two-thirds of the teachers teach industrial arts classes only. However, as the size of the school decreases, industrial arts teachers are called upon to teach nonindustrial arts classes, such as general science, biology, social studies, geometry, and algebra.

Although some public school systems offer work-experience programs, only 15.9 percent of the industrial arts teachers administered the in-school, nonremunerative, general education work-experience programs. Less than 4 percent of all other types of in-school work experience education programs are administered by industrial arts teachers, except in the small school, where this figure rose to about 20 percent.

INDUSTRIAL ARTS—A CURRICULUM AREA OF EDUCATION

It has been stated in the preceding chapter that industrial arts is not a single course nor a subject studied at one grade level. Rather, industrial arts is an area of the curriculum in the public schools beginning in the kindergarten, continuing through all grades, and terminating in programs of adult education. In this respect, industrial arts is comparable to the language arts in our schools.

In this section brief descriptions of industrial arts will be given for the following levels: elementary, middle grades, junior high, senior high, and adult.

1. Industrial Arts in the Elementary School. In 1971 a group of twenty-one educators attended a National Conference on Elementary School Industrial Arts which was made possible by a grant from the Office of Education with the cooperation of the American Council for Elementary School Industrial Arts. The group formulated this definition for elementary school industrial arts:

> Industrial arts at the elementary school level is an essential part of the education of every child. It deals with ways in which man thinks about and applies scientific theory and principles to change his physical environment to meet his aesthetic and utilitarian needs. It provides opportunities for developing concepts through concrete experiences which include manipulation of materials, tools, and processes, and other methods of discovery. It includes knowledge about technology and its processes, personal development of psychomotor skills, and attitudes and understandings of how technology influences society.[3]

Teachers at the elementary level have been using industrial arts activities for a great many years. Some teachers use the term *industrial arts* to describe the efforts and experiences of their pupils, while others use such titles as integrated handicrafts, practical arts, handwork, or construction-type activities. Regardless of the name, however, first-hand experiences with tools, materials, industrial concepts, and processes constitute an important element in the common learnings of all elementary pupils from kindergarten through grade six. Gerbracht and Babcock list the following contributions of elementary school level industrial arts to the learning process:

1. Acquaint children with the industrial-technological aspects of their culture.

---

[3]William R. Hoots, Jr. (ed.), *Industrial Arts in the Elementary School: Education for a Changing Society*, National Conference on Elementary School Industrial Arts (Greenville, N.C.: East Carolina University, 1971), p. 3.

2. Reduce levels of abstraction, develop adequate meanings, and enrich the curriculum.
3. Provide for individual differences.
4. Provide socializing experiences.
5. Supply the beginnings of occupational information.
6. Make school and learning enjoyable.
7. Motivate learning.
8. Teach fundamental manual-mechanical skills.[4]

Figure 3-1 shows three elementary school pupils engaged in constructive activity.

Fig. 3-1. Learning to work with tools and materials is important in building a child's understanding of his world. (Courtesy, William R. Hoots, Jr., East Carolina University)

Currently there are three schools of thought on the place of industrial arts or construction-type activities in the elementary school. One viewpoint holds that industrial arts is a body of subject matter which is to be taught in the same manner as other subjects, such as English, history, mathematics, and so on. It is the opinion of this group that industrial arts has objectives similar to those in the middle school or the junior high school, except, of course, that the informative and manipulative learning experiences are scaled to fit younger pupils.

The second philosophic viewpoint is that industrial arts is an activity

[4]Carl Gerbracht and Robert J. Babcock, *Elementary School Industrial Arts* (New York: The Bruce Publishing Company, 1969), p. 24.

method of teaching. As such, it has no informative content to be taught. There are no preplanned manipulative experiences nor any standard projects to be made. The use of tools and materials is confined to the solving of problems arising from experiences in the "regular" school subjects. In other words, at this level industrial arts is conceived to be a means or method by which the elementary teacher is better able to achieve her goals. Thieme states it this way:

> ... it serves the whole school to do better the things the school is already trying to do by using the activities of elementary industrial arts to enrich, broaden, and illustrate classroom units of instruction.[5]

The third school of thought expresses a compromise point of view in which industrial arts is regarded as both content and method in the elementary school. Proponents of this viewpoint state that industrial arts activities should be carried on in kindergarten through grade three as an integral part of the learning process. No special time would be allotted for industrial arts since construction-type activities would simply be used when appropriate to the learning experiences. In grade four, individual and small-group projects which are integrated with classroom activities would continue. Projects such as dioramas and models would lend realism and vitalize regular classroom experiences. Then as students approached the upper elementary level, increasing amounts of time would be made available for manipulative learning experiences on an individual basis to meet pupil needs and interests. Figure 3–2 shows three upper elementary students researching material which correlates industrial arts and social studies.

While it cannot be denied that diverging philosophies do exist, there seems to be unanimity of opinion that industrial arts does have a vital role to play in the elementary school. Gerbracht and Babcock express it well when they state:

> Industrial arts activities will not answer all the needs of elementary education, but neither can all these needs be met without industrial arts. Industrial arts activities are necessary in a complete program.[6]

The implementation of industrial arts experiences in the elementary school usually is the responsibility of the classroom teacher. Some classrooms are of the self-contained type and are equipped with benches, tools, storage racks, and other equipment necessary to carry on

---

[5]Eberhard Thieme, "Pupil Achievement and Retention in Selected Areas of Grade Five Using Elementary Industrial Arts Activities Integrated with Classroom Units of Work" (Unpublished doctoral dissertation, The Pennsylvania State University, 1965), p. 49.

[6]Carl Gerbracht and Robert J. Babcock, *Industrial Arts for Grades K–6* (Milwaukee, Wis.: The Bruce Publishing Company, 1959), p. 1.

Fig. 3–2. These girls are busily engaged in a multiple motivation project called the "Living Crafts Program" for grades five through eight. (Courtesy, James O. Reynolds, Dayton Public Schools, Dayton, Ohio)

construction-type activities with groups of pupils. This arrangement is ideal because the teacher and one or more pupils may engage in construction activities whenever the need arises without even leaving the classroom. Instruction may be carried on by the classroom teacher who has acquired the necessary competencies for working with tools and materials. Or, the teacher may be assisted by an industrial arts specialist, consultant, or supervisor who makes scheduled and on-call visits to classrooms to assist in the learning process.

A second plan of implementation involves a central laboratory facility within the school which is staffed by an industrial arts teacher. Under this plan students are sent to the laboratory to engage in construction-type activities under the direction of the industrial arts teacher. It is necessary, of course, for the classroom teacher and the industrial arts teacher to work closely together. Obviously, if facilities are limited, some classes may have to wait their turn to use the laboratory. Such delay may well impede the progress of instruction. The solution to this problem is simply to provide additional laboratories to meet the needs of the teachers and their pupils.

The third plan is similar to the second one except that the laboratory is a mobile unit housed in a truck, van, or trailer. On arrival at a given school, the tools, supplies, and equipment may be unloaded and used in an all-purpose room of the school. Or the mobile unit itself may be used as the laboratory. Commercially-produced mobile units are now available which provide a complete array of tools, supplies, and equipment in a self-contained air-conditioned trailer approximately 10 feet by 30 feet. Mobile units are manned by industrial arts specialists who assist the classroom teachers in providing construction-type activities. A mobile laboratory unit is especially valuable in servicing schools in rural areas where small pupil enrollments may prohibit the establishment of more elaborate facilities or where the school district is simply unable to purchase the necessary tools and equipment.

2. Industrial Arts in the Middle School. In recent years a new concept in education called the Middle School has been gaining such prominence that it may well be indicative of an emerging trend in American education. Although the literature abounds with the theory and philosophy of middle grades education, implementation at the local level is proceeding at a somewhat slower pace. Currently the Middle School is receiving much attention from local, state, and national educational agencies as a fertile site for developing programs of career education. Certainly the idea of career exploration fits well into the philosophy of the Middle School. It seems likely at this point that the Middle School with its growing emphasis on career education may well develop into a dominant force in American public school education.

As yet no standard organizational pattern for the Middle School has developed largely because educators are not quite agreed on this point. Some plans call for a Middle School with two grades while others foresee a school with three or even four grades. Educators are agreed, however, that the bounds of the middle grades range from grades 5 through 8; typical grade combinations seem to be 5–6–7, 5–6–7–8, 6–7–8, or 7–8. To help clarify the situation one state department of education has defined a Middle School as any combination of three consecutive grades from 5 through 8. Thus for the educational system in this state, the Middle School may consist of either grades 5–6–7 or grades 6–7–8. Most authorities seem in agreement that a three-year school is needed to facilitate the transition of boys and girls from the elementary grades to the departmentalized organization at the high school level.

The exact role that industrial arts will play in the Middle School is not yet clarified nor defined. Experts in general education and in industrial education are agreed, however, that industrial arts has unique educational contributions to make and must be included in any complete program for the middle grades. Industrial arts educators have devised numerous innovative programs, but no standard pattern has emerged. Currently there are several points of view concerning industrial arts in middle school programs. Some educators recommend that the present junior high school industrial arts program be moved into grades 6–7–8; others state that elementary school industrial arts activities of a more advanced nature should be prepared for these grades. A third group perceives a combination of industrial arts, fine arts, and home economics into a unified arts program. A fourth group envisions industrial arts as one of a number of disciplines contributing to as many as 15 occupational clusters comprising the program of the middle school which emphasizes career education. In this case, industrial arts (like the other contributing disciplines, e.g., fine arts, home economics, et al.) may lose its identity as industrial arts, but would be closely related to one or more occupational clusters, for example, manufacturing, construction, communications, transportation, and so on.

In seeking to provide direction for the industrial arts field, the Council of Industrial Arts Supervisors of the American Industrial Arts Association prepared this statement of philosophy:

> The industrial arts program is concerned with common learning needed by all persons to function effectively in our industrial-technological society; the development of attitudes, interests, abilities, and skills, as well as the acquisition of information about occupations and professions. The program includes problem-solving techniques, an activity approach, the interpretation of humanistic values, appreciations and understandings, and provisions for essential individual success.[7]

In continuing its appraisal of the role of industrial arts in the middle grades, the Council offered the following suggestions for implementing an instructional program:

> Teachers work with youngsters in the laboratory, construct items related to other units of study.
>
> Units of instruction with commonalties, such as design, funnel through the unified arts areas.
>
> Introductory units designed for broad overviews in areas such as industry, research, and technology are used to study history, relationships between subjects, and characteristics of society.
>
> Units of instruction that couple study and application provide information relating to occupations and professions.
>
> Mass-production methods and industrial organization techniques are utilized.

---

[7] *Industrial Arts in the Middle School,* Prepared by the American Council of Industrial Arts Supervisors, American Industrial Arts Association (Washington, D.C.: American Industrial Arts Association, 1971), p. 3.

Separate courses are offered in (1) communications (drawing and design, graphic arts or visual graphics, and photography); (2) construction and manufacturing (wood and noncellulose material and metals); and (3) power (sources, transmission, and utilization).[8]

Whichever direction industrial arts is to take in the middle school, it seems certain that this will become in the near future one of the most exciting areas of education for all teachers. It behooves the industrial arts teacher to accept the challenge of middle grades education and to help these programs develop across the nation.

3. Industrial Arts in the Junior High School. The rise of the junior high school movement in this country and the recognition of the general educational values of industrial arts occurred at approximately the same time, that is, shortly after the turn of the century. Ever since then industrial arts has found a natural home in the junior high school largely because of the common exploratory characteristics which each possesses. At the junior high school level industrial arts is regarded as an essential ingredient in the educational program of every boy and girl. In many states all boys are required to take industrial arts and a growing number of schools are now making industrial arts activities available to girls as well. As mentioned above, the industrial arts program is exploratory in nature in keeping with the general character of the junior high school. Thus, a wide range of learning experiences with the tools, materials, and processes of industry is offered in a well-rounded program. In other words, breadth rather than depth is emphasized at this level.

Since the keynote of the program is exploration, its worth is increased proportionately as its range of activities is broadened. A well-balanced program will include learning experiences in construction, manufacturing technology, visual and electronics communications, and power technology. Figure 3–3 shows several junior high school pupils gaining first-hand experiences in the world of building construction.

From the foregoing it is obvious that industrial arts is a vital component in junior high programs of career education with special contributions to the occupational clusters of manufacturing, construction, communications, and transportation.

4. Industrial Arts in the Senior High School. The role of industrial arts at the senior high school level never has been so clearly defined as it has in the junior high school. Mild uncertainty may exist even among experienced industrial arts teachers as to the nature and purpose of a a program for senior high school students. There are several tenets, however, on which a sound industrial arts program for these grade levels can be predicated. These are: (1) that every senior high school

---

[8] *Ibid.,* p. 6.

Fig. 3-3. These junior high school boys are learning about occupations in the world of construction. (Courtesy, Warden B. Muller, Reynolds Junior High School, Trenton, N.J.)

boy and girl should have the opportunity to elect industrial arts regardless of educational goal, (2) that industrial arts has educational values for every high school student, be he academically talented or less gifted, and (3) that industrial arts courses at the senior high school should neither duplicate nor repeat those in the junior high school.

In times past many high school courses were not unlike those at the junior high school level. Some teachers offered essentially the same course to all students. It is now recognized that a single type of industrial arts course can no longer meet the needs of senior high school pupils. It is important, however, that an opportunity to take industrial arts should be made available to every student regardless of his major area of interest. This means that several different kinds of industrial arts courses, each with distinct and separate objectives, are needed to meet the diverse needs and interests of today's students. For this reason, industrial arts in up-to-date high schools is being developed on a multitrack plan with at least three major emphases. These include preprofessional, pretechnical, and free elective.

1. Preprofessional. Senior high school industrial arts with a preprofessional emphasis is designed for college-bound students who are planning to enter the professions, such as architecture, engineering, journalism, law, medicine, or science. For example, a preengineering student has much need of experience in mechanical drawing and design as well as a working knowledge of industrial materials, tools, machines, and processes. Advanced industrial arts coursework, such as visual communications, manufacturing and construction technology, and power technology, can lay sound foundations for future undergraduate college work in these areas. To illustrate, the skills and knowledges gained in an advanced high school course in architectural drawing will assist the student shown in Fig. 3–4 when he continues this work in college. Certain colleges now offer exemptions from freshman drawing courses if students score well on placement tests; such students begin their college work in engineering graphics at more advanced levels.

Fig. 3–4. Industrial arts drawing helps lay the foundation for a professional career. (Courtesy, Bragg Stockton, Dallas Independent School District, Dallas, Texas)

In addition, the college-bound student, especially the future engineers, scientists, and architects, need opportunities under the guidance of a skillful industrial arts teacher to engage in individual research and experimentation with problems of a technical nature. The industrial arts drawing room and laboratory can serve well as a proving ground for the application of science and mathematics in the student's special area of interest.

One of the greatest services which industrial arts can offer any college-bound student is to assist in the development of his rational powers as an individual. This, it will be recalled from Chapter 1, is the central purpose of American education. The development of the ability to think is not dependent upon the type of subject matter to be studied, but rather upon the method of teaching employed. In other words, critical thinking can be developed by the study of academic, aesthetic, or practical subject matter. Thus, the ability to think can be developed through problem-solving experiences with the tools, materials, and processes of industry. The development of the power of critical thinking will immeasurably assist all college-bound students regardless of their ultimate professional goal.

It is obvious that industrial arts cannot be scheduled for an incidental period or two during the week, if students are to achieve its full and unique benefits. Accordingly, industrial arts classes for preprofessional students should meet daily and receive a full unit of credit in the same manner as academic coursework.

2. Pretechnical. Senior high school industrial arts with a pretechnical emphasis is intended for those students who are not college bound. A few of these students may decide later to begin college; some may enter programs of vocational-industrial-technical programs either at the high school or posthigh-school levels; still others may enroll in different types of formal training. However, most students in this category will seek employment following graduation. Since they are undecided about their future career at this time, these students probably are enrolled in the general curriculum.

The industrial arts experiences for the pretechnical group should be occupationally oriented. Emphasis should be placed on occupational information supported where possible with work-study experiences. The laboratory work should emphasize the development of skill in the use of tools and machines, measuring devices, and testing equipment. Specialized experiences in power technology, construction technology, manufacturing technology, or visual communications may assist the student in reaching a career choice. Such coursework, especially if part of a work-experience program, would be of invaluable assistance in helping pretechnical students make a successful transition from school to vocational-industrial-technical programs to the workaday world.

A special sequence of industrial arts activities should be organized for students who are slow learners and who will probably drop out of school. Occupational training and counseling should be provided for such students in selected lower-level service occupations.

3. Free Elective. In addition to providing special services for preprofessional and pretechnical groups, industrial arts should be available to any high school boy or girl as a free elective. Some students enrolled in nontechnical curricula often find that studies in their major area are demanding in terms of time and effort required. For students like these, the avocational aspects of industrial arts will provide a change of pace, a fresh outlook, and worthwhile relaxation. A wide array of industrial arts offerings should be available to meet individual needs and interests of these pupils. Experiences in industrial crafts and home mechanics may be especially appropriate for some of these pupils.

In describing a free elective industrial arts program, Weber states:

> Whereas one boy or girl might become involved in a highly technical program in electronics or rocketry, another may choose to develop the degree of craftsmanship necessary to do fine silversmithing. The program should be broad enough in concept and extensive enough in facilities to allow for individual study, and group projects; research and experimentation, and skilled craftsmanship; the solution of technical problems, and the creation of beautiful objects; an enlarged understanding of scientific concepts and a heightened appreciation of things aesthetic.
>
> The possibilities of such a program of industrial arts is limited only by the imagination of the teacher. This would truly be industrial arts for general education. Such a program would surely help considerably to develop in each pupil the skills of: (1) flexibility, (2) creativity, (3) making wise choices, (4) learning to know one's self, (5) understanding others, and (6) the skill of learning.[9]

5. Industrial Arts in Adult Programs. It is estimated that nearly 35 million adults enroll annually in organized educational activities; this means that approximately 25 percent of our adult population is involved in some form of continuing education or avocational programs of an educational nature.

Industrial arts activities are among the most popular of all offerings in recreational and avocational programs for adults in the community. Classes are often oversubscribed because adults like to use the well-equipped industrial arts laboratory as their home workshop. Today the typical American believes in "do-it-yourself" activity. He enjoys constructional activities not only for the monetary savings involved, but also for the personal satisfactions gained in creating some desirable or needed object for his home or family.

Adult classes are generally held once or twice weekly in late afternoons or evenings. Usually an adult enrolls in an industrial arts course with a definite purpose, problem, or project already in mind. For exam-

---

[9]Earl M. Weber, *The Role of Industrial Arts in Tomorrow's Schools* (Washington, D.C.: American Industrial Arts Association, Inc., 1966), p. 12.

ple, he may wish to construct a piece of furniture, build a boat, repair his lawnmower, sharpen up a trade skill, or as shown in Fig. 3–5 simply try some new manipulative experience for avocational purposes.

Fig. 3–5. Woodturning is a fascinating pastime which provides this adult much personal pleasure and satisfaction. (Courtesy, William J. Wilkinson, Nether Providence High School, Wallingford, Pa.)

From this discussion it is not meant to imply that only men are interested in adult classes in industrial arts. On the contrary, great numbers of women enroll as well. Popular shop activities for the fair sex are crafts, wood refinishing, and upholstery.

Informal instruction is the general rule in adult classes, except when a beginning group can be treated as a whole, e.g., a class in blueprint reading. The industrial arts teacher will soon discover that the abilities and interests of adults are as varied as his regular pupils. Thus, highly individualized teaching also characterizes adult class meetings. The teacher usually serves as a resource person or consultant to the experienced worker while reserving more formal teaching for those with little or no experience with tools, materials, and processes.

It must be remembered, too, that adult classes constitute an excellent public relations vehicle for the industrial arts program. Adult classes can contribute significantly to upholding strong public support for industrial arts as well as fostering and maintaining active interest in the school itself. More will be said about industrial arts and public relations in a later chapter.

INDUSTRIAL ARTS AT THE COLLEGE LEVEL

This chapter was purposely limited to a consideration of industrial arts in today's public schools. However, passing reference should be made to programs at both the two year community college and the four year college and university levels. Here industrial arts performs the following academic and professional services.

1. Teacher Education

Industrial arts teachers are prepared via four year programs leading to the baccalaureate degree and certification to teach in the public schools. Industrial arts education is also recognized as a major field of graduate study leading to both the master's and the doctor's degrees.

2. Service Courses

Selected industrial arts laboratory and professional courses are offered on a service basis for students enrolled in certain curriculums, such, as, engineering, elementary and special education, art, journalism, and so on. Industrial arts is often made available as a minor field of study for students majoring in academic areas of the liberal arts as well as for majors in education, particularly art, special, or elementary subjects. At some institutions a core of industrial arts coursework is an integral part of the curricular requirements for elementary education and special education majors.

3. Free Electives

A number of colleges and universities offer industrial arts courses as part of general education for any student on a free-elective basis (see Fig. 3–6). These courses may be selected from the regular laboratory and professional offerings of the industrial arts department. Some institutions prepare special courses designed to help students become aware of our industrial and technological society. One institution prepared a special course titled "Modern Technology and Civilization." Initially the course opened with an enrollment of 11 students; five years later the enrollment was 2,000 students in one academic year alone.[10]

Fig. 3-6. These college-level students are becoming acquainted with some of the electronic technology of our society. (Courtesy, Philco Corporation)

THE SUBJECT AREAS OF INDUSTRIAL ARTS

It has been noted previously that (1) the central purpose of education is to develop in students the ability to think and (2) this ability cannot be developed without subject matter which serves as a necessary vehicle of implementation. This section is devoted to a listing of the major subject areas of industrial arts together with brief descriptions of their component subdivisions.

[10]Donald P. Lauda and Robert D. Ryan, "Industry 192," *The Journal of Industrial Arts Education*, vol. XXVIII, no. 3 (January-February 1969), p. 14.

Like the world of industry which it seeks to interpret to youth, industrial arts is comprised of many varied and interesting subject areas. There is more than sufficient content in each area and subdivision for coursework ranging from beginning or exploratory levels through advanced levels. Thus, it is not uncommon to find a sequence of courses in one area, for example, Woodworking I, II, and III. The breadth of industrial arts comes from its varied areas; its depth comes through sequential coursework in a single area.

Traditionally, industrial arts laboratories have been organized on a materials-centered basis, such as, woodworking, sheet metal, plastics, etc. Today there is a trend toward new terminology and the combination of some of these materials into new categories; reference will be made to this point at the end of this section. However, the following seven headings constitute the building blocks from which are drawn the major learning activities in industrial arts education.

1. Woodworking

Historically, instruction in industrial arts centered around woodworking and even today more students are engaged in woodworking activities than in any other subject area in industrial arts. Subdivisions of woodworking include carpentry, cabinet and furniture making, wood turning, wood finishing, pattern making, and upholstery.

2. Drawing

The universal language of industry is drafting. It is not surprising, therefore, to discover that this is one of the most common and most important subject areas in industrial arts. Some form of drawing is utilized in almost every materials area of industrial arts if only to the extent of simple planning, sketching, or blueprint reading. In addition to these informal applications of drawing, there are also a variety of specialized, formal courses including mechanical drawing (engineering graphics), electrical drafting, machine drafting, architectural drawing, sheet metal drafting, topographical drafting and structural drafting.

3. Metalworking

The fact that metalworking is the backbone of American industry simply underscores its importance in industrial arts education. Metalworking courses in industrial arts include student experiences with a wide variety of materials and with many industrial processes. Subdivisions of metalworking include metal machining, welding, casting, heat treating and finishing, sheet metal forming (fabrication and a spinning), and forging.

4. Electricity/Electronics

This area of industrial arts seems to divide itself naturally into two divisions. *Electricity* is concerned with magnets, power sources, circuitry, switch equipment, servicing, and load devices, such as motors, lights, and heaters. *Electronics* is concerned with communications, such as radio and television; computers (digital and analog); electronic controls, including smoke and heat detectors; and other electronic equipment using vacuum tubes and semi-conductors.

5. Graphic Arts

Printing is a basic means of communication, and without it our society would collapse like a house of cards. For this reason printing, or rather graphic arts—a newer and broader term—is an important major area of industrial arts. The subdivisions of graphic arts center around the four basic means of graphic reproduction: letterpress printing, lithography, gravure, and screen process printing. In addition, there are a variety of related activities, such as linoleum or wood-block cutting, intaglio, thermography, rubber stamp making, stencil cutting and duplicating processes, bookbinding, paper-making, and photography.

6. Industrial Crafts

In industrial arts crafts courses, the emphasis should be on the word *industrial* which thus differentiates them from the arts and crafts courses in another discipline in which creativity and aesthetic values predominate. Some of the subdivisions of industrial crafts include ceramics, glass, fibcrglas, plastics, leatherworking, lapidary, model making, jewelry making, and textiles. Any of these subareas may in itself become a course or a series of courses from exploratory through advanced levels. For example, in plastics, students may use a variety of materials including polystyrene, nylon, vinyl, ABS, acrylic, and phenolic, as well as polyester and epoxy resins. These and other materials may be used in industrial processes such as *molding* (extrusion, rotational, and compression); *laminating* (pressure and reinforced plastic); *thermoforming* of sheet material (vacuum forming, plug and ring forming, and blow forming); *foaming processes; coating; casting* and *encapsulating*. It should be obvious from this partial analysis that the subarea of plastics surely contains adequate content of both an informative and a manipulative nature for a beginning and advanced coursework.

7. Power Technology

As an area of industrial arts, power technology is the study of man's attempts to harness wind, water, steam, electrical, solar, and atomic power to his needs. Sometimes the term *transportation* is used in connection with this area. Subdivisions include student experiences with steam, gasoline, and diesel engines; electrical generators and motors; hydraulics and pneumatics, as well as devices to transmit power.

A CHANGE IN TERMINOLOGY

As previously mentioned a change in terminology is becoming evident throughout the field. Instead of woodworking or metalworking, some now use the terms *wood technology* or *metals technology*. A machine shop has become a metals removal laboratory, foundry is called metals casting, and welding is called metals joining.

A more basic change, however, involves the combination of some of the materials areas into new categories with new descriptive titles. Some of these new groupings are (1) construction, (2) manufacturing or production, (3) visual communications, and (4) power technology. Construction and manufacturing (production) include woods, metals, plastics, ceramics, industrial crafts, and other selected materials. Visual communications includes all forms of mechanical drawing, sketching, technical drafting, blueprint reading, photography, and all of the graphic arts. Power technology includes the study of power, generation and transmission, fluidics, power mechanics and transportation, as well as electricity and electronics.

Certain of these new combinations, especially at the junior high school level, involve changes far greater than terminology. The construction laboratory is concerned with the theory and practice of construction in general including small homes, public buildings and community planning as well as highways, bridges, dams, and tunnels. Manipulative experiences in mock-up home construction may include masonry, carpentry, plumbing, electrical installations, and model making. The production laboratory emphasizes the development of concepts and understandings of industrial management and automation as well as first-hand experiences with line production.

Regardless of terminology, approach, or method used, the seven materials areas noted herein comprise the bulk of the subject matter content for industrial arts in our public schools. It is from these areas of industry and technology that the informative and manipulative learning activities evolve. Through problem-solving experiences with the tools, materials, and processes of industry, students will develop their powers of critical thinking which is the central purpose of both industrial arts and American education.

TYPES OF LABORATORY ORGANIZATION

The subject matter content of industrial arts is commonly organized under three main types of shop organizations. These are (1) the unit shop, (2) the compreshensive general shop, and (3) the general unit shop. The plan of organization utilized in a particular school depends on a number of factors including: (1) numbers of students and grade levels to be served, (2) program objectives, (3) staff and laboratory facilities, and (4) financial support available.

1. Unit Shop

A unit shop is one in which the students are engaged with the tools, processes, materials, and informative aspects of a single trade or occupational area. Examples of unit type organization include a machine shop, a sheet metal shop, a carpentry shop, a television repair shop, and so on. Vocational-industrial education has always utilized unit shops because this type of organization lends itself well to vocational methods, procedures, and objectives. Since the range of work is usually limited to one industrial material (e.g., wood, sheet metal, plastic) much duplication of equipment is necessary to provide sufficient work stations for an average-size class. Also, the restrictions of one industrial material limits the possibilities of using the project method as the vehicle of instruction. For unit shop teaching the preparation of the teacher requires concentration in one field. Figure 3-7 shows a unit shop in operation.

It is, of course, impossible to group all unit shops together and attribute to them certain practices or methods of instruction. They differ among themselves almost as widely as they differ from general shops. They share, however, certain features in common which merit their designation as belonging to this type of organization. Admitting variations and exceptions, these common characteristics are as follows:

    1. The course of study is likely to be based on a trade-analysis approach—that is, major stress is likely to be placed on the covering of certain predetermined processes, operations, and information.

    2. Skill is likely to be stressed much more strongly than other objectives.

    3. Prevocational values are apt to predominate.

Fig. 3-7. A unit machine shop. (Courtesy, Clausing Corporation)

4. A uniform course of study for all students is common with teacher reliance on written instruction sheets.

5. A set series of projects is likely to be required from all students.

6. Higher standards of workmanship are likely to prevail than in the case of the general shop, one reason for this being that the teacher is usually an expert in a particular field.

Although historically the first industrial arts shops were of the unit type, this form of organization has gone out of favor today because it seldom adequately fulfills the broader objectives of industrial arts education. Unit shops are still found in certain senior high schools and in large city systems where heavy student enrollment makes it feasible to develop a large number of unit shops with a specialized teacher for each.

2. Comprehensive General Shop

With the evolution of the junior high school there developed a demand for a less specialized organization of industrial arts activities which would meet the exploratory needs of adolescent students. As a result, the comprehensive general shop began to find a place in American education. This type of organization is designed to provide instruction in a variety of different activities carried on in unrelated, separate work areas within the same laboratory. Thus a comprehensive general shop may have a woodworking area, a metals area, an electrical area, a crafts area, a graphic arts area, and so on. Figure 3–8 shows one corner of a general shop in which may be seen metal machining, bench metalwork, materials testing, and electricity/electronics. Student activities go on simultaneously in all areas or in a selected number of areas. The intent of the comprehensive general shop is to provide breadth instead of depth as in the unit shop. This type of organization lends itself well to the research and experimentation approach to industrial arts.

Fig. 3–8. Metals technology and communications are studied in this corner of a general shop. (Courtesy, Clausing Corporation)

Because it is an economical plan for providing a wide number of diversified activities in one room, the comprehensive general shop is well adapted to one-teacher situations, that is, where there is only one industrial arts teacher in the school. Since one teacher directs a number of different activities, he must be broadly educated and technically capable in a variety of fields.

A general shop differs from a unit shop by having activities in two or more materials areas carried on simultaneously. The number and variety of activities may range all the way from a woodworking shop which also offers some mechanical drawing to a laboratory which offers experiences in six or eight different areas. Many combinations of areas occur within the general pattern and wide variation may be noted in matters of method and administration. There are certain characteristics, however, common to most general shops. These include:

1. Activities in two or more industrial areas are evident.
2. A large number of industrial materials are used.
3. The teacher is versatile in many areas.
4. Equipment is diversified, rather than specialized.
5. Breadth of experiences is considered more important than depth in any particular field.

A variation of the comprehensive general shop organization should be mentioned. In schools with two or three industrial arts teachers, the areas usually found in the comprehensive general shop are often divided among the available staff. Thus, in a two-teacher school one instructor might be teaching woodworking, crafts, and metals simultaneously while the other might be handling drawing, electricity, and graphic arts. If there were three teachers in the school, each might be responsible for only two areas or additional areas might be added to the program.

3. General Unit Shop

A third type of shop organization has evolved to meet the need for student experiences which will have some of the exploratory aspects of the comprehensive general shop, but which will, at the same time, be specialized enough to challenge the advanced student. This type of organization is called a general unit shop or a limited general shop; it combines some of the best features of both the comprehensive general shop and the unit shop.

An example of a general unit shop would be a general metals shop where a student might have experiences in sheet metal, art metal, foundry work, forging, welding, spinning, and machine work. Another example would be a general graphic arts laboratory where students carry on such activities as letterpress printing, photo-offset lithography, screen process printing, linoleum or wood block cutting, intaglio, papermaking, rubber stamp making, and bookbinding. Note that while these areas represent several occupational fields they are all related in the sense that they are within the field of graphic arts. General unit shops with their diversity of equipment, range of materials, and the experience possibilities both in breadth and depth are well suited for achieving the objectives of industrial arts. Figure 3–9 illustrates a typical general metals shop.

Fig. 3–9. Looking into the hot metals and sheet metal areas of a typical general metals shop. (Courtesy, Richard M. Birch, California State College, California, Pennsylvania)

The general unit shop organization is in widespread use today. Its rapid growth and development are especially pronounced in newer school systems, particularly in consolidations and jointures, where the services of three to six industrial arts teachers are needed. For example, when there are three industrial arts teachers in a given school, one may find a general woodworking shop, a general metals shop, and a general graphic arts shop. Using the newer terminology of the field, such laboratories would be called construction, manufacturing, and visual communications. Exactly which subareas would be included in each laboratory would depend upon the courses of study as determined by the teachers of the department.

SUMMARY

Selected findings of a national survey by the Office of Education revealed the following:

1. Three-fourths of all American public secondary schools have industrial arts classes.

2. Subject matter content in industrial arts is currently centered in woodworking, drafting, and metalworking.

3. Nearly four million students study industrial arts with the enrollments concentrated in the junior high school and in schools whose enrollment size exceeds 1,000 students.

4. Since 1954 there has been an increase in the compulsory requirements for both boys and girls to take industrial arts.

5. Nearly 95 percent of the 40,000 industrial arts teachers in this country hold a professional teaching certificate. About 60 percent have a bachelor's degree and 35 percent have a master's degree.

Industrial arts is an area of the curriculum in the public schools beginning in the kindergarten, continuing through all grades, and terminating in programs of adult education.

In the elementary school teachers have been using industrial arts activities for many years, although they may have called them practical arts, construction-type activities, or other titles. The three schools of thought on industrial arts in grades K–6 are that industrial arts (1) is an activity method of teaching, (2) has content and should be taught along with other subjects, (3) is both a teaching method and has content of its own. Industrial arts is taught in elementary schools in (1) self-contained classrooms, (2) a central laboratory facility within the school, and (3) mobile laboratory units which visit a number of schools, especially those in remote areas.

Industrial arts fits well into the philosophy and practices of the Middle School. It makes valuable contributions to the curriculum when taught separately or when included in a unified arts program. It is a vital curricular component in those schools where new concepts of career education are being implemented.

In the junior high school industrial arts has long been regarded as an essential part of the general education of every boy and girl. In programs where career education is being emphasized, industrial arts activities comprise at least four of the fifteen occupational clusters in career exploration including (1) manufacturing, (2) construction, (3) communications, and (4) transportation.

At the senior high school level, industrial arts is a multitrack program and serves three groups of students: college-bound, pretechnical, and those students who take industrial arts as a free elective.

Industrial arts activities are popular in community programs for adults. Such classes are excellent public relations vehicles.

Industrial arts has found a place in colleges and universities by providing (1) teacher education programs, (2) service courses for non-majors, and (3) free electives.

Seven subject areas of industrial arts were described including: (1)

woodworking, (2) drawing, (3) metalworking, (4) electricity/electronics, (5) graphic arts, (6) industrial crafts, and (7) power technology. Newer technology is coming to the fore, for example, wood technology, metals technology, construction, production, and visual communications.

Three main types of industrial arts laboratory organization were identified: (1) unit shop, (2) comprehensive general shop, and (3) general unit shop.

DISCUSSION TOPICS AND ASSIGNMENTS

1. In parallel columns contrast the unit shop and the general shop in regard to such features as physical equipment, types of projects, teacher competencies, shop organization, and methods.

2. Which viewpoint do you support concerning the place of industrial arts in grades K–6? Explain.

3. What reasons can you advance for making industrial arts a requirement for all girls at the junior high school level?

4. How can industrial arts assist the college-bound student to develop his power of critical thinking? Be specific.

5. In what ways are adult programs useful vehicles of public relations for industrial arts education?

SELECTED REFERENCES

A Career in Teaching Industrial Arts. Washington, D.C.: American Industrial Arts Association, Inc. (No date).

Anderson, Lowell D. "The Activity-Based Middle School," *Man/Society/Technology*, vol. 31, no. 1 (September-October 1971), pp. 10–12.

Approaches and Procedures in Industrial Arts, Fourteenth Yearbook, American Council on Industrial Arts Teacher Education. Bloomington, Ill.: McKnight and McKnight Publishing Company, 1965.

Brimm, R. P. "Middle School or Junior High?" *Man/Society/Technology*, vol. 31, no. 1 (September-October 1971), pp. 6–8.

Gerbracht, Carl and Robert J. Babcock. *Elementary School Industrial Arts.* New York: The Bruce Publishing Company, 1969.

Gilbert, Harold G. *Children Study American Industry.* Dubuque, Iowa: Wm. C. Brown Company, Publishers, 1966.

Hoots, Jr., William R. (ed.) *Industrial Arts in the Elementary School: Education for a Changing Society.* Greenville, N.C.: East Carolina University, 1971.

Industrial Arts in the Middle School, American Council of Industrial Arts Supervisors, Washington, D.C.: American Industrial Arts Association, Inc., 1971.

Lauda, Donald P. and Robert D. Ryan "Industry 192," *The Journal of Industrial*

Arts Education, vol. XXVIII, no. 3 (January-February 1969), pp. 12-14.

Maley, Donald. "A New Role for Industrial Arts in the Senior High School," *School Shop,* vol. XXIX, no. 7 (March 1970), pp. 41-44, 65.

Mannion, Edmund J. and Alvin W. Spencer. "1980 Projections," *Man/Society/Technology,* vol. 31, no. 1 (September-October 1971), pp. 22-27.

Schmitt, Marshall L. and Albert L. Pelley. *Industrial Arts Education—A Survey of Programs, Teachers, Students, and Curriculum.* U.S. Department of Health, Education, and Welfare, OE-33038, Circular Number 791. Washington, D.C.: Government Printing Office, 1966.

Selective Educational Experiences Diagramed for Progress, Bureau of General and Academic Education. Harrisburg, Pa.: Pennsylvania Department of Education, 1969.

Simonson, Virginia E. "Technology for Children," *School Shop,* vol. XXXI, no. 3 (November 1971), pp. 27-29.

Weber, Earl M. *The Role of Industrial Arts in Tomorrow's Schools,* Bulletin #9. Washington, D.C.: American Industrial Arts Association, Inc., 1966.

Chapter 4

Objectives of Industrial Arts

In Chapter 1 the purposes of general education were discussed at some length. These are the overall aims that should characterize and direct the progress of all phases of education which can be classified as general education. Each branch of learning so included should have its own specific objectives and statements of values to be achieved. These should grow out of, and be in harmony with, the aims of general education. Any branch of learning the aims of which conflict with those of general education or do not grow naturally from them should probably be classified as "specialized" rather than general education.

Industrial arts has long been regarded as an essential part of general education. The close relationship existing between the two was the subject of Chapter 2. It is to be expected, therefore, that the objectives for industrial arts will be closely related to those of general education and may be derived from them.

HOW TO DERIVE OBJECTIVES

One way to derive the objectives for a subdivision of general education, for example, a program or a course, is to analyze the goals of general education to determine what implications, if any, exist for that body of subject matter. The first step in the analysis is to identify the salient features, characteristics, or cardinal points associated with each of the major purposes of general education. It will be recalled that these goals have been summed up previously as: (1) to transmit a way of life, (2) to improve that way of life, and (3) to meet the needs of individuals. The second step is to draw implications for the subdivision, in this case industrial arts education, based on careful study of these characteristics. From these implications a list of objectives can be inferred for the given subdivision of general education.

USE OF OBJECTIVES

With a list of formally stated objectives at hand, the next step is to restate them in terms of expected student outcomes. Careful consideration of these statements of terminal student behavior will bring to mind suggestions for teacher and student activities to achieve the desired changes. Terminal student behavior provides the basis for criterion tests to measure the attainment of that behavior. Thus the cycle is complete. If the students engage in the activities selected, they should attain certain desirable behavior changes which are measurable by means of the criterion tests. The acquisition of the desired behavior changes means student attainment, at least in part, of the program objectives. In turn, achievement of program objectives indicates fulfillment, again at least partially, of the goals of general education.

It is the intent of this chapter to derive the objectives for industrial arts from those of general education. Subsequent chapters will deal with student behavioral changes and suggested learning activities as well as criterion tests for evaluating the attainment of behavioral changes.

IMPORTANCE OF OBJECTIVES

The importance of instructional objectives cannot be overemphasized. Without objectives, students and teachers are helplessly adrift on a sea of meaningless activities without a sense of direction. With them, students and teachers can chart a direct course to desirable goals. In other words, objectives not only give direction to an educational course or program, they provide reasons for its existence and give meaning to the efforts of the teacher and the students. In a sense, objectives are like the letters *alpha* and *omega* of the Greek alphabet. The formulation of course objectives constitutes the alpha or first step in constructing a course of study which is the backbone of any planned program. Objectives provide the basis for selecting the content to be taught and for selecting student activities and experiences. Thus, a teacher may evaluate any educational method, project, or idea to decide whether it merits inclusion in a program simply by appraising it in terms of its contributions to an educational goal (best expressed in terminal behavior).

The objectives for a given course prescribe the physical conditions essential for optimum learning. The layout of a laboratory, the selection of tools, machines, and equipment can be determined successfully only in terms of course objectives. In like manner, course objectives condi-

CHART I
Relationships of Industrial Arts to General Education

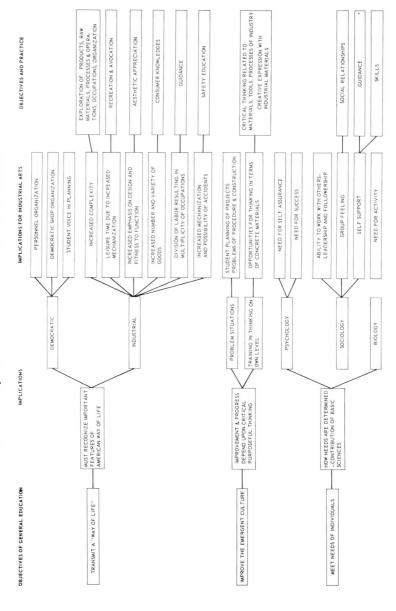

Objectives of Industrial Arts

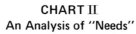

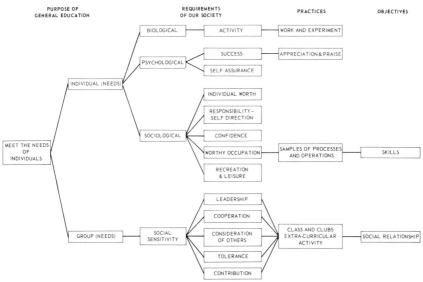

tion both teaching methods and laboratory management practices. Objectives are also in the omega or last phase of any educational program. That is, the evaluation of an educational effort is valid only in terms of previously stated objectives.

ANALYSIS OF THE RELATIONSHIPS OF INDUSTRIAL ARTS TO GENERAL EDUCATION

The analysis presented in Chart I shows graphically how the objectives for industrial arts may be derived from those of general education. Column I lists the basic purposes of general education as presented in Chapter 1; the implications are listed in Column II. Thus it is clear that if the transmission of a way of life is an accepted objective of general education, the important features of that way of life must be recognized and plainly defined. The salient features of present-day American society relate to the fact that it is democratic as well as industrial and technological in nature.

Likewise, if one accepts the improvement of the emergent culture as a defensible aim for general education, then it is essential that the basis for such improvement be clearly evident. The advancement of our society is dependent upon the development and practice of critical thinking by its members. There is no other way to improve our culture.

The importance of critical, purposeful thinking in bringing about societal progress has been noted in Chapter 1. Our schools have the responsibility to face students with problem situations and provide them with specific training in critical thinking *at their own levels.*

The implications evident in meeting the needs of individuals has been shown to be a knowledge and understanding of how such needs are determined. The sciences which contribute most directly to determining the basic needs of individuals are psychology, sociology, and biology. The contributions of these sciences already have been examined. A more detailed analysis of this objective, meeting the needs of individuals, is depicted in Chart II.

IMPLICATIONS FOR INDUSTRIAL ARTS TO TRANSMIT A WAY OF LIFE

Several features of Column III of Chart I merit consideration for the significance they may hold for education and especially for industrial arts.

1. A Democratic Society

Since the American way of life is essentially democratic, one may infer that the classroom itself (in this case the industrial arts laboratory) should be organized on a democratic basis. Training for future participation in a democratic society can be obtained in no more effective way than through living democratically in the real-life situation of the classroom. This suggests that the industrial arts laboratory should lean more toward a learner-centered environment than one teacher directed, especially if the latter is conducted in an authoritarian manner with unilateral decisions. This means that pupils should be given opportunities to help in planning classroom procedures and in laboratory organization; that they should be given responsibilities commensurate with their abilities; and that each, under the guidance of the instructor, should have a voice in determining objectives, procedures, and methods to be used. Further, each student should have the opportunity to develop such personal character traits as leadership, followership, consideration of others, and so on. An effective pupil personnel organization in the industrial arts laboratory is one method for accomplishing this end. Figure 4–1 shows a group of industrial arts students actively planning their shop program.

Fig. 4-1. Students participate in planning the shop program. (Courtesy, State University of New York, Oswego Campus)

2. An Industrial Society

In addition to being democratic, the fact that the American way of life is also highly industrialized and technological, raises many implications for education through the industrial arts program. Some of the characteristics of our industrial society with their implications for industrial arts education include the following:

Increasing Complexity of Industry. An analysis of industry shows it to be increasingly complex from the standpoint of organization, processes and operations, and products. The use of the computer in manufacturing systems has heralded new production heights which are almost impossible for the layman to comprehend. Today's computers not only control the actual production of intricate parts, but also perform a multiplicity of related functions, such as computing operational efficiency, diagnosing machine malfunction and equipment failures, and pinpointing "bottlenecks" in production. By monitoring tool dullness computers provide advance warning when a production tool will need to be changed. Computers also perform valuable record-keeping operations including inventorying, flow of incoming materials, establishing cycle times of production, and recording overtimes (see Fig. 4-2). A computer system in one automotive plant controls 30 machines as well as monitoring all incoming materials for each machine. It automatically controls approximately 350 operations and processes and at the same time reports on general efficiency, slowdowns and machine stoppages, and other important factors in the production cycle.

Fig. 4–2. A fish-eye lens shows a central computerized order processing system housing a core memory file for 200,000 part numbers. A battery of disk file units contains some 90 million characters of data. Execution time of the computer is 900 billionths of a second. This computer keeps track of 15 regional, seven zone, and two master parts depots. (Courtesy, Chrysler Corporation)

In addition to its growing complexity, American industry is characterized by decentralization. In colonial times a boy or girl could observe and, perhaps, participate in the entire process of making cloth from the raising of the sheep and the shearing of the wool to the weaving and shrinking of the finished goods right at home. Today the sheep may be raised in Wyoming, the wool processed in New York, and the cloth woven in New England.

While certain large industries have become geographically decentralized within their own units of organization, many industrial concerns today find it more economical and efficient to practice another form of decentralization by purchasing parts from specialized subcontractors. For example, the Ford Motor Company spends more than three billion dollars annually in purchasing thousands of different items from 17,000 independent suppliers, some large, some small. The number of subcontractors in our space program is phenomenal. For one space capsule alone 4,000 contractors from 42 states were involved in making one or more of the 1,367,059 parts used in the spacecraft. Additionally, over 10,000 parts were manufactured by 1,500 contractors for the launch vehicle.

Truly American industry is both highly complex and decentralized. These factors have completely removed it from the everyday experience of the average boy and girl. The need for our schools to provide insights and understandings of our industrial and technological society requires no further expansion here. Thus, an important objective for industrial arts education can be inferred: *To explore industry and American industrial and technological civilization in terms of its evolution, organization, raw materials, processes and operations, products, and occupations.*

Division of Labor. Another characteristic feature of modern industrial life which results from the increased complexity of industry is the division of labor. Within any large industry the number of different operations may well run into the thousands. Rarely does a single worker perform all the operations which comprise a finished product. More likely that worker performs only one small operation which he repeats over and over again. The division of labor in American industry results in a multiplicity of occupations. Industrial jobs are changing continuously. New inventions, improvements in manufacturing processes, and general economic conditions are among the factors which keep industry and its host of jobs and occupations in a constant state of flux. The average worker today will change jobs several times—perhaps four, five, or even six times in his working lifetime.

An analysis of this aspect of industry presents the schools with another challenge, namely, to provide vocational guidance information and experiences for pupils. Education has long accepted this charge and has recognized vocational guidance as a major aim under the larger goal, "to transmit a way of live." Thus, from one of the goals of general education a vital guidance role for industrial arts can be inferred: *To provide information about, and insofar as possible, experiences in, the basic processes of many industries in order that students may be more competent to choose a future vocation.*

Great Number and Variety of Manufactured Goods. Another feature of modern American industry is the increased number and variety of manufactured goods. At the turn of the century there were available only a few dozen kinds of fabrics; now there are thousands. Today one firm alone manufacturers more than 140,000 different types and sizes of solder. Every year sees thousands of entirely new items added to the list of consumer goods available to the purchaser.

The science of automobile manufacturing has evolved into a state of customized mass production. Today's buyer wants the economies of mass production, but at the same time desires to tailor his car to suit his individual taste. This is entirely possible, however, with the infinite varieties of engines, transmissions, body styles, accessories, and color combinations available. In fact, one Detroit manufacturer states his firm could operate any of several assembly plants for over a year without producing precisely the same car twice.

The tremendous number and variety of manufactured goods provides industrial arts with this consumer education objective: *To increase consumer knowledges to a point where students can select, buy, use, and maintain the products of industry intelligently.*

Leisure Time due to Increased Mechanization. The division of labor and the development and improvement of machine processes to replace hand labor have tended to increase the amount of leisure time for all workers. Although the hardy pioneer toiled from dawn to dark, the tendency today is to reduce still further the shortened work week which now rarely exceeds 40 hours. Some industrial concerns have already embarked on a four-day work week. The need for wholesome recreation to utilize free time thus provided is evident. Adults engage in construction-type activities around the home, at their boats, in sports, and in hobbies related to industry. Figure 4-3 shows an adult in a favorite home activity—refinishing furniture. As a stated goal of general education, the worthy use of leisure time implies a similar goal for industrial arts education: *To develop recreational and avocational activities of a constructional nature.*

Increased Emphasis on Design and Fitness to Function. Another trend evident in modern industry is the emphasis on sound design and fitness to function. Useless decoration is disappearing from manufactured goods and is being replaced by a simplicity of line and form. Each product is fitted to its particular use and at the same time provided with beauty of line and mass without encumbering embellishments that have no structural value.

From this trend one may infer the following objective for industrial arts education: *To increase an appreciation for good craftsmanship and design, both in the products of modern industry and in artifacts from the material cultures of the past.*

Increased Possibility of Accidents. The increased mechanization of industry has increased the possibility of industrial accidents. This is true despite the concerted efforts by industry and governmental agencies to protect the worker against hazards of his occupation and his own carelessness or indifference.

Since our culture is so dominatingly industrial and technological, it behooves the schools to teach pupils how to live and work safely in the environment. The implication for an industrial arts objective is obvious: *To develop safe working practices.*

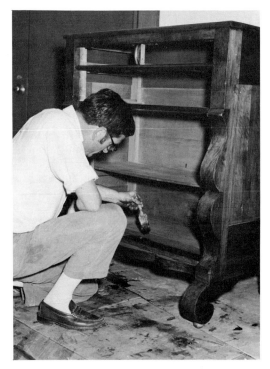

Fig. 4–3. Refinishing furniture is a favorite pastime for many adults. (Courtesy, William J. Wilkinson, Nether Providence High School, Wallingford, Pennsylvania)

IMPLICATIONS FOR INDUSTRIAL ARTS TO IMPROVE THE EMERGENT CULTURE

In addition to transmitting the American way of life, a major goal of general education is to extend and improve that way of life. As stated previously, society can advance and improve itself only through the development of and practice in the process called critical thinking. The Educational Policies Commission recognized the development of the rational powers of each pupil as the central purpose of American education (see Chapter 1). The Commission noted:

> Individual freedom and effectiveness and the progress of society require the development of every citizen's rational powers.[1]

[1] *The Central Purpose of American Education*, Prepared by the Educational Policies Commission, National Education Association of the United States and the Association of School Administrators (Washington, D.C.: National Education Association, 1961), p. 21.

And the society which best develops the rational potentials of its people, along with their intuitive and aesthetic capabilities, will have the best chance of flourishing in the future. To help every person develop those powers is therefore a profoundly important objective and one which increases in importance with the passage of time.[2]

It is not difficult to infer a major objective for industrial arts arising harmoniously from the important goal of general education—to teach students how to think. Since industrial arts is based upon problem-solving experiences with the tools, materials, and processes of industry, it is only natural to recognize its unique contribution to the central purpose of American education. Thus, an important objective for industrial arts becomes: *To develop critical thinking as related to the materials, tools, and processes of industry and to encourage creative expression in terms of industrial materials.*

IMPLICATIONS FOR INDUSTRIAL ARTS TO MEET THE NEEDS OF INDIVIDUALS

The basic needs of the individual seem rooted in the fields of psychology, sociology, and biology. In meeting these needs it is evident that industrial arts can make substantial contributions. Where appropriate, objectives for industrial arts will be inferred.

1. Contributions to Psychological Needs of the Individual

From Column III of Chart I it is noted that from a psychological point of view, it is important for each pupil to develop a feeling of self-assurance and that he experience at least a measure of success.

In industrial arts the students enjoy feelings of real achievement because (1) the results are tangible, (2) a feeling of self-assurance and self-confidence develop, and (3) success is enjoyed. Many cases have been noted where substantial improvement in all school subjects has resulted from initial success attained in industrial arts courses.

Industrial arts can help both the gifted and the less gifted student. Problem-solving experiences with tools, materials, and processes of industry will challenge the minds of even the best academically talented students. In particular, college-bound students need to learn how to think as provided by experiences in industrial arts education. Growth and development of the power of critical thinking will immeasurably

---

[2] *Ibid.,* p. 11.

Objectives of Industrial Arts

assist these students regardless of what their professional goal may be. On the other hand, industrial arts can also benefit those who are less gifted and who are unable to meet even minimum standards in academic studies.

2. Contributions to Sociological Needs of the Individual

Sociologically each pupil should develop as one of the group. He needs to learn to work with others, to develop leadership and followership abilities, and to have some ideas about eventual self-support (see Chart 1).

The informal organization of the industrial arts class makes possible many fine opportunities for working with others without distinction as to race, creed, color, or economic status. Figure 4–4 shows how this may be accomplished in an industrial arts class. Working with others on a common problem or project also helps to develop a sense of belonging as well as a sense of consideration for others. By participating in the student personnel plan, each pupil comes to feel he is part of the group, that his efforts and contributions are accepted and that his worth is recognized. Through such a plan of organization both leadership and followership qualities as well as tolerance and cooperation can be developed.

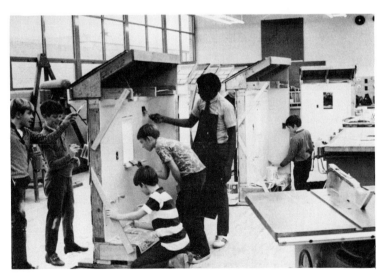

Fig. 4–4. Industrial arts makes possible cooperative effort on a common project. (Courtesy, Warden B. Muller, Reynolds Junior High School, Trenton, New Jersey)

Analysis of the sociological needs of the individual leads to the inference of this objective for industrial arts: *To develop desirable social relationships, such as cooperation, tolerance, leadership and followership, and tact.*

At about the junior high school level or perhaps even the middle school grades for some pupils, the typical student begins to give some thought to his life work. He feels a need of developing toward a position of economic independence. His contacts in the industrial arts laboratory with actual industrial materials and his studies of various occupations tend to meet this need by providing information about the many types of work and by giving experiences which are valuable from a guidance standpoint. Though its true purpose is not to produce skilled workers, industrial arts performs a valuable vocational guidance function. Students learn about industrial occupations, the world of work, and are provided with "hands on" exploratory experiences in a wide variety of trades, occupations, and industrial crafts.

From an analysis of the basic needs of the individual, the guidance objective can be inferred. This was cited previously under the heading "Division of Labor" in this chapter as the following: *To provide information about, and insofar as possible, experiences in, the basic processes of many industries in order that students may be more competent to choose a future vocation.* To this objective the following may be added: *To develop a degree of skill in a number of basic industrial processes.*

3. Contributions to the Biological Needs of the Individual

From the biological standpoint, it has become increasingly evident from many recent studies that all individuals, and especially children, need activity. It is unnatural for them to be quiet for any extended periods of time. Boys and girls like to be active, to move about, and to plan and do things; it is as abnormal for them to be quiet as to go without food. The reason thus becomes clear why learning should be related to activities, whenever possible, rather than to passive study. The enthusiasm with which a pupil leaves his studies for the playground or the shop is a direct result of his natural need for activity. The point needs no further clarification that industrial arts is an action-packed activity requiring each pupil to gain first-hand experiences with the tools, materials, and processes of industry.

It should be noted that in the foregoing discussion of needs, no attempt has been made to discuss all the needs which youth may have; these have been more adequately presented in Chapter 1. In this chap-

Objectives of Industrial Arts

ter only those needs which appear to be in some manner related to practical or industrial arts education have been considered. An equally strong case could doubtless be made for other areas within the educational field.

IMPORTANT OBJECTIVES OF INDUSTRIAL ARTS

Each of the various subject areas within the curriculum of the public school has its own specific objectives. The area of industrial arts is no exception. If industrial arts objectives are defensible, however, it is essential that a direct and readily recognizable relationship exist between them and the aims of general education. Based on the implications outlined in Columns II, III, and IV of Chart I, it was the intent of his chapter to draw certain definite and specific objectives which include a justification for and the purposes of industrial arts. Column V of the chart is an attempt to do this in abbreviated form. Some of the important objectives of industrial arts, as shown by an analysis of the purposes of general education, are (not in rank order of importance):

1. To explore industry and American industrial and technological civilization in terms of its evolution, organization, raw materials, processes and operations, products, and occupations.

2. To develop recreational and avocational activities of a constructional nature.

3. To increase an appreciation for good craftsmanship and design, both in the products of modern industry and in artifacts from the material cultures of the past.

4. To increase consumer knowledges to a point where students can select, buy, use, and maintain the products of industry intelligently.

5. To provide information about, and insofar as possible, experiences in, the basic processes of many industries in order that students may be more competent to choose a future vocation.

6. To develop critical thinking as related to the materials, tools, and processes of industry and to encourage creative expression in terms of industrial materials.

7. To develop desirable social relationships, such as cooperation, tolerance, leadership and followership, and tact.

8. To develop safe working practices.

9. To develop a degree of skill in a number of basic industrial processes.

SUMMARY

The objectives for a given subject area may be derived by analyzing the basic purposes of education to determine what implications, if any, exist for that subject. From these implications may be deduced objectives and practices. Analysis of the aims of general education reveal numerous implications for industrial arts. From these implications the following nine objectives may be derived: (1) exploration of industry, (2) avocation, (3) aesthetic appreciations, (4) consumer education, (5) vocational guidance, (6) critical thinking and creative expression, (7) social relationships, (8) safety education, and (9) skill in basic industrial processes.

DISCUSSION TOPICS AND ASSIGNMENTS

1. Compare and contrast the objectives of industrial arts as listed in this chapter with those in the A.V.A. publication, *Industrial Arts in Education.**

2. How can the objectives listed in this chapter be reconciled with those of Bonser, as outlined in *Industrial Arts for Elementary Schools* by Bonser and Mossman?*

3. Compare the objectives in this chapter with those outlined by Hostetler.*

4. Study Chart I and list any additional objectives which you feel could be justified for industrial arts.

5. Starting with the three basic purposes of general education, indicate how the objectives for any selected subject might be developed.

SELECTED REFERENCES

Bonser, Frederick G., and Lois Coffey Mossman. *Industrial Arts for Elementary Schools*. New York, N.Y.: The Macmillan Company, 1923.

Fuzak, J. A. "Reflective Thinking as an Aim in Industrial Arts Education," *Education*, vol. 65, no. 10 (June 1945), pp. 583–588.

Guidance in Industrial Arts Education for the '70s." American Council of Industrial Arts Supervisors. Washington, D.C.: American Industrial Arts Association, 1971.

Hostetler, Ivan. "What Objectives Should Be Emphasized in Industrial Arts?"

* See *Selected References*.

Improving Industrial Arts Teaching. Conference Report, June 1960. Washington, D.C.: U.S. Department of Health, Education, and Welfare, OE-33022, Circular no. 656, pp. 11–22.

Industrial Arts in Education. Washington, D.C.: American Vocational Association, 1965.

Pendered, Norman C. "Should the Objectives of Industrial Arts Change?" *Industrial Arts and Vocational Education*, vol. 50, no. 9 (November 1961), pp. 18–20, 54.

This We Believe. Washington, D.C.: American Industrial Arts Association, Inc., 1965.

Wilber, Gordon O. "Industrial Arts and the Pedagogical Lag," *School Shop*, vol. 12, no. 1 (September 1952), pp. 9–10, 32.

Chapter 5

Analysis of Objectives

The previous chapter outlined specific objectives for industrial arts and showed how they were derived from the aims of general education. It is the purpose of this chapter to show how these objectives may be used as a basis for instruction in industrial arts education.

It is evident that if a teacher has objectives they should be used. It is not sufficient to subscribe to a set of aims and then forget about them or keep them in the desk drawer to be shown to those who may be interested. Rather, these objectives must be used and kept in mind constantly for they constitute the very foundation of the entire program. Before objectives can be used, however, they must be analyzed carefully in terms of the outcomes expected.

BEHAVIOR CHANGES AS OUTCOMES

Thoughtful analysis of educational objectives will indicate that behavior changes in students are what are really desired. While the behaviorial-change approach is receiving current emphasis in educational circles, it is not a new idea. John Ruskin once said:

> Education does not mean teaching people to know what they do not know. It means teaching them to behave as they do not behave . . .

In retrospect, it is interesting to recall that the 1948 edition of this book featured an analysis of industrial arts objectives in terms of expected student behavior changes. In this respect the text was unique because at that time the outcomes in fields of education were not commonly expressed in behavioral terms. Since then, however, the pendulum began to swing ever so slowly in this direction. Subsequent editions of this text in 1954 and 1967 followed the same pattern of expressing objectives in student behavioral terms. This is true, also, of this edition for it is the authors' belief that the analysis of objectives in

Analysis of Objectives 81

terms of student behavior is the proper approach and is a powerful tool in the improvement of instruction. Perhaps the attention given to teaching machines and programed instruction in the last decade or so has been responsible, at least in part, for the heavy emphasis presently being given to expressing objectives in terms of student behavior. Certainly the mode today in all areas of education is to express educational objectives in student behavioral format.

The following illustrations will support the viewpoint that expressing the outcomes of education in terms of student behavior has found ready acceptance by many in the field.

The 1950 edition of the *Encyclopedia of Educational Research* noted that:

> The purposes of general education are better understood in terms of performance or behavior rather than more narrowly in terms of knowledge.[1]

In 1953 a nationwide study of the behavioral outcomes of elementary education was undertaken by the Educational Testing Service, the Russell Sage Foundation, the Department of Health, Education, and Welfare, and the Department of Elementary School Principals of the National Education Association. These sponsoring agencies formed a Mid-Century Committee on Outcomes in Elementary Education. Their report outlines in detail specific behavior patterns for elementary school children.[2]

This elementary school study was deemed so useful that four years later a similar study was made for the secondary schools. In this research the *Purposes of Education in American Democracy* (see Chapter 1) were analyzed in terms of those behavior outcomes which ought to be evident if the objectives were being achieved.[3] The report of this investigation appears under the title: *Behavioral Goals of General Education in High School.*[4]

The Industrial Arts Policy and Planning Committee concurs with the viewpoint that education is really a matter of changing the behavior of students by observing that:

---

[1]T. R. McConnell et al., "General Education," Walter S. Monroe (ed), in *Encyclopedia of Educational Research* (New York, N.Y.: The Macmillan Company, 1950), p. 489.

[2]Nolan C. Kearney, *Elementary School Objectives* (New York, N.Y.: Russell Sage Foundation), 1953.

[3]*The Purposes of Education in American Democracy*, Prepared by the Educational Policies Commission, National Education Association of the United States and the Association of School Administrators (Washington, D. C.: National Education Association, 1938), pp. 47–108.

[4]Will French et al., *Behavioral Goals of General Education in High School* (New York, N.Y.: Russell Sage Foundation, 1957).

> The school is essentially a behavior-changing institution. That is, school experiences make pupils different than they would be if they lacked these experiences.[5]

The remainder of the AVA booklet from which this quotation was taken is devoted to statements of desired behavior changes for nine suggested objectives of industrial arts.

In 1963 Pace noted:

> Evaluation studies have long reflected the belief that objectives should be defined behaviorally and that the effectiveness of programs and institutions should be viewed in relation to all of their objectives.[6]

In their *Study of American Industry—A Guide for Secondary Schools* (1965), industrial arts personnel in the State of Maine identified expected pupil behavioral changes for each objective listed for grades seven through twelve. This analysis covered pupil experiences in the manufacturing and construction industries, power and transportation industries, and service industries.[7] See Chapter 7 for additional information.

Unless one defines and delineates educational objectives in terms of terminal behavior of students, there is apt to be much confusion and uncertainty as to just what is meant. Mager puts it this way in his book, *Preparing Instructional Objectives:*

> An objective is an intent communicated by a statement describing a proposed change in a learner—a statement of what the learner is to be like when he has successfully completed a learning experience. It is a description of a pattern of behavior (performance) we want the learner to be able to demonstrate.[8]

For example, if the *development of safe working practices* is accepted as an objective of industrial arts, then the student who has studied industrial arts will be expected to behave differently when working with tools and machinery from one who has not had such experiences. Specifically, the industrial arts pupil will observe all safety rules, will wear clothing appropriate to the work being done, including safety glasses and other protective devices, will use safety guards, will

---

[5] *A Guide to Improving Instruction in Industrial Arts* (Washington, D.C.: American Vocational Association, 1953), p. 13.

[6] C. Robert Pace, CUES (College and University Environment Scales) Preliminary Technical Manual (Princeton, N.J.: Educational Testing Service, 1963), p. 5

[7] Maine State Department of Education, *Industrial Arts Technology—A Guide for Secondary Schools in Maine* prepared by Department of Education, Vocational Division. (Augusta, Maine: State Department of Education, 1965).

[8] Robert. F. Mager, *Preparing Instructional Objectives* (Palo Alto: Calif.: Fearon Publishers, 1962), p. 3.

Analysis of Objectives

use only safe tools or machines, and will make certain that others cannot be hurt by the work he is doing. These actions on the part of the student indicate that learning has taken place and that behavior changes have occurred. Thus, a student's behavior after he has finished an industrial arts course should be different from that when he started. If this is not the case, probably learning has not taken place.

It must be recognized that student behavioral changes are not induced in a vacuum. They develop and will be retained longer through first-hand experiences between the student and his environment. As in all learning, the pupil should be actively involved in the learning process. The role of the teacher is to identify desirable behavior changes and to help students acquire them through experience activities.

TYPICAL BEHAVIOR CHANGES

Since behavior changes are the desired outcomes, each objective for industrial arts must be analyzed in terms of those changes which appear desirable. As stated previously, it is not enough that the teacher subscribe to the idea of "consumer education" as an objective for industrial arts, he must inquire further: "Just what is meant by consumer education?" A generalized concept of what each objective means is not acceptable. Specific outcomes are the only measurable results of any type of instruction and should be the immediate goal. In other words, the teacher should examine each objective thoughtfully and ask himself: "Just what behavior changes do I expect from my students as evidence that this objective has been attained?" and "How do I expect my students to be different when the course is completed?" This analysis step cannot be ignored if concrete and tangible results are desired from each of the accepted objectives of industrial arts.

Sometimes teachers appear to confuse the terms *ideals* and *objectives*. It has been said that ideals are like the stars in the sky; we never reach them, but like the mariner, use them as a guiding light. Some teachers seem to feel that this description fits objectives too. They regard objectives as vague, nebulous, visionary in nature, and consequently unattainable. This is not so! Objectives are attainable, especially if they are stated in terms of student behavior changes.

In the past, industrial arts aims and purposes were vague and unattainable because their true significance was not disclosed by searching study of what was expected by way of behavior changes in students.

It may be assumed that no two persons attempting to analyze the objectives of industrial arts in terms of behavior changes would make identical lists. It is probable, however, that most successful teachers would arrive at lists which would include many common elements. It is also likely that any group of individuals interested and educated in the industrial arts field could take such lists and come to a common agreement on the most desirable behavior changes for the students of a given grade, school, or community. Many such groups of teachers or supervisors have met to determine objectives; but few have taken the additional and most important step of identifying the specific behavior changes which they wish to bring about. *Such group action appears to be one of the most vital needs facing the industrial arts teaching profession today.*

Naturally the desired behavior changes will differ between grade levels and among the different types of industrial organizations. Chart III is intended to show what is meant by an analysis of objectives in terms of desired behavior changes. The chart is a follow-up of Chart I (see Chapter 4) in which the objectives of industrial arts were derived from an analysis of the goals of general education. The analysis (Chart III) is based on a program for the junior high school level which is assumed will be taught in a comprehensive general shop or its equivalent.

In the following sections of this chapter each objective of industrial arts is followed by suggested behavior changes expected in students. It is important for the teacher to acquaint all pupils with the objectives of the course and to indicate what behavioral changes are expected.

Objective 1—Exploration of Industry

To explore industry and American industrial and technological civilization in terms of its evolution, organization, raw materials, processes and operations, products, and occupations.

Expected behavior changes from students:

1. They will be able to list the major industries of our country, locate them geographically, and discuss principal products and manufacturing processes.

2. They will recognize the scientific principles underlying industrial operations and processes.

3. They will visit industries whenever possible to learn about methods, processes, products, and the like.

4. They will recognize industrial processes and methods and will attempt to relate them to the industrial arts laboratory.

5. They will be able to understand new industrial processes by comparing them to more familiar ones.

6. They will be able to define common industrial terms (see Fig. 5–1).

CHART III
Sample Behavior Changes from an Analysis of One Objective

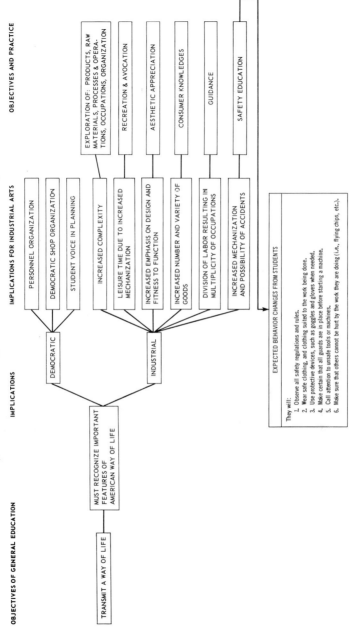

Fig. 5-1. Learning the names of tools is fun for these elementary school pupils. (Courtesy, California State Department of Education)

7. They will be familiar with industrial organizations and will relate them to personnel organization of the industrial arts laboratory. Their cooperation in the pupil personnel system will increase.

8. They will be familiar with the sources of some of the raw materials of industry and will be able to discuss their transportation, processes, and industrial uses.

9. They will read about industry, its historical development, current trends, and the lives of important inventors and industrial leaders.

10. They will seek information about new inventions and developments in industry and technology.

11. They will be able to interpret the language of industry—industrial and technical drawing (see Fig. 5-2).

12. They will study about and help conserve our Nation's natural resources.

13. They will learn more about and contribute time and effort to those concerned with pollution control of our environment.

Analysis of Objectives

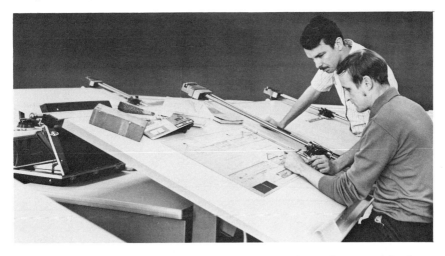

Fig. 5-2. Industrial arts provides many opportunities for students to study the language of industry. (Courtesy, John Mitchell, Gorham State College of the University of Maine, Gorham, Maine)

14. They will understand some of the fundamentals of economics by which industry operates.

15. They will read about the evolution of labor unions and will be able to interpret current problems of capital, management, and labor.

Objective 2—Avocation

To develop recreational and avocational activities of a constructional nature.

Expected behavior changes from students:

1. They will read such magazines as *Popular Mechanics, Home Craftsman,* and *Popular Science.*

2. They will ask advice on how to carry on constructive activities out of school.

3. They will become interested in, and will engage in, one or more constructional hobbies.

4. They will spend spare time in the shop either in school or at home (see Fig. 5-3).

5. They will ask questions and talk about their hobbies.

6. They will consult catalogs for information about their hobbies.

7. They will contribute to class discussions with information gained from reading along lines of their interests.

Fig. 5–3. A home workshop can provide many interesting hours of wholesome recreation.

8. They will take the initiative in visiting industries along the lines of their interests.

9. They will make the acquaintance of, and form friendships with, others having similar interests.

10. They will develop home workshops.

11. They will design and construct projects in their laboratory work which are related to their hobby interests.

Objective 3—Aesthetic Appreciations

To increase an appreciation for good craftsmanship and design both in the products of modern industry and in artifacts from the material cultures of the past.

Expected behavior changes from students.

1. They will know the names of outstanding craftsmen and designers of the past and present and will be able to recognize the work for which they are known.

2. They will be able to identify pieces of period furniture as well as contemporary styles and trends.

3. They will appreciate good design in artifacts and will show such appreciation in speech and actions. (see Fig. 5–4).

4. They will understand the principles of good design and will be able to recognize them in the products of industry.

Analysis of Objectives 89

Fig. 5–4. Appreciation of past cultures is reflected in the fine work of this high school student from Horsham, Pennsylvania. (Courtesy, Stanley Tools, New Britain, Connecticut)

5. They will be able to apply the principles of good design in the construction of their projects.

6. They will recognize and avoid poor design and overdecoration.

7. They will understand some of the physical properties of certain industrial materials and will select materials which are appropriate to the design and intended use.

8. They will appreciate good craftsmanship and construction in industrial products.

9. They will be able to criticize constructively the designs of others.

Objective 4—Consumer Education

To increase consumer knowledges to a point where students can select, buy, use, and maintain the products of industry intelligently.

Expected behavior changes from students:

1. They will examine the products of industry carefully and will judge their value before buying.

2. They will understand the common processes by which materials are shaped, formed, and assembled in industry.

3. They will look for constructional features in determining the quality of a product. (see Fig. 5–5).

Fig. 5–5. Industrial arts promotes consumer knowledge. (Courtesy, Bragg Stockton, Dallas Independent School District, Dallas, Texas)

4. They will become acquainted with trade names and will look for proven brands when purchasing.

5. They will recognize quality and will buy accordingly.

6. They will buy on the basis of their needs, rather than entirely on the basis of price or salesmanship.

7. They will maintain and use manufactured goods in such a way as to prolong their life and usefulness.

8. They will study sales materials carefully and sift out propaganda and false advertising from honest sales efforts.

Objective 5—Vocational Guidance

To provide information about, and insofar as possible, experiences in the basic processes of many industries in order that students may be more competent to choose a future vocation.

Expected behavior changes from students:

1. They will read about and will discuss occupations and the world of work with their friends, teachers, and parents.

2. They will choose elective courses in school which provide additional information about occupations.

3. They will visit industries or watch with interest motion pictures showing workmen at various trades and occupations.

4. They will talk with workers in various trades and occupations concerning the work in which they are engaged.

5. They will be able to cite wages, entrance requirements, working conditions, and opportunities for advancement in various occupations (see Fig. 5–6).

6. They will be familiar with local, state, and federal laws as these relate to industrial employees.

7. They will be familiar with the social security benefits for industrial workers.

8. They will be familiar with the health risks and dangers of various jobs and occupations in several industries.

9. They will decide that they are not fitted for or are not interested in certain vocations. They will make a tentative choice of a vocation or family of occupations.

Objective 6—Critical Thinking and Creative Expression

To develop critical thinking as related to the materials, tools, and processes of industry and to encourage creative expression in terms of industrial materials.

Expected behavior changes from students:

1. They will be able to recall and apply the steps in critical thinking, that is, the scientific method of problem solving.

2. They will increasingly attempt to solve their own problems.

3. They will think through the correct procedure for making a project and will then follow their plan.

4. They will realize that it is as important to be able to plan a project as to construct it (see Fig. 5–7).

5. They will be able to solve simple problems involving the tools, materials, and processes of industry in construction and repair problems at home and at school.

6. They will be able to formulate and test hypotheses for given problem situations.

7. They will experiment with new ways to solve construction problems and will make improvements on the basis of their findings.

8. They will evaluate the work of others in accordance with the principles of scientific thinking.

Fig. 5–6. These boys are learning about the world of work through "real-life" construction activities. (Courtesy, Pittsburgh Public Schools and the *Pennsylvania Industrial Arts News*)

9. They will be able to apply knowledge of the physical properties of industrial materials in the creation of new ideas, designs, and in the construction of products.

10. They will be able to take ideas from different sources and create new designs or products combining elements of creativity and utility.

11. They will be able to analyze tool and machine failures and identify causes for such breakdowns.

Analysis of Objectives

Fig. 5–7. This student is learning that planning must always precede construction. (Courtesy, California State Department of Education)

Objective 7—Social Relationships

To develop desirable social relationships, such as cooperation, tolerance, leadership and followership, and tact.

Expected behavior changes from students:

1. They will develop group spirit and loyalty.
2. They will cooperate with others in group activity.
3. They will assume and discharge leadership responsibilities in the student personnel organization.
4. They will organize or participate in club activities.
5. They will accept leadership responsibilities in club organizations.
6. They will give help and advice willingly; they will seek help from others when in need.
7. They will accept assignments given them by leaders in the personnel organization and will recognize the leadership of others.
8. They will work willingly with individuals who may be of a different race, creed, color, or economic status (see Fig. 5–8).

Fig. 5–8. In industrial arts boys learn to work with boys regardless of race, color, or creed. (Courtesy, Donald Prescott, Snow Hill Junior High School, Snow Hill, North Carolina. Photo by Robert McElroy)

9. They will respect the work, efforts, and rights of others.

10. They will seek to understand those whose views may differ from their own.

Objective 8—Safety Education

To develop safe working practices.

Expected behavior changes from students:

1. They will develop a high degree of safety consciousness.

2. They will be able to cite and will observe all safety regulations and rules.

3. They will wear safe clothing and clothing suited to the work being done.

4. They will use protective devices, such as safety glasses and gloves, when needed (see Fig. 5–9).

5. They will make certain that all guards are in place before starting a machine.

Analysis of Objectives

Fig. 5–9. Learning to work safely is part of industrial arts education. Notice safety glasses and face shield. The shavings appear to be on the boy's shirt, but the camera actually caught them in mid-air. (Courtesy, Willson Products Division, Reading, Pennsylvania)

6. They will call attention to unsafe tools or machines and will warn others who may be working in an unsafe manner.

7. They will promote safe practices in the laboratory and will be familiar with emergency procedures.

8. They will make sure that others cannot be hurt by the work which they are doing (e.g., flying chips, etc.).

9. They will be familiar with causes and effects of common industrial accidents.

Objective 9—Skill in Basic Industrial Processes

To develop a degree of skill in a number of basic industrial processes.

Expected behavior changes from students:

1. They will be able to select the correct hand tool or machine to perform a given operation.

2. They will be able to use common hand tools and machines safely and effectively.

3. They will perform tool processes with an increasing degree of accuracy.

4. They will continuously improve the quality of workmanship in their projects (see Fig. 5–10).

Fig. 5–10. An example of fine craftsmanship. This Hepplewhite armchair was designed and built by a high school senior in Everett, Massachusetts. It was a grand-prize winner in national competition. (Courtesy, Stanley Tools, New Britain, Connecticut)

5. They will develop pride in their craftsmanship.

6. They will grow in self-assurance as indicated by a willingness to attempt more difficult projects.

7. They will practice difficult operations in order to perfect their skills.

WHAT OBJECTIVE IS OF MOST WORTH?

The industrial arts teacher must be concerned with all of the objectives of industrial arts. This means that he should endeavor to bring about as many desirable behavior changes as possible for each objective. It is recognized that some students will progress much further than others in the number of desirable behavior changes which will be achieved. This is another way of saying that individual differences must be recognized in industrial arts as in any other subject. Other things being equal, however, the more of the desirable behavior changes that can be observed in any class, the more nearly have the objectives of the course been attained.

It must be realized, however, that neither the teacher nor his students will have the time or energy to engage in all the activities which will fully develop these behavior changes. Somewhere along the line some choices will have to be made. Some things will have to be done; some things will have to be left undone. A problem of immediate concern is what best can be done in industrial arts in the time available.

It is acknowledged that the school cannot equip a student with all that he needs to know to last him a lifetime. Tyler states it this way:

> With the rapid acquisition of new knowledge, it is no longer possible to give the student in school an adequate command of the facts in each major subject which will serve him throughout the balance of his life. The school can only start him on a life-long career of continued learning.[9]

It has been said that one-half of what we know today technically will be obsolete in ten years from today and that one-half of what we will need to know technically has not been invented yet.[10]

Like the school itself, the industrial arts program cannot equip its students with all the skills and knowledges they will need to last them a lifetime in our industrial and technological society.

The age-old question is still with us: What should the schools teach? How best can the school equip today's students for tomorrow's world? Realizing the immense complexity of this task, the Educational Policies Commission identified the development of the ability to think as the

---

[9] Ralph W. Tyler, "Frontiers in Industrial Arts Education," *Frontiers in Industrial Arts Education, 28th Annual Convention Proceedings.* (Washington, D.C.: American Industrial Arts Association, 1966), p. 18.

[10] Eric Walker, President, The Pennsylvania State University, University Park, Pennsylvania (circa 1966).

central purpose of education.[11] This matter was discussed more fully in Chapters 1 and 2 and the suggestion was advanced that *the development of the ability to think may well be regarded as the central purpose of industrial arts education*. This objective was derived earlier in this chapter from one of the three major goals of education, namely, to improve the emergent culture. The objective in terms of industrial arts education was stated as follows:

> To develop critical thinking as related to the materials, tools, and processes of industry and to encourage creative expression in terms of industrial materials.

If this is accepted as the cardinal aim of industrial arts, then the teacher may well give special preference to the development of critical thinking by means of problem-solving experiences with their resultant behavior changes.

SUMMARY

Objectives are vital to an educational program. They provide direction, guidance, and enable evaluation to take place. However, a mere statement of objectives is not enough; objectives must be translated into statements of desirable behavior patterns. These describe the typical behavior of a student after he has finished an educational program or a course. It is not expected that two teachers will prepare identical lists of behavior patterns upon analysis of the same objective. Nor is it expected that the students or teachers will be able to engage in all of the activities deemed necessary to bring about all the behavior patterns for the nine objectives of industrial arts. Ideally, students and teachers will strive for the attainment of as many desirable behavior patterns as possible in the limits of the time available.

The development of the ability to think is the central purpose of industrial arts education. For this reason the teacher and students should devote special attention to behavior changes that foster the growth and development of this ability.

DISCUSSION TOPICS AND ASSIGNMENTS

1. Select any objective of industrial arts and list desired behavior changes on (1) the elementary level, (2) the junior high school level, and (3) the senior high school level.

[11] *The Central Purpose of American Education,* Prepared by the Educational Policies Commission, National Education Association of the United States and the Association of School Administrators (Washington, D.C.: National Education Association, 1961), p. 12.

2. Under each objective for industrial arts, list as many additional desirable behavior changes as you can think of.

3. After an objective has been selected, what further steps should be taken before it can be used?

4. Do you think Mager's approach* to stating objectives has any implications for industrial arts education? Justify your answer.

5. Can the objectives of industrial arts education be fitted into the *Taxonomy of Educational Objectives?** Explain in detail.

6. Identify several behavioral outcomes described in French's *Behavioral Goals of General Education** to which industrial arts can make a significant contribution.

SELECTED REFERENCES

A Guide to Improving Instruction, Industrial Arts Policy and Planning Committee of the American Vocational Association. Washington, D.C.: American Vocational Association, 1953.

Bloom, Benjamin S. (ed.). *Taxonomy of Educational Objectives.* Handbook I—Cognitive Domain. New York, N.Y.: David McKay Company, Inc., 1956.

Darm, Adam E. "Using Behavioral Objectives and Problem-Solving in Kinematics," *School Shop,* vol. XXIX, no. 7 (March 1970), pp. 60–62.

Denver Public Schools, *Industrial Arts Education in the Denver Public Schools.* Denver, Colo.: Board of Education, 1952.

French, Will and associates. *Behavioral Goals of General Education in High School.* New York, N.Y.: Russell Sage Foundation, 1957.

Krathwohl, David R. "Stating Objectives Appropriately for Program, for Curriculum, and for Instructional Materials Development," *Journal of Teacher Education,* vol. 16, no. 1 (March 1965), pp. 83–92.

Mager, Robert F. *Preparing Instructional Objectives.* Palo Alto, Calif.: Fearon Publishers, Inc., 1962.

Maine State Department of Education. *Industrial Arts Technology: A Study of American Industry.* A Guide for Secondary Schools in Maine prepared by Department of Education, Vocational Division. Augusta, Maine: State Department of Education, 1965.

Morrison, Edward J. "The Use of Behavioral Objectives in Instructional Materials Development," *American Vocational Journal,* vol. 45, no. 2 (February 1970), pp. 46–48.

Pace, C. Robert. *College and University Environment Scales* (CUES). Preliminary Technical Manual. Princeton, N.J.: Educational Testing Service, 1963.

The Central Purpose of American Education. Educational Policies Commission, National Education Association of the United States and the American

*See *Selected References.*

Association of School Administrators. Washington, D.C.: National Education Association, 1961.

Tischler, Morris. "Conceptual Thinking in a Technical Society," *Industrial Arts and Vocational Education,* vol. 55, no. 6 (June 1966) pp. 26–28.

Wilber, Gordon O. "Method for Selection of Industrial Arts Activities," *The Industrial Arts Teacher,* vol. 4, no. 3 (March 1945), pp. 1, 4.

Chapter 6

Learning Activities in Industrial Arts

The learning activities in industrial arts education are of two general types: (1) manipulative and (2) informative. The manipulative includes all of the constructive activities carried on with the tools, machines, materials, and processes of industry. These include the projects which the student produces, the problems he solves, and the experiments he performs.

The second type of learning activity includes the informative lessons and concomitant learnings which take place in the industrial arts laboratory and which cannot be classified as manipulative. Older terms which have been used to describe this kind of learning include *related, related technical,* or *related information.* Under this heading are classified such lessons as, for example, the names of tools and materials, the production of steel, the story of mass production, industrial development and the conservation of natural resources, criteria for selecting an occupation, and countless others.

The problem immediately confronting the teacher is, "How am I to choose the proper learning activities for my classes?" "How am I to know what projects my students should make and what lessons I should teach?" It is the purpose of this and succeeding chapters to outline certain principles which underlie such choices.

BASIS FOR SELECTING LEARNING ACTIVITIES

There is one primary purpose for learning activities. That purpose is to achieve the objectives of the particular course in question. This fact holds true whether the subject is Latin, mathematics, Chinese, or industrial arts. While it is true that outcomes which are aside from the stated objectives, may be achieved, they are incidental and learning activities should not be chosen with such casual ends in view. A given learning activity, either informative or manipulative, should be selected on the sole criterion of whether or not it contributes to the specific behavioral outcomes which the teacher has in mind for a particular group of students. These expected behavioral patterns are, of course, based on objectives which are in harmony with the needs and interests of the pupils as well as those of the community and society in general.

It has been established in previous chapters that objectives really mean changes in behavior on the part of students. The test for judging whether any specific learning activity should be included in a given course is, therefore, to ask the question: "Does it contribute significantly toward bringing about one or more of the desired behavior changes?" If the answer is "yes," then that item may well become a part of the course. If, on the other hand, the answer is "no," then—regardless of how interesting or desirable that particular item may be—it should be rejected. This criterion holds true whether one is considering a new project, a field trip, a film, or whatever.

LEARNING ACTIVITIES FOR INDUSTRIAL ARTS

The expected outcomes of a general shop program were listed in the preceding chapter. Once these are established, it is relatively easy to determine what learning activities should be selected. If one knows his destination, it is a simple matter to determine a route which will bring him there. The teacher has only to consider the various behavior changes which have been selected, and then to list the student and teacher activities, such as, industrial visits, lessons, demonstrations, and projects, which appear to have the greatest likelihood of bringing them about. Chart IV illustrates how this might be done for one of the objectives of industrial arts—namely, safety education. The chart is a follow-up of Charts I and III. It will be recalled that Chart I (see Chapter 4) graphically depicts the derivation of the objectives for industrial arts. Chart III (see Chapter 5) continues with an analysis of these objectives in terms of expected behavior changes from students when these objectives are attained. Now, Chart IV completes the analysis by identifying certain student and teacher activities deemed appropriate to effect the desired behavior changes.

It should be recognized, however, that while the teacher must have objectives and must have a clear concept of the behavior changes which he expects to bring about, identical methods for the achievement of such aims cannot be used for all students in the class. Each student is a unique individual. His interests, attitudes, and above all, his needs will determine the exact methods which the teacher will use in achieving

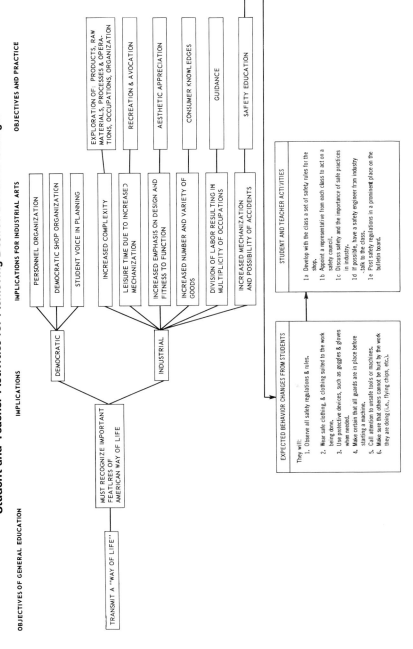# CHART IV
Student and Teacher Activities for Achieving Desired Behavior Changes

the desired behavior changes. Each teacher must study his students and must teach on the basis of their present needs and interests. Since no two members of the class will be exactly alike, a preconceived series of experiences will not be applicable to every member of the class. It is well for the teacher to have in mind a large number of activities that might be used and to realize that new methods and devices may need to be developed to meet unusual situations.

The following sections of this chapter offer specific examples to assist a teacher in selecting learning activities for an industrial arts course. Each expected behavior change from students is accompanied by a list of suggested student and teacher activities which might be utilized to bring about these behavior changes.

Objective 1

To explore industry and American industrial civilization in terms of its evolution, organization, raw materials, processes and operations, products, and occupations.

Expected Behavior Changes From Students	Student and Teacher Activities
1. They will be able to list the major industries of our country, locate them geographically, and discuss principal products and manufacturing processes.	a. Assign readings and reports and have class discussions. b. Organize committees to study particular industries and present findings. c. Draw maps of industrial locations. d. Use films from industrial firms. e. Have optional reading list.
2. They will recognize scientific principles underlying industrial operations and processes.	a. Relate scientific principles involved in demonstrations and experiments. b. Invite science teacher to talk to the class.
3. They will visit industries whenever possible to learn about methods, processes, and products.	a. Plan industrial visits and stress the importance of looking for definite things. b. Have reports and follow-up class discussions on visits. c. Use motion pictures to supplement trips or where trips are not feasible. d. Encourage students to make their own trips.

Learning Activities in Industrial Arts

4. They will recognize industrial processes and methods and will attempt to relate them to the industrial arts laboratory (see Fig. 6–1).	a. Plan visits to observe industrial processes and methods. b. In demonstrating show similarities and differences between school shop methods and industrial methods. c. Organize line-production projects; make comparisons with industrial methods. d. Encourage students to use industrial processes and methods wherever possible in their projects. e. Use motion pictures, filmstrips, and film loops to illustrate industrial processes.
5. They will be able to understand new industrial processes by comparing them to more familiar ones.	a. Recommend and assign readings which describe new or unusual processes of industry. b. Have discussions on how improved processes relate to older ones.
6. They will be able to define common industrial terms.	a. Give homework assignments, e.g., assignment sheets. b. Use vocabulary drill. c. Include terminology in quizzes.
7. They will be familiar with the organization of industry and relate the personnel organization of the industrial arts shop to similar systems in industry. Their cooperation in the personnel system will increase.	a. Hold discussion lessons on general organization of industry. Explain duties and functions of Board of Directors, stockholders, etc. Show lines of authority with a chart. b. Develop personnel organization with class; point out similarities and differences in comparison with industry. c. Visit a typical industry and talk with superintendent and others about their work and responsibilities. d. Have meetings with officers of school personnel organization and discuss problems of interest.
8. They will be familiar with the sources of some of the raw materi-	a. Exhibit various raw materials and talk about them with the class.

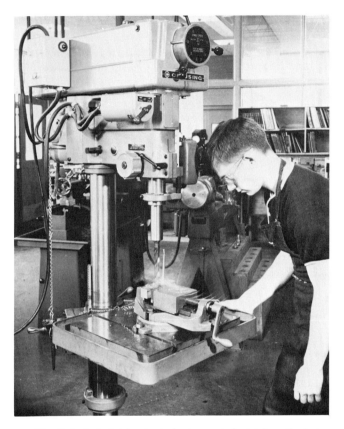

Fig. 6-1. Industrial arts students use industrial methods in the school shop. (Courtesy, Clausing Corporation)

als of industry and will be able to discuss their transportation, processing, and industrial uses.

b. Assign readings and reports. Have students make maps.
c. Use motion pictures, slides, study prints, and other visual aids to illustrate transportation and uses of raw materials.
d. Develop a collection of raw materials in the laboratory.
e. Encourage the use of a variety of raw and semifinished materials in student projects.
f. Have students collect samples of industrial materials.

9. They will read about industry, its historical development, current

a. Assign readings about selected industries and their development.

Learning Activities in Industrial Arts 107

trends, and the lives of important inventors and industrial leaders.

    Choose those, at first, which are naturally interesting to youths, such as, the aerospace industry, automobile production, etc.
  b. Have short reports on readings.
  c. Encourage students to bring in newspapers or magazine articles dealing with current industrial trends.
  d. Assign papers on the lives of inventors and have oral reports.
  e. Listen to recordings depicting reenacting of the birth of great inventions.
  f. Have students construct models of famous inventions.

10. They will seek information about new inventions and developments in industry and technology.

  a. Use bulletin board and displays to stimulate interest in new developments.
  b. Use motion pictures of new processes and materials.
  c. Call attention to new developments during industrial visits.
  d. Encourage students to bring in clippings and magazine articles on new industrial developments.
  e. Use newer materials in demonstrations, if available.
  f. Prepare information sheets.

11. They will be able to interpret the language of industry—industrial and technical drawing.

  a. Demonstrate sketching and elements of drafting.
  b. Show films and film loops.
  c. Have students make sketches of their projects.
  d. Make blueprint reading assignments.
  e. Give blueprint reading quizzes.
  f. Have resource person, e.g., engineer, architect, or contractor, speak to class.
  g. Let pupils examine "real" industrial drawings.

12. They will study about and help conserve our nation's natural resources.

  a. Assign readings and have class discussions.
  b. Invite social studies teacher to talk to class.

13. They will learn more about and contribute time and effort to those concerned with pollution control of our environment.

   c. Use films and recordings.
   d. Practice economy in use of laboratory materials.
   a. Use readings, reports, and discussions.
   b. Invite ecology expert to speak to class.
   c. Hold community-wide seminar on pollution control.
   d. Have pupils identify service project in community and work toward its completion.

14. They will understand some of the fundamentals of economics by which industry operates.

   a. Use assignment sheets based on library readings.
   b. Have reports and class discussion.
   c. Invite economics teacher to speak to class.

15. They will read about the evolution of labor unions and will be able to interpret current problems of capital, management, and labor.

   a. Assign readings and reports on various phases of management-labor problems.
   b. Present both sides of labor-management issues in class discussions.
   c. Invite local resource people (labor steward and management representative) to present their views. If unable to schedule, consider video-tape or telectures.
   d. Include aspects of capital, management, and labor in setting up line-production projects.

Objective 2

To develop recreational and avocational activities of a constructional nature.

Expected Behavior Changes From Students	Student and Teacher Activities
1. They will read such magazines as *Popular Mechanics, Home Craftsman,* and *Popular Science.*	a. Make these and similar magazines available in the laboratory. b. Call special attention to outstanding articles, projects, and pictures to be found in them. c. Use projects from such sources as

Learning Activities in Industrial Arts

	a basis for demonstrations and experiments.
	d. Send students to such sources for project ideas.
2. They will ask advice on how to carry on constructive activities out of school.	a. Let the students know that you are interested in their extra-curricular constructive activities.
	b. Talk with each student about his out-of-school interests.
	c. Take time to answer all questions concerning such activities fully and accurately.
	d. Encourage students to bring their problems to class with them.
3. They will become interested in, and will engage in, one or more constructional hobbies.	a. Discuss hobbies and explain their value. Cite cases where hobbies have contributed to progress and success.
	b. Give demonstrations so as to indicate the recreational aspect of handicraft.
	c. Stage a hobby show.
	d. Promote contests along hobby lines.
	e. Promote projects which have high hobby possibilities.
4. They will spend spare time in the shop either in school or at home,	a. Arrange to make the laboratory available at certain times for those who wish to carry on extra work.
	b. Use the bulletin board to suggest interesting things that students may make.
	c. Indicate by example the pleasure of working on constructive projects.
	d. Display projects which will help students enjoy such hobbies as athletics, fishing, hiking, and scouting.
	e. Cooperate with organizations like the Boy Scouts by allowing students to work out merit badge requirements.
5. They will ask questions and talk about their hobbies.	a. Have individual and group conferences with students concern-

	ing their hobbies. Let them know that you are interested. b. Have class members report on their hobbies. c. Have students bring to class examples of work done on a hobby basis and display them for others to see and talk about.
6. They will consult catalogs for information about their hobbies.	a. Have catalogs of craft supplies and equipment available in the laboratory. b. Refer students to catalogs for answers to their questions. c. Call attention to the fact that catalogs frequently contain ideas for projects and information on processes. d. Display attractive catalogs and leaflets on the bulletin board.
7. They will contribute to class discussions with information gained from reading along lines of their interests.	a. Assign topics concerned with hobbies and let the students report on them. b. Suggest books and articles for the students to read. c. Call on certain students for information about hobbies in which you know they are interested.
8. They will take the initiative in visiting industries along the lines of their interests.	a. Post a list of local industries where visitors are welcome. b. Show your interest by indicating the type of information that can be secured from industrial visits. c. Have students tell the class about visits they have made. d. Display pictures of local plants and industries.
9. They will make the acquaintance of, and form friendships with, others having similar interests.	a. Initiate informal clubs for students having similar interests. b. Bring together students from different classes who have common interests. c. Urge the formation of hobby clubs outside of school. d. Make available movies and other visual aids for interested groups.

Learning Activities in Industrial Arts

10. They will develop home workshops.	a. Display pictures of home workshops.
	b. Demonstrate processes and projects suitable to the home workshop.
	c. Promote an exhibit of work done outside of school.
	a. Encourage students in school to make equipment and tools which can be used in the home workshop.
	e. Have each student select a possible location for a home workshop; develop a floor plan.
	f. Visit home workshops and advise students concerning improvements.
	g. Feature home workshop projects in the school display case.
11. They will design and construct projects in their laboratory work which are related to their hobby interests.	a. Encourage students to select projects that are along the line of their hobby interests.
	b. Display models of projects which are related to popular hobbies.
	c. Help students to think along lines related to their hobbies when choosing and planning projects.
	d. Discuss hobbies in class and suggest projects that can be made in the laboratory.

Objective 3

To increase an appreciation for good craftsmanship and design, both in the products of modern industry and in artifacts from the material cultures of the past.

Expected Behavior Changes From Students	Student and Teacher Activities
1. They will know the names of outstanding craftsmen and designers of the past and the present and will be able to recognize the work for which they are known.	a. Assign reports on early craftsmen, such as Chippendale, Sheraton, Hepplewhite, Cellini, and Revere.
	b. Use flash cards of designers and examples of their work.

2. They will be able to recognize pieces of period furniture as well as contemporary styles and trends.

    c. Have speakers on various types of early handicrafts.
    a. Make a study of period styles of period furniture.
    b. Make a trip to a museum to study period products.
    c. Visit stores or a home where period furnishings can be studied.
    d. Make comparison chart of period styles.

3. They will appreciate good design in artifacts and will show such appreciations in speech and actions.

    a. Discuss the principles of design which are unchanging.
    b. Show examples of good design in woods, metals, etc.
    c. Display pictures of early craftsmanship.
    d. Visit a museum to study examples of early handicrafts.

4. They will understand the principles of good design and will be able to recognize them in the products of industry.

    a. Discuss principles of good design and their application to industrial arts projects.
    b. Use visual aids to develop an appreciation of what constitutes good design.
    c. Show how fitness to function and beauty are related.
    d. Criticize plan sheets in terms of design.

5. They will be able to apply the principles of good design in the construction of their projects.

    a. Help students redesign their project ideas.
    b. Correlate work with that of the art department.
    c. Promote contests for the best suggestions for redesigning a project that is not entirely satisfactory for the purpose.

6. They will recognize and avoid poor design and overdecoration.

    a. Discuss purpose of decoration. Show examples of poor design and nonfunctional decoration.
    b. Help students develop project designs which are in good taste.
    c. Provide opportunities for students to choose between good and poor designs.

Learning Activities in Industrial Arts

	d. Show how overdecoration tends to hide beauty of line and proportion.
7. They will understand some of the physical properties of certain industrial materials and will select materials which are appropriate to the design and intended use.	a. Discuss limitations and characteristics of various materials. Show that a design which is good for one material may be impractical for another. b. Criticize designs in terms of the suitability of materials. c. Show the relationship between designs of early craftsmen and materials which were available. d. Show how the development of new materials and machines has made possible changes and improvements in design. Show characteristics of, and possibilities in, new materials.
8. They will appreciate good craftsmanship and construction in industrial products.	a. Encourage students to reproduce simple antiques. b. Discuss principles of sound construction.
9. They will be able to criticize constructively the designs of others.	a. Give class a poor design and ask them to criticize it. b. Have students critique each other's designs. c. Display work done by members of the class and have discussions of the designs used. d. Have the class choose the best designs and tell why they are good.

Objective 4

To increase consumer knowledge to a point where students can select, buy, use, and maintain the products of industry intelligently.

Expected Behavior Changes From Students	Student and Teacher Activities
1. They will examine the products of industry carefully and will judge their value before buying.	a. Show that two articles which look alike may not be of equal value, e.g., a standard hammer and a dime-store product. b. Emphasize the reasons for weigh-

	ing values when making a purchase.
	c. Ask students to list the things they would look for in buying, for example, a jackknife, a bicycle, or a motor bike.
	d. Outline the principles of comparative shopping.
	e. Have students "go shopping" in the latest edition of *Buying Guide of Consumer Reports*, for say, a colored television.
	f. Visit a local store and report on various grades and prices of articles, e.g., toasters, patio furniture, and the like.
2. They will understand the common processes by which materials are shaped, formed, and assembled in industry (see Fig. 6–2).	a. Use films and film loops.
	b. Analyze industrial processes in terms of principles.
	c. Provide as much "hands on" experience as possible.
3. They will look for constructional features in judging the quality of a product.	a. Demonstrate the difference which construction makes in the strength and appearance of an article, i.e., mortise and tenon versus butt joints, welding versus soldering, etc.
	b. Promote projects with different types of construction; point out the special uses and applications of each.
	c. Show examples of poor construction in articles built to sell at a competitive price.
4. They will become acquainted with trade names and will look for proven brands when purchasing.	a. Discuss merits as well as disadvantages of buying brand-name merchandise.
	b. Make lists of advertised brands of various products.
	e. Devise simple tests of quality and test both marked and unmarked products.
	d. Discuss the importance of recognizing the stamp of approval of organizations, such as National Board of Fire Underwriters,

Learning Activities in Industrial Arts   115

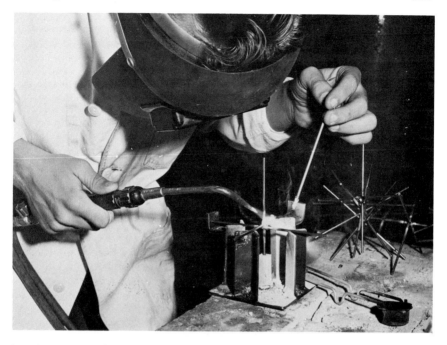

Fig. 6–2. Knowing how to weld will help make this student a better consumer of industrial products. (Courtesy, California State Department of Education)

5. They will recognize quality and will buy accordingly.

6. They will buy on the basis of their needs, rather than entirely on the basis of price or salesmanship.

7. They will maintain and use manufactured articles in such a way as to prolong their life and usefulness.

Good Housekeeping, U.S. Coast Guard, etc.

a. Have students examine samples of various articles and tell why they would choose one or the other.
b. Show relationship between quality and value. Point out that a low price does not always indicate a bargain.

a. Have class give examples of purchases which were made without regard to need.
b. Have students bring in examples of advertisements which aim to stimulate buying, regardless of need.

a. Discuss general maintenance procedures, e.g., lubrication, painting, and inspection.
b. Discuss and demonstrate specific

	maintenance practices for articles which students use, e.g., bicycles, fishing tackle, and home workshop tools.
	c. Teach maintenance of laboratory equipment. Make this part of the personnel organization.
	d. Encourage students to bring in maintenance problems from home, e.g., electric cords, bicycles.
	e. Discuss types of maintenance which should NOT be undertaken at home.
8. They will study sales materials carefully and sift out propaganda and false advertising from honest sales efforts.	a. Ask students to look up articles and books which show how easily one may be tricked into buying worthless goods.
	b. Discuss sales psychology and sales propaganda. Show that an attempt is continuously being made to persuade people to buy, regardless of need, or ability to pay.

Objective 5

To provide information about, and insofar as possible, experience in, the basic processes of many industries in order that students may be more competent to choose a future vocation.

Expected Behavior Changes From Students	Student and Teacher Activities
1. They will read about and will discuss occupations and the world of work with their friends, teachers, and parents (see Fig. 6–3).	a. Outline major occupational groups.
	b. Discuss important points to consider in studying an occupation.
	c. Assign readings on various occupations for class discussion.
	d. Invite guidance counselor to speak to class.
	e. Use guidance films and filmstrips.
	f. Use assignment sheets based on the *Dictionary of Occupational Titles.*
	g. Have students prepare a guidance folder covering intensive study of an occupation.

Learning Activities in Industrial Arts 117

Fig. 6–3. These students and teacher are discussing occupations in the world of work. (Courtesy, California State Department of Education)

2. They will choose elective courses in school which provide additional information about occupations.

   a. Discuss choices of elective courses.
   b. Point out courses that give additional occupational information, e.g., mechanical drawing (draftsman, engineer); chemistry (pharmacist).

3. They will visit industries or watch with interest motion pictures showing workmen at various trades and occupations.

   a. Make industrial visits; use motion pictures when visits are impracticable.
   b. Discuss values of concentrating closely on working conditions, skills, etc. of workers shown in motion pictures.
   c. Encourage students to visit industries on their own.
   d. Have reports from students who have visited interesting industries.

	e. Show films and film loops on various industrial processes. Discuss or test the class on their observations.
4. They will talk with workers in various trades and occupations concerning the work in which they are engaged.	a. Assign students or committees to talk with workers in various occupations. b. Have workers from industry visit the laboratory and talk to the class. c. Encourage students to talk with family relatives about their work and report to the class.
5. They will be able to cite wages, entrance requirements, working conditions, and opportunities for advancement for various occupations.	a. Assign readings, oral reports, and class discussions. b. Visit industries and talk with workers. c. Have students prepare reports on future employment prospects in various occupations and report to class. d. Show motion pictures of occupations. e. Prepare information sheets and assignment sheets. f. Have books and magazines available for supplemental reading.
6. They will be familiar with local, state, and federal laws as these relate to industrial employees.	a. Class reports and discussions. b. Invite employment counselor to speak to class.
7. They will be familiar with the social security benefits for industrial workers.	a. Invite local social security representative to speak to class. b. Assign committee reports.
8. They will be familiar with the health risks and dangers of various jobs and occupations in several industries.	a. Invite safety engineer or representative of local safety council to speak to class. b. Readings, reports, and class discussion. c. Use safety education films.
9. They will decide that they are not fitted for or are not interested in certain vocations. They will make a tentative choice of a vocation or family of occupations (see Fig. 6–4).	a. Give tryout experiences in as broad a sampling of industries as possible. b. Provide information about a large number of jobs and occupations.

Learning Activities in Industrial Arts 119

    c. Use aptitude and interest tests; discuss results with students.
    d. Be on the alert for aptitudes and advise students accordingly.
    e. Talk with individuals about their vocational plans for the future.
    f. Have students indicate present plans for future vocation and list reasons. Discuss choices with individuals.
    g. Have students list career choices in order of preference.

Objective 6

To develop critical thinking as related to the materials, tools, and processes of industry and to encourage creative expression in terms of industrial materials.

Expected Behavior Changes From Students	Student and Teacher Activities
1. They will be able to recall and apply the steps in critical thinking, that is, the scientific method of problem solving.	a. Teach steps in scientific problem solving; give examples; and test for recall.
2. They will increasingly attempt to solve their own problems.	a. Withdraw help gradually from students in the preparation of plan sheets as they grow in ability to prepare them.
	b. Refer students to accessible sources of information whenever possible to do so.
	c. Discuss sources of information where answers to their questions may be found.
	d. Help students enjoy success in problem solving, especially in their initial efforts.
3. They will think through the correct procedure for making a project and will then follow their plan.	a. Demonstrate method of preparing a plan sheet.
	b. Check work in progress to see that plan is being followed.
	c. Display well-developed plan sheets.

Fig. 6-4. This industrial arts student is participating in a 10th grade line-production project prior to out-of-school work experiences in the 11th and 12th grades—all of which is intended to help him make an intelligent occupational choice. (Courtesy, James O. Reynolds, Dayton Public Schools, Dayton, Ohio)

4. They will realize that it is as important to be able to plan a project as to construct it.

  a. Require all students to submit plan sheets with procedure outlined before permitting work to begin.
  b. Impress students with importance of planning by exercising great care in checking plan sheets.

5. They will be able to solve simple problems involving the tools, materials, and processes of industry in construction and repair problems at school and at home.

  a. Provide "hands on" experiences with tools and materials.
  b. Let student do thinking about solutions to problems, not the teacher.

	c. Advise students of kinds of repair jobs NOT to be attempted at home.
	d. Use repair jobs only if they have educational value.
6. They will be able to formulate and test hypotheses for given problem situations (see Fig. 6–5).	a. Discuss purpose of hypothesis and give examples.
	b. Give common problem, have pupils formulate hypothesis; discuss best solutions.
	c. Present problem situations of increasing difficulty and call on pupils to give tentative soluions.
	d. Have variety of problems at several ability levels so each pupil may enjoy a degree of success.

Fig. 6–5. These junior high school students are testing a hypothesis related to rocket-fin design. (Courtesy, George M. Haney and Alan D. Brown, Earle B. Wood Junior High School, Rockville, Maryland)

7. They will experiment with new ways to solve construction problems and will make improvements on the basis of their findings.

   a. Encourage students to think of new ways of solving their problems.
   b. Demonstrate different ways of doing a given construction problem.
   c. Call attention of class to any unusual method or technique developed by one of its members.
   d. Praise students for creative thinking.

8. They will evaluate the work of others in accordance with the principles of scientific thinking.

   a. Show how to evaluate using sample problems.
   b. Have pupils critique each other's solutions to problems.
   c. Use check lists to assist evaluation.

9. They will be able to apply knowledge of the physical properties of industrial materials in the creation of new ideas, designs, and in the construction of projects.

   a. Discuss properties of materials and their limitations in terms of design.
   b. Show by demonstration and example the correct use of materials.
   c. Demonstrate what happens when the wrong material is used for a given purpose.
   d. Have class study uses made by industry of important materials. Relate these applications to the laboratory.
   e. Check with students on their choice of materials.
   f. Check plan sheets carefully for the correct use of materials.
   g. Set up design problems in increasing order of difficulty.
   h. Prepare a test with pictures of various projects and let class suggest most suitable materials to use in making them.

10. They will be able to take ideas from different sources and create new designs or products combining elements of creativity and beauty.

    a. Have available many books, pictures, magazines, and drawings from which students may secure ideas for new designs.
    b. Demonstrate methods of changing a design to make it fit a different purpose.

| | c. Check plan sheets for opportunities to improve designs.
d. Give practice in changing and adapting designs.
e. Give individual help in the development of designs. |
|---|---|
| 11. They will be able to analyze tool and machine failures and identify causes for such breakdowns. | a. Discuss cause and effect relationships.
b. Present trouble-shooting techniques.
c. Set up problem situations with tools and machines and have pupils analyze causes of failures. |

Objective 7

To develop desirable social relationships, such as cooperation, tolerance, leadership and followership, and tact.

Expected Behavior Changes From Students	Student and Teacher Activities
1. They will develop group spirit and loyalty.	a. Organize class as a group with self-elected leaders.
b. Keep records of group accomplishment.
2. They will cooperate with others in group activity.	a. Discuss need for cooperation in order to promote group progress.
b. Give personal praise for cooperation.
c. Be on alert for cases of noncooperation and discuss the problem with those concerned.
d. Provide group problems where cooperation is needed for a solution, e.g., a production project, keeping the laboratory clean, etc.
3. They will assume and discharge leadership responsibilities in the student personnel organization (see Fig. 6-6).	a. Post list of responsibilities.
b. Meet with leaders to discuss policy.
c. Have individual conferences with those who fail to assume responsibilities.
d. Give support to those charged with responsibilities.

Fig. 6–6. Learning to clean up the laboratory develops followership qualities; maybe next week this boy will assume a leadership role in the personnel organization. (Courtesy, California State Department of Education)

4. They will organize or participate in club activities.
   a. Discuss types of clubs related to industrial arts.
   b. Meet with any group that wishes to discuss the formation of a club.
   c. Act as sponsor of one or more clubs and help develop the program.

5. They will accept leadership responsibilities in club organizations.
   a. Work with leaders in promoting live programs.
   b. Discuss function of leadership and need for it in clubs.
   c. Have individual conferences with those who seem to have leadership ability.
   d. See that all pertinent comments and suggestions by club members are given due consideration.

6. They will give help and advice willingly. They will seek help from others when in need.
   a. Assign advanced students to help those who need assistance.
   b. Give suitable praise to those who help others with their problems.

7. They will accept assignments given them by leaders in the personnel organization and will recognize the leadership of others.	a. Discuss the function of leadership in industrial organization and show how it is related to the student personnel plan. b. Indicate the need for cooperation in order to accomplish results. Give full support to all officers. c. Have personal conferences with students who fail to cooperate or with leaders who are unreasonable in their requests.
8. They will work willingly with individuals who may be of a different race, creed, or color.	a. Be impartial to all members of the class. b. Give members of minority groups opportunity to display their abilities. c. Assign to work stations and duties without regard to personalities. d. Confer personally with any student who shows evidence of intolerance. e. Study the contributions of many nations to handicraft and skill. Show examples.
9. They will respect the work, efforts, and rights of others.	a. Permit no horseplay or shouting in the laboratory. b. Teach respect by being respectful of others yourself.
10. They will seek to understand those whose views may differ from their own.	a. Teach tolerance by being tolerant yourself. b. Give every class member a right to be heard.

Objective 8

To develop safe working practices.

Expected Behavior Changes From Students	Student and Teacher Activities
1. They will develop a high degree of safety consciousness.	a. This is a composite of understandings, attitudes, and appreciations developed by activities suggested below.

2. They will be able to cite and will observe all safety rules and regulations.

a. Develop with the class a set of safety rules for the laboratory.
b. Appoint a representative from each class to act on a safety council.
c. Discuss safety and the importance of safe practices in industry.
d. Invite a safety engineer from industry to talk to the class.
e. Post safety rules in prominent places in laboratory.
f. Observe students at work; insist on strict observance of safe practices.
g. Observe all safety rules yourself.
h. Administer tests on safety rules.

3. They will wear safe clothing and clothing suited to the work being done (see Fig. 6–7).

a. Wear the type of clothing you expect students to wear.
b. Observe all safe practices when demonstrating, e.g., rolled sleeves, necktie tucked in, etc.
c. Show motion pictures and call attention to safe clothing or to unsafe practices.
d. Use personnel organization as a means of checking students for unsafe dress.

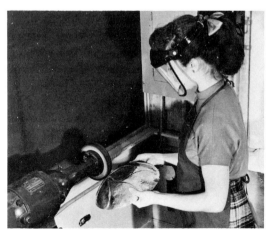

Fig. 6–7. This industrial arts student has given proper attention to clothing and eye protection for the type of work she is doing. (Courtesy, California State Department of Education)

Learning Activities in Industrial Arts

4. They will use protective devices, such as goggles and gloves, when needed.

   a. Use proper safety devices when demonstrating.
   b. Use posters, slides, or still pictures to show correct use of safety devices. Figure 6-7 shows a student correctly protected for buffing.
   c. Discuss importance of using all possible protective devices.
   d. Take pictures of students who are properly protected and post photos on bulletin board.
   e. Provide safety glasses for each student.
   f. Insist that all visitors wear safety glasses in danger zones of laboratory.
   g. Provide some means of sanitizing safety glasses.

5. They will make certain that all guards are in place before starting a machine.

   a. Use proper guards when demonstrating.
   b. Have guards properly color coded.
   c. Check students for proper use of guards before allowing use of dangerous machines.
   d. Show motion pictures, slides, or still pictures of safe practices and call attention to use of guards.
   e. Use safety posters near machines.

6. They will call attention to unsafe tools or machines and will warn others who may be working in an unsafe manner.

   a. Discuss responsibility of each person to call attention to unsafe tools or machines or to unsafe usage.
   b. Have "out-of-order" signs ready to hang on unsafe machines.
   c. Include safety reporting in student personnel organization.
   d. Give recognition to all students reporting unsafe tools, machines, equipment.

7. They will promote safe practices in the laboratory and will be familiar with emergency procedures.

   a. Review and practice emergency procedures for leaving laboratory.
   b. Establish procedure for reporting all accidents, no matter how trivial.

	c. Discuss locations and use of emergency equipment, e.g., panic buttons and fire extinguishers.
	d. Discuss and practice emergency procedures, e.g., in finishing room, in metals casting, etc.
8. They will make certain that others cannot be hurt by the work they are doing (i.e., flying chips).	a. Call attention during demonstrations to care being taken to protect members of class from harm.
	b. Check with students who are starting a new type of work and warn them of possible hazards to others.
	c. Use motion pictures, slides, or still pictures to impress need for being alert for possible danger to others.
9. They will be familiar with causes and effects of common industrial accidents.	a. Teach cause and effect relationships.
	b. Assign readings and reports for class discussion on industrial accidents.
	c. Have safety representative speak on accidents and their prevention.
	d. Show motion pictures.
	e. Set up simulated accident situations and have students provide causes.

Objective 9

To develop a degree of skill in a number of basic industrial processes.

Expected Behavior Changes From Students	Student and Teacher Activities
1. They will be able to select the correct hand tool or machine to perform a given operation.	a. Discuss and demonstrate correct uses of tools and machines which pupils may use.
	b. Check student plan sheets for intended use of tools and machines.
	c. Set up problems whose solutions involve correct selection of tools or machines.

Learning Activities in Industrial Arts

2. They will be able to use common hand tools and machines safely and effectively.
   a. Teach safe practices in tool and machine usage.
   b. Give as many experiences with common tools and machine operations as desirable.
   c. Observe students at work; make corrections and suggestions for improving tool and machine operation.
   d. Give performance tests.

3. They will perform tool processes with an increasing degree of accuracy.
   a. Demonstrate tool processes, holding to a high standard of accuracy.
   b. Set up with the class standards for accuracy which are to be expected for each area of work.
   c. Check projects frequently during the process of construction.

4. They will continuously improve the quality of workmanship in their projects.
   a. Accept only work that represents what may reasonably be expected from each student.
   b. Compare early and later projects and point out where improvement has taken place as well as where further effort should be directed.
   c. Visit an industry where precision work is being done and point out examples of accuracy and fine workmanship.

5. They will develop pride in their craftsmanship.
   a. Exhibit examples of fine workmanship by members of the class.
   b. Praise individuals who do good work on some particular project or part.
   c. Award prizes for good workmanship.

6. They will grow in self-assurance as indicated by a willingness to attempt more difficult projects.
   a. Demonstrate increasingly more difficult techniques.
   b. Encourage students to make each project more difficult than the one before.
   c. Keep the parents informed concerning the progress of students.

7. They will practice difficult operations in order to perfect their skills.	a. Encourage students to practice difficult operations on scrap stock, e.g., mortise and tenon joint, turning a burr, spinning a bead, etc. b. Give frequent performance tests covering difficult skills.

Again it should be pointed out that the list of student and teacher activities outlined in this chapter is not intended as an exact guide or a course of study. No two teachers will desire exactly the same behavior changes nor use the same techniques to bring them about. Neither is it expected that all the suggestions mentioned in this chapter could be attempted within the compass of a single course. The example presents, however, a *method* for attacking the problem and suggested learning activities from which the teacher may choose those items that will most effectively meet his purposes.

SUMMARY

Every industrial arts teacher must answer the important question: "What shall I teach?" The answer lies in the selection of lessons, demonstrations, and other student and teacher activities which will promote the desired behavior changes. It has been the purpose of this chapter to show in detail a method which may be used to assemble learning activities for the industrial arts program.

DISCUSSION TOPICS AND ASSIGNMENTS

1. Contrast and compare the method for the selection of learning activities as outlined in this chapter with the trade-analysis technique used in vocational education (see Selected References).

2. Is the trade-analysis technique satisfactory for selecting learning activities in industrial arts? Why?

3. Select several objectives in this chapter and divide the student and teacher activities into two groups representing, respectively, manipulative activities and nonmanipulative activities. Does one group seem to dominate?

4. Make additions to "Student and Teacher Activities" which you think will promote desirable behavior changes in students for any of the objectives listed in this chapter.

5. Criticize or defend this statement: All subject matter is a means to an end.

6. Is it educationally sound to develop a standard course of study which is applicable to all students of a given grade level? Why?

7. Enumerate "pros" and "cons" for the topic: A National Curriculum for Industrial Arts.

SELECTED REFERENCES

Bollinger, Elroy W., and Gilbert O. Weaver. *Trade Analysis and Course Organization for Shop Teachers.* New York, N.Y.: Pitman Publishing Corporation, 1955.

Feirer, John. "A National Curriculum for Industrial Arts," *Industrial Arts and Vocational Education,* vol. 50, no. 9 (November 1961), p. 17.

―――. "More on a National Curriculum for Industrial Arts," *Industrial Arts and Vocational Education,* vol. 52, no. 8 (October 1963), p. 15.

Friese, John F. "Analysis of Course-of-Study Materials for Industrial Arts," *Industrial Arts and Vocational Education,* vol. 42, no. 7 (September 1953), pp. 208–211.

Fryklund, Verne C. *Analysis Technique for Instructors.* Milwaukee, Wis.: The Bruce Publishing Company, 1965.

Wilber Gordon O. "A Method for the Selection of Industrial Arts Activities," *The Education Digest,* vol. 10 (April 1945), p. 53.

Chapter **7**

Organizing Learning Activities

General procedures were outlined in the previous chapter for identifying learning activities for industrial arts classes. Unanswered, however, were many specific questions, such as, "Which lessons shall be selected and when shall they be taught?" and "What types of construction activities or projects shall be used?" The answers to these and other questions lie in the organization of these learning activities on some logical basis. In this respect, much depends on the judgment of the individual teacher as he applies general principles to his particular situation. Practices and techniques which have been found successful may be helpful. This chapter will outline a plan and suggest principles for the organization of learning activities in industrial arts.

ANALYZE OBJECTIVES IN TERMS OF STUDENT BEHAVIORAL OUTCOMES

Certainly the first step in organizing content for a given course is to determine the objectives for that course. It should be equally clear that the second step is the analysis of these objectives in terms of desired behavior changes. Objectives and behavior changes have been discussed at length in Chapters 4, 5, and 6. Only after these two steps have been taken can the teacher begin to organize the student and teacher activities for a course. The lessons, demonstrations, projects, and activities which are utilized to bring about the desired behavior changes are the raw material from which the course is made.

ORGANIZATION OF LEARNING ACTIVITIES

The average industrial arts teacher finds it more difficult to organize subject content and learning activities into a planned course of study than to teach the skills required for the manipulative activities. One of the reasons for this is that traditionally industrial arts has emphasized the "doing" over the "knowing" aspects. A contributing reason for this may be that physical conditions in the laboratory are not always conducive to formal instruction. That is, even today many shops do not contain an area where an entire class can gather for an informative lesson. Besides, students are anxious and eager to begin to use tools, materials, and machines immediately upon their arrival in the laboratory; they attend only half-heartedly to the so-called informative aspects of the work. For these and other reasons teachers tend to neglect the informative lessons and to concentrate on the manipulative aspects. This problem is not a new one in industrial arts because more than sixty-five years ago Charles R. Richards wrote

> . . . We are facing the question now as we devote our attention to miscellaneous and more or less meaningless projects, or whether we shall seek in an orderly way to develop insight into the basic industries of our time and a knowledge of some of the steps through which these have reached their present form.[1]

It is difficult to understand how an industrial arts course can be justified on the basis of manipulative activities alone. As one reviews any list of desired behavior changes, it will quickly become evident that less than half of them relate to the mastery of tool skills and techniques. *It appears, therefore, that the informative aspects of industrial arts are fully as important (if not more so) than the tool processes.* Some years ago the American Vocational Association said "The general education commitment precludes limiting industrial arts instruction to manipulative-construction activities alone.[2] Statements like this and others have contributed to what appears to be a slowly growing acceptance of this point of view throughout the field.

The question which immediately faces today's teacher becomes, then: "What lessons shall I teach and in what order shall I teach them?" Clearly it is psychologically unsound simply to select lessons at random from a list of activities designed to effect certain behavior changes and to teach such lessons as isolated items of information, unrelated to anything taught before or to follow. To do this would be like teaching various historical facts without any consideration for their chronological sequence. Obviously, it would be impossible to remember facts so taught or to organize them into any reasonable

---

[1]Charles R. Richards, "A New Name," *Manual Training Magazine*, vol. 6, no. 1 (October 1904), pp. 32–33.
[2]*A Guide to Improving Instruction in Industrial Arts* (Washington, D.C.: American Vocational Association, 1953), p. 11.

pattern. It is equally apparent that items of information concerning American industrial society cannot be assimilated or remembered unless they are organized into some kind of integrated pattern. *There is a need for a central core or theme around which all such information can be related.*

THE UNIT METHOD OF TEACHING

In attempting to organize course materials around a unifying core, industrial arts teachers may well study the theory and practice of unit teaching as exemplified by the more progressive teachers at both the elementary and secondary school levels.

1. What Is Unit Teaching?

Unit teaching involves learning situations in which the pupils participate individually and collectively in a variety of experiences that are continuous, related, and unified around a central theme. This core may be a problem, project, or topic (see Fig. 7–1). The purpose of a unit is to guide the growth and development of pupils toward desirable behavioral patterns. Usually a series of related units comprise the instruction for a given period of time, for example, a semester or a year.

The *Dictionary of Education* defines a unit as:

> An organization of various activities, experiences, and types of learning around a central theme, problem, or purpose, developed cooperatively by a group of pupils under teacher leadership; involves planning, execution of plans, and evaluation of results.[3]

Mitchell relates the term more specifically to industrial arts education when he describes the unit method of teaching as follows:

> It is an organizational pattern in which the work of a semester or a year is divided into a series of interrelated, flexible units, each having a unifying element or central theme toward which the activities or learning experiences are directed.
>
> The unit cuts across the arbitrary dividing lines which usually exist between the areas organized on a materials basis. In addition, the piecemeal learning of unrelated, isolated skills or facts can be avoided. The activities which are an essential part of the unit may lead the pupil into any or all of the areas of the lab, and if carefully planned, will provide learning experiences in which the slow learner can find him-

---

[3]Carter V. Good, *Dictionary of Education* (New York, N.Y.: McGraw-Hill Book Company, Inc., 1959), p. 587.

Organizing Learning Activities

Fig. 7-1. This seventh-grade student has constructed a model of a Conestoga wagon and has prepared a written report as part of his contributions to a unit on transportation. (Courtesy, George M. Haney and Alan D. Brown, Earle B. Wood Junior High School, Rockville, Maryland)

self, on the one hand, and the most erudite pupil will be challenged, on the other.

Thus, individual differences in pupil interests, mental and physical aptitudes and capacities can be provided for; the aims of the program can be achieved; and perhaps of increasing importance, the challenges of an ever-changing technology can become real for both pupil and teacher.[4]

The significant contributions of unit teaching have been noted by many educators in recent years. For example, Burton says, "Teaching has improved greatly wherever effort has been made to use modern unit organization."[5]

On this same point Logan and Logan report that

---

[4]John Mitchell, "What Is the Unit Method of Teaching Industrial Arts?" paper read before the Maine Vocational Association, Farmington State Teachers College, Farmington, Maine, August 23, 1963.

[5]William H. Burton, *The Guidance of Learning Activities*, 3d ed. (New York, N.Y.: Appleton-Century-Crofts, 1962), p. 328.

One of the most widely used and most effective methods of securing integration, individualization, critical thinking, creative expression, and enrichment of learning experiences is the unit of work.[6]

In discussing the future role of the industrial arts project, Mitchell writes he is firmly convinced that

> ... the unitary approach provides the more effective answer to this question, particularly in the general shop. The central theme which has its context from the technology, forms a nucleus around which all the activities and learning experiences are evolved.[7]

2. Advantages of Unit Teaching

As previously noted, the principal advantage of unit teaching is that student growth and development occur through continuous, integrated experiences which are focused around a central problem or theme. Thus, the work is not geared to a lesson which begins and ends within a class period. Daily lessons of this nature become so fragmented that the students easily lose sight of course aims. Besides, not all lessons can be fitted into the same block of class time each day. With the unit plan, however, the pupils merely continue their activities from one period to the next until the unit is completed.

Some other advantages claimed for unit teaching include the following:

1. The division of content into teachable units causes the teacher to weigh carefully the selection of units in relation to the behavior changes that are desired and to the ultimate purposes of education.

2. Both independent study by individuals as well as group cooperative activities are possible. The teacher is in a position to make changes in balancing these as the unit unfolds.

3. It is easier to provide for individual differences. Units can be planned to challenge the superior student and at the same time provide a degree of success for the less gifted. The unit method enables the teacher to do remedial teaching.

4. Students usually enjoy this method because units have beginnings and endings and provide job-completion satisfactions. This is in contrast to a single long course that to pupils never seems to end. Then too, some units involve pupil planning of activities and experiences which kindles interest and enthusiasm.

---

[6]Lillian M. Logan, and Virgil G. Logan, *Teaching the Elementary School Child* (Boston, Mass.: Houghton-Mifflin Company, 1961), p. 167.

[7]John Mitchell, "The Future Role of the Project in Industrial Arts," *Industrial Arts and Vocational Education*, vol. 49, no. 8 (October 1960), p. 26.

3. Types of Units

There are several kinds of units, but those in which the industrial arts teacher probably will be most interested are (1) subject-matter unit, (2) activity (experience) unit, and (3) resource unit.

In the subject-matter unit the emphasis is placed on the content to be learned. The teacher determines the unit objectives and selects the learning activities. Details on this type of unit will be presented later in this chapter.

In an activity unit, or as it is sometimes called, an experience unit, the pupils participate in determining objectives, content to be included, and procedures. The unit emphasizes both social and educational development of individuals. An activity unit on transportation for the middle grades may be focused on the activity of constructing models of rafts, wagons, ships, cars, space vehicles, and other modes of transportation, but at the same time science, history, English, and mathematics are studied concurrently with the construction of the projects.

A resource unit is not used with pupils; it is a sourcebook of ideas for the teacher to build units of instruction. It is a collection of (1) introductory and orientation material, (2) objectives, (3) suggested learning activities, (4) community resources, (5) educational aids, (6) suggestions for evaluating outcomes, (7) bibliographies for both teacher and pupils, and (8) guides to related units. Each of the preceding is, of course, directly related to the general title of the resource unit. Often the materials presented encompass a wide range of age or grade levels from which the teacher may select those appropriate for his pupils. Some resource units contain background reading material for the teacher. It is not expected that the teacher will utilize all of the ideas, suggestions, and recommendations contained in the resource unit. His role is to select those items which seem to be most relevant for his students. Thus, the purpose of a resource unit is to serve as a guide for teachers who are developing new units or those desiring to upgrade older ones. Resource units are often prepared by teachers of the same subject in a school system, by teacher education institutions, by professional associations, or by state departments of education.

4. Length of a Unit

A unit may be planned to cover any period of time. Its length depends on the nature of the subject matter, the types of behavioral changes to be effected, the interest span and maturity level of the pupils, and "how" the unit was received by pupils as it began to unfold. Too many units in a course tend toward fragmentation; a single long one defeats the purpose of the unit method. Some teachers feel that a unit should last from three to nine weeks; an introductory unit in a course may last a week or less. Some teachers intentionally vary the length of units for a "change of pace" and to add variety to the course. The length of a unit is of less consequence than the contributions it makes to desirable behavior changes and its relevancy to other units.

GENERAL PHASES OF UNIT TEACHING

A unit passes through several distinct stages as it unfolds in the laboratory. As mentioned previously, each unit is independent and complete in itself, but at the same time is related to all other units in the course. The sum total of all units together constitutes the course of study.

1. Preplanning Phase

Before a particular unit of instruction comes into being, the teacher must determine course objectives and identify titles for the several units which are to bring about the desired behavior changes. After this has been done, the planning of an individual unit begins with the formulation of unit objectives in terms of desired terminal behavior. This is followed by the preparation of a course outline and the identification of learning activities to be employed and a list of lessons to be taught. After plans for evaluation have been made, the preplanning phase ends with the preparation of bibliographies and lists of instructional aids. Details on these steps will be given later in the chapter.

2. Orientation Phase

So far as students are concerned (usually) the first phase of the unit is the introduction or orientation stage. This is the motivating period during which the teacher arouses the interest and enthusiasm of a class. Depending upon the age and pupil maturity level involved, a unit may be introduced with casual questions by the teacher, by an industrial visit, or by a motion picture. For older students the teacher may simply announce the title of the next unit to be studied.

3. Activity Phase

The third phase is the implementation stage or the actual conduct of the unit. This is the data-gathering or "doing" stage. Here the stu-

Organizing Learning Activities

dents will engage in whatever activities have been decided will contribute to the attainment of the goals of the unit. Such activity includes both individual and group or committee action. Typical learning activities might include assigned and free-choice readings, industrial trips, resource persons (industry and community), motion pictures, research-type activities or experiments, demonstrations, and project or construction work. Chapter 6 provided long lists of suggested student and teacher activities. Figures 7–2 and 7–3 show seventh-grade students in the data-gathering and planning stages as well as in the construction phases of a unit on transportation.

Fig. 7–2. Seventh-grade students researching their subtopic selections as part of the total unit. (Courtesy, George M. Haney and Alan D. Brown, Earle B. Wood Junior High School, Rockville, Maryland)

4. Culminating Phase

The unit ends with a final or culminating activity. This is the summarizing stage when the various segments of the unit are pulled together. Culminating activities should be simple, interesting, and involve many

Fig. 7–3. This student is busily engaged in the activity or construction phase of a unit on transportation. (Courtesy, George M. Haney and Alan D. Brown, Earle B. Wood Junior High School, Rockville, Maryland)

class members. Culminating activities are closely related to the interests and maturity level of the pupils. Typical culminating activities might include: a display of project work, committee reports, a class discussion or seminar, student critique, or a unit test. Figure 7–4 illustrates a seventh-grade class engaged in a seminar presentation which concluded a unit on communication and transportation and their contribution to the growth of civilization.

5. Evaluation

An important feature of any unit is the evaluation of pupil growth and development. Evaluation is not begun after all the steps in unit development have been completed; rather, evaluation begins as soon as the unit begins. It involves continuous constructive review by the teacher of student and class efforts. Evaluation should stimulate class members and the group as a whole to even greater effort.

HOW TO PLAN A TEACHING UNIT

Like any other good plan, the unit has several distinct parts or subdivisions. It is the intent of this section to consider each of these and to relate some illustrative examples. Most plans for a teaching unit

Organizing Learning Activities

Fig. 7-4. Culminating activity of a seventh-grade class studying a unit on communication and transportation. (Courtesy, George M. Haney and Alan D. Brown, Earle B. Wood Junior High School, Rockville, Maryland)

probably will include these subdivisions: (1) title, (2) objectives, (3) outline of content, (4) student and teacher activities, (5) lessons to be taught, (6) culminating activities, (7) evaluation, (8) bibliographies, and (9) instructional aids.

1. Title

The title of a unit is a guide for the teacher. In general terms it tells what the unit is all about; some titles indicate the emphasis to be given to the unit. A title should be specific. For example, the title, *A Study of Industry*, obviously is so broad and comprehensive that its bounds defy identification. A better title might be: *A Study of Mass Production in American Industry* (assuming this was the purpose of the unit). If possible, the title should appear interesting to the students and seek to motivate them to want to begin the unit. A title may be phrased either in statement or in question form. Some examples of titles for suggested units are: (1) How Is American Industry Financed and Organized? (2) The Secrets of Mass Production, (3) Styles in Period Furniture, and (4) Labor Unions: Past, Present, and Future.

2. Objectives

Statements of unit objectives are important because (1) they provide direction for both teacher and pupils, (2) they serve as a basis for including or excluding learning activities, and (3) they provide a means for evaluating pupil achievement in terms of behavioral changes.

To illustrate sample objectives, suppose a teacher were interested in developing a unit for junior high school pupils titled: Are Industrial Occupations For Me? Such a unit would be in keeping with the vocational guidance objectives of industrial arts (see Objective 5, Chapters 4, 5, and 6). It will be recalled that this objective seeks to provide information about, and insofar as possible, experiences in, the basic processes of many industries in order that students may be more competent to choose a future vocation. The following represent two sample objectives for the proposed unit.

Upon completion of this unit the pupil should be able to:

1. Cite wages, entrance requirements, working conditions, and opportunities for advancement for several industrial occupations.

2. Mention health risks and dangers of various jobs and occupations in several industries.

The teacher will, of course, determine how many unit objectives will be needed as well as tailor them to fit the needs, interests, and maturity level of the class.

3. Outline of Content

An outline of subject-matter content is requisite for a teaching unit. Not only is it a guide to keep the unit moving in the desired direction, but also it provides a beginning and an ending for the unit. As its name implies the content outline bounds the area of study. Of course, the outline must be in complete harmony with the unit objectives. The degree of detail in the outlining is the prerogative of the teacher. However, much of the success of the unit hinges on the outline, so the teacher will do well to give it the careful attention it deserves. It is to be noted that some of the newer textbooks are organized on the unit basis; this makes them helpful to teachers desiring to prepare new units or to upgrade older ones.

4. Student and Teacher Activities

With the unit title, the objectives, and the content outline firmly in mind, the teacher is now ready to select the learning activities which, it is hoped, will bring about the desired terminal behaviors. Depending

Organizing Learning Activities 143

on the teacher's decision, the pupils may play a major, a minor, or no role in planning the learning activities for the unit.

Some key points to keep in mind while selecting learning activities are:

1. The emphasis should be on pupil activity and not on teacher activity.

2. Keep individual differences in mind and plan for activities covering a wide range of needs, interests, and abilities (see Fig. 7–5).

Fig. 7–5. During the activity phase of this unit, the teacher has carefully planned a variety of interesting activities to meet the range of interests, needs, and abilities of these elementary-school pupils. (Courtesy, California State Department of Education)

3. Select activities which encourage student initiative and the abilities to think, plan, and use time wisely.

4. In writing statements of learning activities, begin each with an action verb, such as, assign, construct, make, draw, et al.

Again using the unit title, Are Industrial Occupations For Me?, suggestions for student and teacher activities may include these three:

1. Have local social security representative talk to class about industrial employee benefits.

2. Provide tryout or "hands on" experiences in as many trades and occupations as possible.

3. Make field trips to industries to observe men and machines at work.

It is understood, of course, that the above are merely examples of the possible learning activities for the proposed unit. As the teacher gives thought to this matter and especially after he teaches such a unit a time or two, many fine learning activities doubtless will come to mind. In addition to the preceding, Chapter 6 contained additional suggestions for student and teacher activities which may be helpful in eliciting desirable behavior changes for a vocational-guidance unit.

5. Lessons to Be Taught

Under this heading should be listed the titles of all lessons planned for the unit. This includes lectures, demonstrations, experiments, and the like. Details for these lessons will be developed later as lesson plans.

6. Culminating Activities

The purpose of these activities is to terminate the unit in a manner that summarizes and emphasizes the important aspects. Pupil age, interests, and maturity level determine the nature of the culminating activity. Some groups may prepare a display of project work, present a series of demonstrations or oral reports, hold a class discussion, or stage a conference. For subject-matter units a test may be used to conclude the unit. In this case it may be desirable for the test questions to follow the sequence in which the unit topics were presented.

7. Evaluation

The evaluative process goes on continually during a unit. It begins when the unit begins and ends when the unit ends. Since behavior changes are desired, the teacher will need to observe pupils closely during the unit as behaviors begin to change. All that can be done in the planning of a teaching unit is to indicate the types of evaluation that will be utilized during the course of the units as well as how and when they will be used.

8. Bibliographies

Two bibliographies are often provided in a teaching unit. One is for the teacher while the other is for student use. Only those books and materials available to the teacher and to the students should be included in the bibliography.

9. Instructional Aids

A teaching unit is usually concluded with a list of useful teaching aids. Again, only those audio-visuals which are easily available to the teacher should be included.

UNIT TEACHING IN THE FIELD

Reference is made here to three current applications of the unit method of teaching in industrial arts education. These include the *Maine State Plan,* the *Maryland Plan,* and a *1968 NDEA Institute.*

The Maine State Plan

In 1965 the State Department of Education published a plan for industrial arts education, grades seven through twelve.[8] This five-year effort was spearheaded by state department staff and industrial arts personnel at Gorham State College of the University of Maine. Involved were scores of industrial arts teachers who met in curriculum groups and workshop sessions.

After objectives had been identified and stated in behavioral format, 21 units were developed for the six-year program. These are summarized as follows:

Grade 7—Manufacturing Industries

Unit titles: *Technology and Civilization* (1 week); *Household Accessories* (8–9 weeks); *Personal Accessories* (8–9 weeks).

Grade 8—Manufacturing Industries

Unit titles: *Camping Equipment* (6–9 weeks); *Hunting and Fishing Equipment* (5–7 weeks); *Communication Equipment* (5–8 weeks).

Grade 9—Manufacturing Industries

Unit titles: *Tools and Home Workshop Equipment* (9 weeks); *Small Furniture for the Home* (10 weeks); *Production Industries—Mass Production* (9 weeks); *Model Power Products* (8 weeks).

Grade 10—Manufacturing and Construction Industries

Unit titles: *Tool and Machine Industries* (12–15 weeks); *Residential Construction* (18–20 weeks); *Transportation Construction* (3–4 weeks).

Grade 11—Power and Transportation Industries; Electrical-Electronics Industries

Unit titles: *Portable Power Plant Industries* (6–9 weeks); *Transporta-*

---

[8] *Industrial Arts Technology: A Study of American Industry,* Guide for Secondary Schools in Maine prepared by Department of Education, Vocational Division (Augusta, Maine: State Department of Education, 1965).

tion by Automobile (6–9 weeks); *Residential Wiring* (3–6 weeks); *Wire and Wireless Communication* (12–18 weeks).

Grade 12—Service Industries

Unit titles: *Small Service Business Management* (2 weeks); *Appliance Servicing* (7–11 weeks); *Automotive Servicing* (6–9 weeks); *Repair and Refinishing Industries* (3–5 weeks); *Area of specialization—Vocational Orientation* (18 weeks). Or, an optional grade 12 is *Area of Specialization—Vocational Orientation* (36 weeks).

The Maryland Plan

In the late 1950's, Maley of the University of Maryland began to formulate what is now known as the *Maryland Plan*.[9] Today the plan embraces three grade levels and includes (1) research and experimentation (ninth grade), (2) study of modern industrial organization and production processes (eighth grade), and (3) an anthropological study of man's technological achievements (seventh grade). This discussion is concerned with the seventh grade program because here the unit method of teaching is employed.

In addressing an industrial arts group on unit teaching, Haney stated:

> One way of introducing children to industrial arts is through the "Unit Method." The "Unit Method" of teaching industrial arts, as distinguished from the traditional "unit shop concept," utilizes a distinctive teaching methodology which is common to the study of broad and comprehensive *units*. These units of study attempt to provide for the student better understandings and appreciations of the growth and development of civilization in general and Western civilization in particular. At the seventh-grade level, the unit method uses an anthropological or historical approach to the study of man's mastery over the raw materials of nature and his progress in changing these materials to conform to his own needs and wants. In essence, its content is concerned with how and why man learned to *make* and *use* tools and machines, harness power and energy sources, and build transportation and communication devices—all of which enabled him to improve his standard of living and have a more enjoyable material way of life. The emphasis of industrial arts subject matter is technology at this level.
>
> At the Earle B. Wood Junior High School, as well as other Montgomery County, Maryland, junior high schools, the "Unit Method" has been implemented as part of a new comprehensive industrial arts cur-

[9]Donald Maley, "Research and Experimentation in the Junior High School," *The Industrial Arts Teacher*, vol. XVIII, no. 4 (March-April, 1959), pp. 12–17.

riculum. Seventh grades have engaged in the study of industrial arts based upon the following major Units or Unit Topics:

1. The Development of Tools and Machines and Their Contribution to the Growth of Civilization.

2. The Development of Power and Energy Sources and Their Contribution to the Growth of Civilization.

3. The Development of Communication and Transportation and Their Contribution to the Growth of Civilization.

Each unit represents a broad and comprehensive area of study. Within each unit there is contained certain basic elements or subdivisions of subject matter that relate to the total unit topic.[10]

NDEA Institute

In 1969 the Office of Education (DHEW, Washington, D.C.) sponsored an NDEA Institute entitled *Units for the Laboratories of Industries*.[11] Twenty teachers from fourteen states gathered at Gorham State College of the University of Maine. Their purpose was to develop and evaluate teaching units dealing with manufacturing for the junior high school level. These units were tested experimentally on 60 boys and girls. Video tapes were made of the units being taught and were evaluated by institute participants. A copy of one of the units developed during the institute, *Unit 1—Introduction to Technology and Industrial Arts*, is contained in Appendix 1.

LESSON PLANNING

Good lessons begin with good teacher preparation. Some kind of planning is essential to successful teaching and there is no substitute for it. Struck puts it this way:

> ... the man of public affairs would not choose to face his tasks without previous thought, so teachers ... are committed to planning their work.[12]

Sometimes preservice or beginning teachers express the feeling that as soon as they gain experience in teaching that they will no longer

---

[10]George M. Haney, "A Study of Man's Technological Achievements Through 'The Unit Method,'" paper read before the industrial arts section meeting of the Maryland Vocational Association, College Park, Maryland, March 12, 1966.

[11]*Units for the Laboratories of Industries*, Office of Education (DHEW) Washington, D.C. Gorham State College, Maine. John Mitchell, Institute Director, August 1968, 486 pp. Microfiche ED–031–544.

[12]F. Theodore Struck, *Creative Teaching* (New York, N.Y.: John Wiley and Sons, Inc., 1938), p. 118.

need to plan their lessons. This could not be further from the truth! In a recent research investigation Conrad found that experienced teachers did more planning than less experienced teachers.[13] Also, he discovered that teachers who emphasized the qualitative aspects of planning involved their students in responses requiring independent judgment, opinion, prediction, and conclusion more frequently than did other teachers.

The industrial arts teacher engages in several kinds of planning. For example, his long-range plan for the semester or year is his course of study. In turn, this consists of a series of unit plans. Units are composed of lessons. In other words, lessons are to units as units are to a course of study.

A lesson is considered to be that portion of an instructional period which is devoted to the development of one or more concepts and/or motor skills. Each lesson should have a plan which is nothing more than a teacher's outline of the main points of the lesson in the order in which they are to be presented. It is simply a teacher's guide for a lesson. Thus, in a given unit numerous lessons will be given, such as demonstrations, informative lessons, talks by resource people, industrial visits, and the like. A lesson is not to be confused with a class period; conceivably more than one lesson could occur in a given class period.

Lesson planning is an important responsibility for all teachers. All lessons will need to be planned to some degree. Some plans will need to be written; others will not. Some teachers like to make greatly detailed plans while others prefer an abbreviated style. Ericson and Seefeld note that the extent of detail in a lesson plan will depend on the teacher's experience in this area, the degree of his technical skill, and his ability to think in an abstract way through a series of steps without the use of a written analysis.[14] In this connection Struck notes:

> Just as the greatest leaders in political, economic, social, and business affairs differ in dependence upon notes when they appear at important meetings—some use them extensively, some sparingly, and some not at all—so teachers too differ in their ability to do their work effectively without access to written plans.[15]

Insofar as lessons are concerned, the industrial arts teacher engages in two kinds of planning. The first of these is a plan for an informative lesson and the second is for a demonstration. In each case careful plan-

---

[13] Robert James Conrad, "A Study of the Relationship Between Lesson Planning and Teacher Behavior in the Secondary Classroom" (unpublished doctoral dissertation, University of Utah, 1969).

[14] Emanuel E. Ericson and Kermit Seefeld, *Teaching the Industrial Arts* (Peoria, Ill.: Charles A. Bennett Company, Inc., 1960), p. 65.

[15] Struck, *loc. cit.*

Organizing Learning Activities

ning is essential to the achievement of the lesson aims and unit objectives. It is highly recommended that a lesson plan be written for each demonstration and informative lesson to be given during the unit. These can be filed where easily available. Before giving one of these lessons, it is a simple matter for the teacher to refresh his memory by quickly reviewing the content and also to make certain that all tools, supplies, equipment, and instructional aids will be on hand when needed.

The format in which lesson plans are prepared varies greatly. Teachers should experiment with a variety of styles and formats in an effort to find one which especially suits their manner of teaching. There are several divisions or headings which are probably common to most good comprehensive lesson plans. These include the following:

1. Identification Data. This information usually includes the school and teacher's name, industrial arts course or number, grade level, section number (if needed), unit number and title, and lesson number and title. In other words, the teacher records in this section pertinent information relating this lesson to the course.

2. Objectives. It is important to state the aims of the lesson in behavioral terms, that is, what behavior changes are expected from the students as a result of this lesson? This practice is merely an application to lesson planning of the analysis techniques for course and unit objectives that were presented in this chapter and in Chapters 5 and 6.

3. Teaching Aids/Materials/Tools and Equipment. Although grouped here for discussional purposes, each of these categories merits a separate heading on the lesson plan. Supplies and materials as well as tools, machines, and equipment should be listed. All aids to instruction, such as overhead transparencies, slides, chalkboard sketches to be used, and other instructional media should be listed. It is desirable, but not necessary, to record each list in the order in which the items will be used in the lesson. These headings should appear toward the beginning of the lesson plan rather than at the end so that the teacher can quickly review what is needed for the lesson.

4. Introduction. In this section the teacher impresses the students with the lesson aims and their importance. He also seeks to introduce the lesson in an interesting manner to motivate and stimulate the pupils.

5. Instructional Steps. Here the lesson content is presented, usually in outline form, in whatever detail is deemed desirable by the teacher. Some teachers like key sentences written out, while others prefer only a skeleton outline.

6. Termination. The teacher should plan a means to end the lesson formally. There are several ways in which this may be done. Under a heading called *Summary of Main Points,* the teacher may provide a

Midland Senior High School
Ninth Grade Industrial Arts

Course: Manufacturing II
Unit: 8 — Joining Metals
Lesson: 4 — Informative

SILVER BRAZING ALLOYS AND FLUX

Objectives: As a result of this lesson, student will be able to:
    1. Relate the purpose and the two principles underlying the use of flux in silver brazing (silver soldering).
    2. Recall the basic ingredients in flux.
    3. Recall the three elements contained in silver solders and which is varied in order to control melting temperature.
    4. Recall the three basic types of silver solders and the range of their flow temperatures.
    5. Associate the four basic metal heat colors with the temperature ranges they represent.

Teaching Aids:

    1. Chalkboard and colored chalk
    2. Silver Brazing Alloy Composition Chart
    3. Heat Colors Chart
    4. Silver Solder Identification Chart

Materials:

    1. Household Borax
    2. Handy & Harmon Flux
    3. Samples of silver solder: sheet, wire, and strip.

Introduction:

    Recently I demonstrated to you the process of silver brazing by brazing a simple "tee" joint. What could we do if we had several joints to braze, but they were on opposite sides of the main piece? Might not the first one melt and break loose while we were working on the second? Experienced silver brazers use brazing alloys (silver solders) of different melting points. Let's investigate what these different alloys are and how they are used.

Instructional Steps:
    I. Composition of silver solders
        A. Elements of alloys
            1. silver — proportions vary from 2/3 to 3/4
            2. copper — proportions vary around 1/5
            3. zinc — proportions vary from 1/8 to less
        B. Melting temperatures vary

Show:
Silver Alloy
Composition Chart

            1. zinc controlling element — adding zinc lowers melting point — 1/8 maximum.
            2. Easy-Flow contains 15%
            3. Hard-Flow contains 3%
    II. Types and Forms of Silver Solder Available
        A. Forms:
            1. Sheet   — 26-28 gauge most practical
            2. Wire   — 12 to 14 gauge commonly used by jewelers
            3. Strip   — most popular with silversmiths; sold by the foot
                a. 1/32" x 1/16" for light objects
                b. 1/16" x 1/8" for heavy objects
        B. Types:
            1. Common types used — marked for identification

Show:
Silver Solder
Identification Chart

                a. "Easy Flowing" — straight end
                b. "Medium Flowing" — twist end
                c. "Hard Flowing" — loop end
            2. Special purpose
                a. Low temperature (1175°) for jewelry findings
                b. High temperature (1460°) for objects to be enameled

Fig. 7–6. Left and right: A sample plan for an informative lesson. (Courtesy, John M. Shemick, The Pennsylvania State University)

III. Purpose and Kinds of Flux
    A. Why is a flux needed?
        1. Heated metals combine with oxygen in air to form a coating
        2. Coating called an oxide which prevents solder from adhering
        3. Flux prevents oxides by:
            a. forming film over metal, keeps air away from metal
            b. dissolves small amounts of oxide that may form
    B. Kinds of Flux
        1. Shop-made water-base paste

Show:
Box of borax

            a. Household borax, oldest type, still preferred by many metalsmiths
            b. Boric acid added for high temperature work
                75% borax + 25% boric acid
        2. Commercial preparations

Show:
Jar of Handy &
Harmon Flux

            a. Available in liquid or paste forms
            b. Handy & Harmon paste, liquifies at low temperatures, excellent for low temperature solders ($1100°F$)

IV. Metal Heat Color as Approximate Temperature Indicator
    A. First visible red, $900°F$, bottom of silver soldering range

Show:
Heat Colors
Chart

    B. Dull red, $1100°F$, low temperature silver soldering of jewelry findings
    C. Cherry red, $1400°F$, between dull and cherry red common types of silver solder alloys melt and flow
    D. Salmon red, $1640°F$, CAUTION sterling silver melts

<u>Termination:</u>

    A. Summary of main points
        1. Composition of silver solder alloys
            a. silver
            b. copper
            c. zinc — key element affecting melting point
        2. Types and forms of silver solder available
            a. sheet
            b. wire
            c. strip, most popular with metalsmiths
            d. types
                (1) Easy-Flow, plain end
                (2) Medium-Flow, twist end
                (3) Hard-Flow, loop end
        3. Purpose and kinds of flux
            a. Shop-made, borax and water; add boric acid for high temperature work
            b. Commercial, Handy & Harmon — excellent for low temperature silver solder alloys
        4. Metal heat color
            a. First red — $900°F$
            b. Dull red — $1100°F$
            c. Cherry red — $1400°F$
            d. Salmon red — $1640°F$ — DANGER — sterling silver melts
    B. Checking understanding
        1. If we had hard-flow silver solder alloy, what could we add to it which would cause it to melt at a lower temperature?
        2. What two things might cause silver solder to ball up and not flow throughout the joint?

References:
    Walker, John R. *Modern Metalworking*, Goodheart-Willcox, 1965, pp. 29-5 to 29-8
    Bovin, Murray. *Silversmithing and Art Metal*, 1963, pp. 19-23

Midland Senior High School
Ninth Grade Industrial Arts

Course: Manufacturing II
Unit: 8 — Joining Metals
Lesson: 3 — Demonstration
Time: 15 minutes

HOW TO SILVER BRAZE

Objectives: The student will be able to:

1. Prepare properly, the metal parts to be brazed by
   a. cleaning surfaces, bright and dirt free;
   b. fitting pieces together with 0.001 to 0.003 inches clearance;
   c. applying flux evenly, and
   d. assembling and firmly supporting pieces to be soldered.
2. Heat properly and flow filler alloy by
   a. using neutral or reducing torch flame, and
   b. heating pieces uniformly.
3. Completely clean the finished brazed joint with water.

Tools and Equipment:

1. Brazing torch outfit
2. Welding table
3. Hold-down bar
4. Wire brush
5. Safety goggles

Materials:

1. Brazing flux with brush
2. 2 pcs. 1/8" x 1" x 2" C.R. steel
3. Filler alloy (Easy-Flo)
4. Emery cloth
5. Steel wool

Introduction:

By this time you know how to soft solder. When greater strength and/or heat resistance is needed, the process of silver brazing is often used. Although silver brazing may appear to be similar to welding, it is different in that unlike metals can be joined by the process.

Instructional Steps:

I. Prepare the metal to be joined
   A. Good fit and proper clearance are important
      1. Allow opening between 0.001" to 0.003" because filler metal flows by capillary action, and thin film of filler metal makes stronger joint.
   B. Clean metal surfaces to be joined
      1. Mechanically with
         a. steel wool
         b. emery cloth
         c. fine file
      2. Chemically with acid pickle
      3. Solvents for grease and oil
   C. Apply flux
      1. Apply evenly to all surfaces to be joined
      2. Aply just before brazing

Fig. 7–7. **Left and right:** A sample lesson plan for a demonstration. (Courtesy, John M. Shemick, The Pennsylvania State University)

Organizing Learning Activities

> D. Assemble and support
> 　　1. Soon after the flux is applied
> 　　2. Fit together for proper clearance
> 　　3. Support to prevent shifting during heating because of expansion of metal as it is being heated, and the surface tension effect of flowing filler metal
> II. Heat and flow filler alloy
> 　　1. Adjust torch
> 　　　　a. Neutral to reducing flame
> 　　　　b. Greenish "feather" extending out from inner cone
> 　　2. Apply uniform heat
> 　　　　a. Keep torch moving; avoid hot spots
> 　　　　b. Concentrate flame on heaviest metal parts
> 　　3. Apply filler alloy
> 　　　　a. Temperature test: drop a little alloy on the joint. If it "balls-up", it is not hot enough; if it "lays out", temperature is right.
> 　　　　b. Alloy will flow completely via joint-cappillary action
> III. Final cleaning
> 　　1. Hot water adequate (normally)
> 　　2. Contaminated flux, add acid to water, or
> 　　3. Wire brush off flux
>
> Termination:
>
> I. Summarize main points
> 　　A. Preparing surfaces
> 　　　　1. Clearance (0.001 to 0.003")
> 　　　　2. Clean
> 　　　　3. Flux
> 　　　　4. Support
> 　　B. Heating
> 　　　　1. Neutral flame
> 　　　　2. Uniform heating
> 　　C. Cleaning with hot water
> II. Checking understanding
> 　　A. Why may brazing be preferred to soft soldering?
> 　　B. What advantage does brazing have when compared to welding?
>
> ---
>
> Reference:
> 　　Walker, John R. *Modern Metalworking*, Goodheart-Willcox, 1965, pp. 29-1 to 29-8.

brief review of the highlights of the lesson. Or, the lesson may be terminated with key questions grouped under a subheading called *Checking Understanding*. Thought-provoking questions may be asked orally. Or, a series of objective-type questions may be projected on the screen via the overhead projector. Still another means to bring the lesson to a conclusion is with an *Application* subheading. Here the teacher describes plans or suggests opportunities for students to use what they have learned. Application is without question one of the most important parts of any lesson. Depending on circumstances, the teacher may use one or all of these ways in the same lesson to terminate it.

7. References. Here are listed instructional materials for use by both teacher and students. Only materials immediately available should be included.

8. Assignment. In addition to the foregoing divisions, some lesson plans also have an *Assignment* heading under which is provided directions for further student application. An assignment should not be busy work, but should be especially selected to promote better understanding of content or the development of a skill. It may involve manipulative activity in the laboratory or a homework assignment to answer selected questions.

9. Other Useful Data. It must be kept in mind that the headings and information contained on a lesson plan are there solely to help the teacher in planning and in presenting the lesson. Thus, a teacher should include those items of information he finds useful and in a format that he likes. Some teachers like to have a lesson plan heading called *Questions for Later Use on Tests*. Obviously a good time to prepare test items is about the same time the lesson is being planned or given. These questions may be placed on the back of the lesson plan (or on subsequent sheets) so that additions and deletions may be made easily.

Some teachers use a lesson plan heading called *Critique* under which they jot down constructive comments after the lesson has been presented. The intent of this section is to help the teacher upgrade and improve the lesson next time it is given. Useful notations which may be included are: (1) omissions of subject content, (2) suggestions for a different approach, (3) notes on time (too little or too much) spent on certain portions or on the entire lesson, (4) student questions, (5) observations of class reaction, and so on.

It should be obvious that not all lessons will require planning under each of the headings previously suggested. A lesson plan should be simple and well suited to the teacher's manner of teaching. It should be useful but under no circumstances wasteful of an instructor's time. Figure 7–6 illustrates a sample plan for an informative lesson, while Fig. 7–7 shows a lesson plan for a demonstration.

As previously stated, lesson plans are to help the teacher organize the material to be presented and to facilitate the actual presentation of the lesson. The preparation of a lesson plan is a useful exercise in itself if for no other reason than that it enables the teacher to think through what is to be accomplished by the lesson and how it is to be done.

Just exactly how a given teacher will use the plan during presentation of the lesson is a matter of professional competency and personal preference. Some teachers like to work from the lesson plan itself while others prefer to speak from key points outlined on 3×5 cards. In either case it is probably good practice even for an experienced teacher to review the lesson plan and/or cards prior to presentation of the lesson before the class.

COURSE OF STUDY CONSTRUCTION

The importance of planning has been emphasized throughout this chapter. It was pointed out, too, that the industrial arts teacher must be concerned with both long- and short-range planning. Lesson planning is an example of short-range planning while a course of study represents long-range planning. In previous sections both lesson and unit planning were presented. The task now at hand is to assemble these into a course of study.

1. What Is A Course of Study?

In the *Dictionary of Education* a course of study is defined as:

> A guide prepared by administrators, supervisors, and teachers of a particular school or school system as an aid to teaching a given subject or area of study for a given grade or other instruction group.[16]

The importance of the task of course construction is noted by Friese who states:

> The aims we establish or accept influence the subject matter selected. The subject matter directs pupils' thoughts along one path rather than another. The methods selected for presenting subject matter directly affect the attitude toward and method by which pupils approach the solution of problems. Our national life is affected each time that a course of study is constructed.[17]

If one subscribes to the unit method of teaching, then the backbone of a course of study consists of a series of related units. To this core may be added the goals of general education, the objectives of industrial arts, descriptions of the school population and the learning laboratory, appendices, and other desired information.

2. Steps In Building A Course of Study

Much of the material needed to assemble a course of study has already been discussed in detail in this and in previous chapters. The task now remaining is to select appropriate content and arrange it in a suitable format. Suggested steps in building a course of study are as follows:

---

[16] Carter V. Good, *Dictionary of Education* (New York, N.Y.: McGraw-Hill Book Company, Inc., 1959), p. 143.

[17] John F. Friese, *Course Making in Industrial Education* (Peoria, Ill.: Charles A. Bennett Company, Inc., 1966), p. 11.

I. Prepare statements of objectives
   A. State goals of general education (see Chapter 1).
   B. State objectives of industrial arts (see Chapter 4).
   C. Select course aims.
II. Describe student population for whom course is intended
   Include age, sex, and grade levels. Describe socioeconomic status and other characteristics and abilities.
III. Select titles for teaching units (see Chapter 7)
   Have as many units as needed for course. Determine time to be spent on each unit. For each unit, do the following:
   A. State unit objectives in terms of expected behavior changes.
   B. Prepare a content outline for unit.
   C. List learning activities, both student and teacher (see Chapter 6).
   D. Select culminating activities.
   E. Outline evaluation plans. Describe techniques to be used and tell when. If unit tests are to be employed, prepare samples.
   F. Prepare bibliographic readings for both teacher and students.
   G. List audio-visual aids.
   H. Prepare lesson plans for unit (not necessary to prepare all lesson plans to be used, just samples)
IV. Arrange units in order of presentation
V. Describe learning environment (see Chapters 19–22)
   Indicate materials areas represented and make sketch of layout.
VI. Describe laboratory management (see Chapters 16–17)
   Explain personnel system, safety program, preventative maintenance program, ordering and inventorying procedures, materials and tool handling methods, daily class routine, records system, etc.
VII. Plan final course evaluation (see Chapter 18)
   Outline appraisal techniques, grading system, sample final examination questions.
VIII. Prepare appendices as needed
   Bibliography of supplementary resource material
   Sources of resource units
IX. Prepare a "Table of Contents"

SUMMARY

This chapter sought to answer the important question, "How shall the teacher organize learning activities in industrial arts?" The need for organizing instructional matter around a central theme was emphasized. The unit method of teaching can successfully meet this need. The term was defined, five advantages of unit teaching were given, the four phases through which a unit progresses were presented, and three types of units were cited. The steps in preparing a unit include: (1) select title, (2) state objectives in behavioral format, (3) outline content to be covered, (4) select learning activities, (5) list lessons to be taught, (6) plan culminating activity, (7) outline plans for evaluating behavior changes, (8) prepare bibliographies, and (9) list instructional aids. Three examples of unit teaching were cited including the Maine State Plan, the Maryland Plan, and an NDEA Institute.

The importance of planning was stressed and the following points in lesson planning were discussed: (1) identification data, (2) objectives, (3) teaching aids/materials/tools and equipment, (4) introduction, (5) instructional steps, (6) termination, (7) references, and (8) assignment. Two detailed examples of lesson plans were offered. After defining the term *course of study* and highlighting its importance, the chapter concluded with an outline of the steps to be followed in preparing a course of study.

DISCUSSION TOPICS AND ASSIGNMENTS

1. Does the unit method of teaching hold any implications for the physical features or layout of an industrial arts laboratory? Explain your answer.

2. Suggest titles for several units that might be included in a ninth-grade course in basic electricity.

3. Select a teaching unit of your choice and make a plan for it.

4. Prepare a lesson plan for a demonstration to a seventh-grade class on the topic, *How to Use a Crosscut Saw*.

5. Make a lesson plan for an information lesson to an eighth-grade class on the topic, *The Operation of a Blast Furnace*.

SELECTED REFERENCES

Billett, Roy O., Donald Maley, and James J. Hammond. *The Unit Method.* Washington, D.C.: American Industrial Arts Association, Inc., 1960.

Boatwright, Ronald L. "The Course of Study," *Journal of Industrial Arts Education,* vol. XXIV, no. 5 (May-June 1965), pp. 50–51.

Callahan, Sterling G. *Successful Teaching in Secondary Schools.* Chicago, Ill.: Scott, Foresman and Company, 1966.

Conrad, Robert James. "A Study of the Relationships Between Lesson Planning and Teacher Behavior in the Secondary Classroom." Unpublished doctoral dissertation, University of Utah, 1969.

Drew, Alfred S. "Resource Units in Industrial Education," *Industrial Arts and Vocational Education,* vol. 53, no. 6 (June 1964), pp. 30–31, 52.

Friese, John F. *Course Making in Industrial Education.* Peoria, Ill.: Charles A. Bennett Company, Inc., 1966.

Kukula, Stanley J. "The Orientation Unit," *Industrial Arts and Vocational Education,* vol. 54, no. 1 (January 1965), pp. 23–24.

Haney, George M. "A Study of Man's Technological Achievements Through 'The Unit Method.'" A paper read before the industrial arts section of the Maryland Vocational Association, College Park, Maryland, March 12, 1966.

Maine State Department of Education. *Industrial Arts Technology: A Study of American Industry.* Guide for Secondary Schools in Maine prepared by Department of Education, Vocational Division. Augusta, Maine.: State Department of Education, 1965.

Maley, Donald. "Research and Experimentation in the Junior High School," *The Industrial Arts Teacher,* vol. XVIII, no. 4 (March-April 1959), pp. 12–17.

Mitchell, John. "What is the Unit Method of Teaching in Industrial Arts?" A paper read to the Maine Vocational Association, Farmington State Teachers College, Farmington, Maine, August 23, 1963.

———. "The Future Role of the Project in Industrial Arts," *Industrial Arts and Vocational Education,* vol. 49, no. 8 (October 1960), pp. 24–26.

Office of Education (DHEW). *Units for the Laboratories of Industries.* Report of an NDEA Institute conducted at Gorham State College of the University of Maine. Microfiche ED-031-554. Washington, D.C.: August 1968.

Ragan, William B., and John D. McAulay. *Social Studies for Today's Children.* New York, N.Y.: Appleton-Century-Crofts, 1964.

Chapter 8

The Industrial Arts Project

It has been stated previously that problem solving or the ability to think is not only the central purpose of industrial arts, but of general education as well (see Chapter 2). Industrial arts seeks to promote the development of critical thinking in students through problem-solving experiences with the materials, tools, and processes of American industry. These experiences become educationally beneficial through the project via the project method of teaching. A *true* industrial arts project is actually a high form of problem solving. It follows that the *true* project is a learning tool in the development of the rational powers of the individual.

WHAT IS A PROJECT?

The *Dictionary of Education* defines a project as follows:

> a significant, practical unit of activity having educational value and aimed at one or more goals of understanding; it involves investigation and solutions of problems and, frequently, the use and manipulation of physical materials; planned and carried to completion by the pupils and teacher in a natural manner.[1]

Thus a project is a worthwhile educational activity that is planned by both students and teacher to reach a given educational objective. As an activity unit of instruction, each project is complete in itself, but should be related to other informative units or projects within the course. A *true* project is a problem-solving experience involving purpose, planning, execution, and evaluation.

In industrial arts education the execution phase may involve manipulation or use of industrial materials in servicing, construction, or

---

[1] Carter V. Good, *Dictionary of Education* (New York, N.Y.: McGraw-Hill Book Company, 1959), p. 421.

experimentation. Typically, the project in industrial arts results in some tangible object or is a specific job to be done. For example, construction of a table lamp, a screwdriver, or a transistor radio might be industrial arts projects (see Fig. 8–1). Grinding the valves of an automobile, repairing a small engine (see Fig. 8–2) or installing an electric-light fixture are jobs that also may be considered as industrial arts projects.

Fig. 8–1. An outstanding example of an industrial arts project. (Courtesy, Stanley Tools, New Britain, Connecticut)

The project method of teaching is simply the method of instruction whereby a *true* educational project is utilized. In this method it is not the end product that determines whether or not the project method has been employed; one cannot tell merely by examining the completed object or product. Rather, the origin and purpose, the planning involved, the relationship between teacher and student, and the evaluative procedure must be considered in deciding whether the project method has been employed. *The mere construction of an object by a student does not necessarily involve the project method of teaching.*

In industrial arts the *true* project involves a manipulative experience of important educational value with teacher responsibility and supervision of student activity in planning, execution, and evaluation of

Fig. 8–2. Not all industrial arts projects are of a constructional nature nor of a take-home variety. This boy's project is to find out why the engine will not run, and to repair it. (Courtesy, Westerly Park Junior High School, State College, Pennsylvania)

an object, job, or other purposeful activity for which the student has a felt need. Through this method students may work individually or in cooperative groups on selected problems. Opportunity is provided for students to use initiative, resourcefulness, and problem-solving abilities in arriving at logical conclusions or solutions to problems.

The steps involved in critical thinking and in the project method of teaching have a high degree of sameness as is depicted in Chart V. Note in the chart that both a *true* project and a problem to be solved arise in the mind of a student based upon a felt need or a real interest. Upon recognizing the problem, the student then begins to formulate a tentative solution or hypothesis based upon available information, observations, and previous experiences. The student now proceeds to test his hypothesis by collecting data or by manipulating, experimenting, and constructing. Upon completion of the constructive or summary and conclusions stage, the student evaluates his solution or finished product in terms of his original problem.

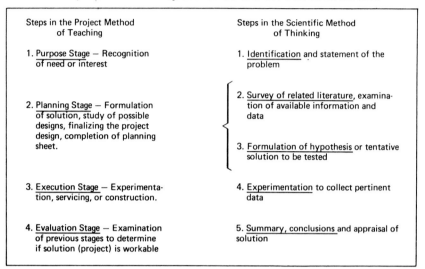

CHART V

Comparison of the Project Method and the Scientific Method

Steps in the Project Method of Teaching	Steps in the Scientific Method of Thinking
1. Purpose Stage — Recognition of need or interest	1. Identification and statement of the problem
2. Planning Stage — Formulation of solution, study of possible designs, finalizing the project design, completion of planning sheet.	2. Survey of related literature, examination of available information and data
	3. Formulation of hypothesis or tentative solution to be tested
3. Execution Stage — Experimentation, servicing, or construction.	4. Experimentation to collect pertinent data
4. Evaluation Stage — Examination of previous stages to determine if solution (project) is workable	5. Summary, conclusions and appraisal of solution

SOME HISTORICAL ASPECTS

The use of projects for industrial arts purposes has been a gradual development over a long period of time. It seems to have been introduced from two main sources, the Swedish sloyd and the Arts and Crafts movement. The Russian system of tool instruction which had greatly influenced the forerunner of modern industrial arts, had concentrated on exercises, such as the making of joints. The prime purpose of the exercise was to develop skill. Upon completion the exercise was discarded because it had no pupil interest nor utilitarian value. With the introduction of Swedish sloyd in this country, with its emphasis on the making of useful articles, there was begun a new trend toward reducing emphasis on exercise work by allowing students to make one or more pieces of furniture after a suitable number of exercises had been completed. At the same time the influence of the Arts and Crafts idea resulted in a movement for better design and finish. Thus, these three influences—Russian hand work, Swedish sloyd, and Arts and Crafts—merged together to become Manual Arts. The basic purpose of this new trend was the making of useful and well designed articles or projects.

When the manual arts were broadened and enriched, in keeping with the principles of industrial arts which were developing during the early decades of the present century, the project idea was carried over into the new program. While the influence of the exercise is still felt in some schools, the project is now recognized as the basic manipulative vehicle of industrial arts education.

PRIMARY FUNCTION OF THE PROJECT

The prime purpose for the use of projects in industrial arts is to bring about desirable behavior changes in pupils and thus help them realize selected educational goals. This is the only justification for the use of any project in any industrial arts program.

The teacher should regard the project as an opportunity—an opportunity for his students to attain the objectives of the course. To him the project per se is only of secondary, perhaps even minor, importance. He regards the project not as an end in itself, but merely as a means to an end. It is a vehicle of instruction through which certain behavior changes can be effected. A project is not educationally acceptable unless the teacher is thoroughly convinced that by making it the student will acquire certain behavior changes considered to be desirable. Thus the choice among many possible projects becomes critical in terms of whether it meets the needs of a particular student and will help him attain the objectives of the course.

From the point of view of some students, the project may well be the most important feature of an industrial arts course. For them the construction of projects appears to be the prime purpose of the entire program. Making a successful project provides many students with an opportunity to feel the pride of accomplishment with a consequent rise in self-esteem which is often difficult for them to attain in other school activities. The fact that the project constitutes tangible evidence which can be proudly shown to family and friends is also important. The teacher should not discount the value the student may attach to even a poorly-made project which is the result of his own labors. Pride in craftsmanship is a legitimate expected behavior change. However, while these values of an educational project are worthy of mention, they do not in themselves constitute sufficient justification for the making of a project.

To parents and school administrators, industrial arts projects frequently stand as concrete evidence of pupil accomplishment and as a return on their educational investment. Often they judge the worth of a program by the number and quality of projects completed by students. The industrial arts teacher should never underestimate the impact of well-designed, well-executed, take-home projects. At the same time, however, he should make special effort to communicate the nature, purposes, and expected real outcome of his program. He should not sustain the image of a supervisor of project construction in his school.

STEPS IN THE PROJECT METHOD

The four steps in the project method of teaching are: (1) purpose, (2) planning, (3) execution, and (4) evaluation. It is interesting to observe that each of these steps must be taken by the pupil. This does not mean that the teacher is relieved of any activity or responsibility, but it does mean that the student occupies the center of the learning stage with the teacher standing in the background, but always ready to assist in the performance. Obviously, the project method is highly student-centered rather than teacher-centered.

1. Purpose

Here the student recognizes his problem, assays his needs, or expresses his interest. Initially, the teacher's role is to kindle interest and, if necessary, motivate the individual. Later he must temper the student's choice of problem with other factors, such as his estimation of the student's ability including interest span, the behavior changes to be expected, and the time and laboratory resources available. This step is probably the most difficult one for the teacher because some pupils will want to attempt things beyond their capabilities. The teacher must resolve this issue diplomatically and expeditiously. He must remain in control of the situation in a firm, friendly manner realizing at all times that the educational responsibility is his alone.

2. Planning

In this stage the pupil studies his problem or need and begins to gather data from available sources as well as from his own experiences and those of the teacher. Soon tentative solutions should come to his mind. He may reject some solutions and improve others, but finally he selects one worthy of testing (constructing). During this time, as depicted in Fig. 8–3, the teacher serves as a resource person or guide ready to share his expertise. He must remember that the ability to think can be developed in students only through their practice and success in solving problems. Although he must take whatever steps he can to help ensure the pupil a successful problem-solving experience, he must resist the temptation to do the thinking for the pupil. He must give the student every opportunity to solve his own problems.

Fig. 8–3. During the planning stage of a project, the teacher serves as a resource person. However, he must guard against doing the thinking for the student. (Courtesy, California State Department of Education)

The planning stage will culminate in a plan of action, a procedure for the investigation, or the completion of a project planning sheet. The latter will include among other things a sketch, working drawing, bill of materials, and a sequential procedure. This is one of the most important phases in the project method of teaching. It has been stated on more than one occasion in this text that *it is more important from an educational point of view for a student to be able to plan his work well than it is for him to execute his plan skillfully.*

3. Execution

This is the doing, servicing, experimenting, testing, or constructing phase. Here the teacher brings to bear his skills and knowledge of the tools, equipment, and processes of industry. He helps direct the manipulative and construction work by showing, telling, and demonstrating as needed. The execution phase ends with the completion of the finished project. Again, the teacher must resist the temptation to work on the student's project. Only where the project is likely to be ruined and materials wasted or where the student's self-confidence is endangered, is the teacher justified in working on the student's project.

4. Evaluation

The purpose of the evaluation stage is to determine if the hypothesis was workable, that is, was the solution to the problem a satisfactory one. The student returns to his original need or problem and considers the planning phase, evaluates the execution phase, and makes his decision. He considers, also, how he might have improved the solution or end product by changes in any of the three preceding steps. In this evaluative process he is aided by the teacher.

CRITERIA FOR PROJECT SELECTION

An industrial arts project should be selected for inclusion in a course on the sole basis of its contribution to desired behavior changes. Unfortunately, most of the project literature currently in the field completely ignores desirable behavior changes or the educational benefits which may accrue if a student completes a given project.

All industrial arts teachers are anxious to find and to use good projects. Not all teachers, however, have definite criteria for determining what constitutes a good or a poor example. However, the teacher can gain a measure of project acceptability by means of the Kauffman scale of learning values which is reproduced as Chart VI.[2] This scale is in terms of common types of projects, the learning involved, and the resultant degree of acceptability. As such it constitutes selected criteria for evaluating the degree of problem solving in a given project.

In addition, careful study of the desired behavior changes will suggest other important criteria for selecting an industrial arts project. These criteria are in the form of questions which the teacher should ask himself.

1. Does the Project Promote the Type of Behavior Change Desired? Unless a project contributes significantly to the attainment of at least

[2]Henry J. Kauffman, "Restating the Concept of Projects," *School Shop*, vol. XVIII, no. 3 (November 1958), p. 34.

CHART VI
Learning Values for Types of Projects *

Type of Project	Learning Involved	Acceptability
Project made and assembled by manufacturer and needing only superficial finishing by student.	Learning is held to a minimum.	Only for extremely dull students.
Project parts made by manufacturer or teacher and assembled and finished by the student.	Some problem solving but the most important has been done. A low degree of satisfaction may result.	Not desirable, but occasionally justified when shop facilities are limited.
Student copies plan for project from book or job sheet without any personal adaptation.	Little problem solving; mostly organization procedures.	Not desirable but useful for unimaginative students.
Student studies a variety of plans and makes variations or combines ideas for personal need.	Considerable problem solving might be involved.	Acceptable procedure for many students.
Student studies standard plans and evolves a new design or function for an object. Uses reference books for study of design and materials.	An alert mind is needed for such activity. The result may lack some desirable qualities, but much learning has occurred.	Only a small percentage can profit by this approach.
Student has need which cannot easily be solved by standard references. He studies the pros and cons; investigates historical precedent in the problem; studies materials and designs; perhaps visits industry; finally, builds a masterpiece.	Only a few can operate on this level. Learning and gratification are at a maximum.	Very acceptable.

* Courtesy of Henry J. Kauffman and *School Shop*.

one of the objectives, through one or more behavior changes, there can be no justification for its construction. Whatever other merits the project may have or no matter how many other criteria it may meet, it should not be used if it cannot stand this test. The first and most important question which the teacher should ask in considering a new project is, "Will it help the student attain the objectives which have been set?"

2. Does the Project Involve Processes Which Are Related to Industrial Methods and Techniques? Since a prime objective of industrial arts relates to the exploration of industry, it is essential that a major part of the project work involve processes which illustrate or exemplify methods of industry. The field of industry is so broad and varied that time can ill be spent on projects which do not contribute to the exploration of some process new to the learner. Even a purely craft project may illustrate in an elementary way some method used by industry for working similar materials. Tactful guidance by the instructor will help to meet this criterion.

3. Is the Project in Keeping with the Grade Level and Ability of the Student? Nothing can be more discouraging to the student than to begin a project and, after it is well started, find that he has misjudged his ability or interest span and that he is unable to complete his undertaking successfully. It will be recalled from the discussion on critical thinking (see Chapter 2) that a student must enjoy success, especially in his initial problem-solving efforts. The teacher should take every precaution to guard against failure and disappointment. A project which is too large or requires too high a degree of skill and accuracy should be avoided. In many cases projects of this type can be redesigned and simplified to a point where they become practicable.

4. Does the Project Have Real Interest for the Student? A good project should have high "boy interest." Interest in a project is of two general types. The boy may wish to make a project because he wants it for his own purposes. Examples of this type of interest might include a gun rack, fishing-tackle box, or a model of a jet plane. The second type of interest is that which prompts a boy to build a record cabinet for use in his home. He may not wish to use this cabinet himself, but he is anxious to receive the praise and commendation which his parents and neighbors will bestow. In the past most industrial arts projects have contributed to the latter type of interest. Only in occasional cases have projects contributed to meeting the personal needs of students. *More emphasis should be directed toward developing and using projects which have a primary appeal to the boy because he has a personal need and use for them.* A teacher who can stimulate students to express their own real interests in the choice of projects will have the satisfaction of knowing that he has done more than kept them busy.

5. Does the Project Present a Challenge? A project can be too easy as well as too difficult. Unless both the planning and execution of the project present a problem to the student, it is doubtful whether he will learn anything new from its construction.

6. Can the Project Be Completed within a Reasonable Length of Time? The interest span of the average student is apt to be limited. A project which takes too long to complete will likely lead to boredom. The teacher will need to judge carefully the interest span of each individual student. If intense interest is present, some very young students have been known to work for long periods of time on a single large project.

7. Is the Project Well Designed? The ability to analyze a project critically from the standpoint of design should be stressed. Plans that indicate poor design should be reworked until satisfactory proportions, materials, and methods of construction are attained. Such design modifications offer challenging experiences in problem solving.

8. Is the Project Economical of Materials? Students should appreciate the value of materials. Large projects which have few educational outcomes should be avoided. Usually it is possible to accomplish equally good results (in terms of behavior changes) by making small projects as through large ones. Considerable tact and diplomacy may be required for the teacher to persuade a student that he should substitute a smaller, but educationally richer, project for a large one. However, this should be done when possible.

If each proposed project is checked against these criteria and a majority of affirmative answers result, the teacher may feel relatively certain that the project is educationally acceptable.

METHODS FOR USING PROJECTS

Teachers of industrial arts have widely different methods for using projects. Each will have his own particular technique for choosing, assigning, and checking on projects, but most methods will fall into one of the following categories:

1. All Projects Assigned

In cases where the teacher is mainly interested in making certain that all students cover a predetermined series of processes and operations, a set of assigned projects is frequently used. Models and project sheets are commonly the bases for assignments in such cases. Clearly, this is the easiest method for the teacher. Once the series has been established, little preparation by the teacher will be required; and as each member of the class will be doing approximately the same thing, the problem of timing lessons and demonstrations is simplified. This method does not lend itself readily to student planning because all plan sheets would be similar and would tend to become standardized. Neither does a set series of projects meet individual needs or make provision for individual differences. Pupil interests appear to be of secondary importance and, hence, are not considered. *The conscientious industrial arts teacher will consider seriously the limitations of arbitrarily assigned projects in light of his objectives before adopting this technique.*

2. Choice within Groups

In an attempt to avoid the strict regimentation of rigidly assigned projects, many teachers have developed a system of choice within groups. Various groups of projects within each of the areas represented in the shop are prepared. Each group contains several projects or jobs which are similar in regard to the processes and the degree of difficulty involved. Usually, students are required to select one or more projects from each group. In some cases drawings or project sheets are available for each project, while in other cases students are expected to find or develop their own plans. This method has the obvious advantage of giving the student some latitude in the choice of projects he selects, and also, provides an opportunity for student planning. It tends also to ensure the covering of selected processes which the instructor has decided are important. If students' interests and needs are recognized by allowing the substitution of other projects similar to those listed, this technique for assigning projects becomes a workable one.

3. Free Choice

In some industrial arts classes students are given complete freedom in the choice of their projects. It would seem at first glance that this would be the perfect system for meeting individual needs and interests. It would exemplify the project method of teaching at its best. In actual practice, however, not all industrial arts teachers seem able to direct a program of this type so as to attain the desired objectives. Nor do all students possess the degree of critical thinking required. Without some direction students tend to follow whims or interests of the moment in starting projects that are not within their ability. Sometimes material and labor are wasted, to say nothing of the disappointments which result from repeated failures. It is possible for the clever teacher, through the careful use of planning sheets, to direct a program of free selection successfully. A carefully selected display of suitable projects will help the student to select wisely. However, the teacher will need to keep the desired objectives clearly in mind and to work with each student individually toward the achievement of these aims.

4. A Workable System

If a *true* project has evolved through the project method of teaching, it is evident that the student will have played a dominant role in the selection of the project. This practice will work well with students who have already had previous successful experience in problem solving. Such a technique may be applicable for certain groups of individuals, but for the majority of beginning students who have not yet experienced success in problem-solving activities, a combination of the three methods previously mentioned may make a more practical approach for the teacher with large classes.

A practical system for beginning classes would seem to include (1) an assigned project or projects to get the beginning class started and underway, (2) groups of projects from which the student may choose, and (3) the opportunity for students to substitute other projects that seem to cover approximately the same activities as those previously selected. After beginning students have gained competencies through successful problem-solving experiences via required and limited choice projects, the teacher may then utilize free choice in project selection as appropriate to certain classes and individual students. The ultimate goal of the teacher and the students should be, of course, the use of a *true* project using the project method of teaching. At the same time, however, the teacher must realize that this goal may never be realized by every one of his students.

TYPES OF PROJECTS

The project method of teaching is well suited to both individual and to group cooperative efforts.

1. Individual Projects

The great majority of all project work in industrial arts is on an individual basis. In fact, the emphasis on individual instruction is a salient feature of industrial arts education. The use of the project method with individuals is self-evident and needs no further amplification here. It must be kept in mind, however, that not all individual projects need to be of the take-home variety. Figure 8-4 shows a student whose project has altruistic value, in this case for his school. Some areas of industrial arts, for example, graphic arts, abound in similar opportunities for projectwork combining educational value and school service.

2. Class Projects

In addition to individual projects occasional opportunity should be provided for students to work together toward some common goal. The use of class projects enables the teacher to help students realize an important objective of industrial arts—the development of desirable social relationships (see Objective 7, Chapter 5). As outlined under this objective, expected behavioral outcomes may include the development of group spirit and loyalty, leadership and followership qualities, cooperation, tolerance, and tact.

Class projects may have civic or school value. Such projects may

Fig. 8–4. An industrial arts project of value to the school. (Courtesy, Franklin L. Busch, Dover High School, Dover, Pennsylvania)

include scenery for the school play, a bicycle rack for the school grounds, hurdles for the track team, a piece of equipment for the shop, a score board for the football field or gym, a slide for the playground, a float for the annual parade, repairs of Christmas toys for needy children, or the production of school newspaper, magazine, or yearbook. Through production-type experiences other class projects may help students explore industry in terms of its organization, materials, and processes (see Objective 1, Chapter 5).

The class project is well adapted to the unit method of teaching which was described in Chapter 7. By careful planning on the part of both students and the teacher, class projects can add interesting variety or "change of pace" to the class routine.

Enterprising industrial arts teachers have modified the idea of class projects so that several interesting varieties now exist. These include (1) production activities which result in many products, and (2) group effort which results in a single large project.

1. Production Activities Resulting in Many Products. Numerous references in the literature of the field attest to the popularity of production activities in the industrial arts laboratory. Some of this may be due, perhaps, to the growing number of institutions that are now requiring coursework in production methods in the industrial arts teacher education curriculum.

It is difficult to understand how teachers can achieve the objectives of industrial arts without resorting in some form or other to production activities. Some teachers refer to class production activities as *mass production* while others call them *line production.* The latter term is preferable because a production line or assembly line is usually involved.

A typical production activity generally involves the formation of a company including a Board of Directors and the selling of stock certificates to generate capital for the production of a specific product. The class then plans an industry-like organization with various divisions, such as engineering, production, and business. Role playing is an important feature of a school production activity and one which arouses student interest and enthusiasm. Sometimes the class sets up a personnel department or employment office where interviews are conducted for vacancies in the organization. Or the class may decide simply to elect key officials, but rarely does the teacher make any appointments.

Students on the engineering staff design the product to be manufactured, prepare working drawings, complete the prototype, and arrange for "tooling up" including the design of jigs, fixtures, and templates. The production department arranges the production line, trains workers, makes safety inspections, initiates production, maintains quality control, conducts time and motion studies, and packages the finished product. Meanwhile, the business department has been engaged in material procurement, sales, marketing, advertising, and consumer relations as well as cost accounting and delivery schedules. After final accounting, the Board of Directors (assuming the company has been successful!) authorizes redemption of the stock certificates, the issuance of dividends, if any, and the dissolution of the company.

Figure 8–5 shows a line production activity at the elementary school level. These pupils were studying a unit on Japan and decided to make a pair of "real" Japanese sandals for each member of the class. The finished sandals are shown in Fig. 8–6.

2. Group Effort Resulting in a Single Large Project. In the production activities previously described, the class effort resulted in a number of finished products. In the following two variations to be described, class effort resulted in a single large project. Some teachers stimulate class spirit and enthusiasm by causing two different class sections to engage in friendly competition with each other on similar projects.

The first variation involves students working cooperatively to produce a project which depicts a typical major American industry, such as steel, coal, automotive, petroleum, etc. Usually the completed pro-

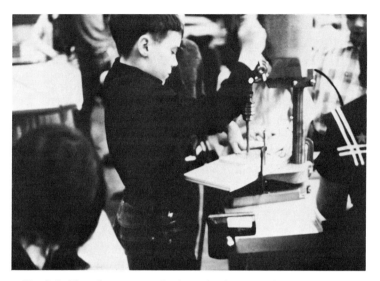

Fig. 8–5. This elementary school pupil is learning about industry by means of a line production activity. (Courtesy, Bald Eagle Elementary School, Wingate, Pennsylvania. Photo by Franklyn C. Ingram)

ject depicts the organization and operation of some processing industry or manufacturing plant. In form the project may be a large working model, a table or wall display, or a diorama.

Through role playing activities the students study the various occupational areas of the industry. These may include research and engineering, personnel training, safety education, plant management, union organization, and business or cost accounting. In addition, each student is expected to assume responsibility not only for the design and construction of some component of the project, but for its assembly to the overall project as well.

In using this approach, Brown and Haney add the following requirement for their junior high school students.

> The student also chooses a specific topic related to the selected industry, researches the topic, and presents to the class a written and oral report on his findings. Such topics might include: (1) an introductory overview of those general features of the industry's productive system related to raw material processing, product or component manufacture, inspection and quality control standards, performance testing, packaging, warehousing, shipping, and product distribution; (2) origin, history, and geographic location of firms within the industry; (3) new technological developments within the industry; (4) profits and financial outlook of firms within the industry; (5) labor-management prob-

The Industrial Arts Project 175

Fig. 8–6. These sandals were one of the construction activities engaged in by an elementary school class studying a unit on Japan. (Courtesy, Bald Eagle Elementary School, Wingate, Pennsylvania. Photo by Franklyn C. Ingram)

lems within the industry; and (6) competition with other firms and industries.[3]

A group cooperative effort can also be used to produce a project which fulfills a need of the school. For example, in one high school a group of 13 seniors designed and constructed a metals band saw for the shop. In addition to producing a useful piece of equipment, the students felt they had been greatly challenged by this problem-solving experience and received a good deal of practical experience on a variety of machines.[4]

The second variation also results in a single, large project but this time involves entire classes from several shops within the same industrial arts department. In other words, the total industrial arts facility (faculty and students) of the school combine their talents and interests into one grand, cooperative effort.

Dello-Russo reports great success in utilizing this technique with senior high school students. In one instance 50 students under the direction of three teachers in four different shops cooperatively pro-

[3]Alan D. Brown and George M. Haney, *Industrial Arts Programs* (Rockville, Md.: Earle B. Wood Junior High School, 1970), p. 3.
[4]James H. Bishop, "Team-Producing a Metal Band Saw," *School Shop*, vol. XXIX, no. 5 (January 1970), p. 40.

duced a 1903 Delton automobile in 12 weeks of working time (see Fig. 8–7). In describing this problem-solving adventure Dello-Russo writes, "There was a place in this project for talented college preparatory students."[5]

Fig. 8–7. The 1903 Delton on exhibition at the New Jersey Vocational and Industrial Arts Convention. (Courtesy, Robert Dello-Russo, Westfield Senior High School, Westfield, New Jersey)

Classes from each of the school's four industrial arts laboratories cooperated in this venture. Since there never was a 1903 Delton automobile, the drafting class created the original design and made the working drawings, templates, and patterns. Basic mechanical design and assembly were developed by the automotive class. The woodworking class manufactured the plywood body, upholstered the seat in black leather, and applied a hand-rubbed finish. The machine shop did all the machining and fabrication of metal parts including king pins, steering linkage, bushings, nuts, shafts, centrifugal clutch, differential, brake drums, and leaf springs.

After many engineering problems had been researched and solved by the students, Dello-Russo writes:

[5]Robert Dello-Russo, *Group Project Technique—Summary* (unpublished), Westfield Senior High School, Westfield, New Jersey, (n.d.).

... Twelve weeks later we rolled the Delton out of the shop. Its bright wine-colored lacquer finish pinstriped in gold was a beautiful sight. The engine started on the first pull and the Delton chugged off on its maiden trip.

The response was good. All controls functioned well; the braking was smooth and effective; the gear ratio was perfect.

And, the students were jubilant![6]

This group project enjoyed immediate success. The Delton won numerous awards, citations, and prizes at local and state meetings and earned national recognition for its builders. Blueprints of the student-designed car subsequently appeared in *Popular Science* magazine.[7]

This first multiclass group project was deemed so educationally beneficial that another one was attempted the following year. For this project the students chose a ground effects machine. Actually a novel automobile powered by a two-cylinder engine, the ground effects machine would lift itself about three inches off the ground and move along at ten miles per hour. Thirty-seven students in the same four-shop industrial arts department cooperated in the venture. This group project won several awards in statewide competition at science and industrial arts fairs.

A third group project resulted in the production of a gyro glider. This project occupied every spare minute of five boys for a period of seven weeks. When airborne the glider would rise to about 150 feet and glide at 20 miles per hour to a convenient landing site.

After the success of the 1903 Delton, the community of Westfield became so conscious of the outstanding cooperative work being done in the industrial arts department at their local high school that several anonymous donors financed other group projects.

Another example of a large group project occurred in Bay View High School (Milwaukee) under the direction of Anderson and Tibbetts. Here students produced a sport biplane. More than 75 students were involved at one time or another during the five semesters required for the airplane's completion. FAA inspectors followed all phases of the construction before permitting the first test flight. School authorities gave credit to the project for retaining certain students in school who, otherwise, would have been dropouts.[8]

[6] *Ibid.*
[7] Herbert R. Pfister, "New Cars in the Old Time Style," *Popular Science*, vol. 174, no. 6 (June 1959), pp. 125–131.
[8] Agner Anderson and Marlyn Tibbetts, "Sport Biplane," *Industrial Arts and Vocational Education*, vol. 55, no. 9 (November 1966), p. 46.

SOURCES FOR PROJECT IDEAS

Seemingly one of the most persistent problems which face the industrial arts teacher is the selection of suitable projects for his classes. This appears to be the case whether the instructor encourages student selection or all projects are assigned.

Students should be encouraged to choose and design their own projects. However, experienced teachers know that such latitude is not possible with all students (especially at first); nor is it likely that even the most apt student will be ready with a new idea as soon as he has finished a given project. It is suggested, therefore, that the instructor have on hand a list of "type" projects, such as sports equipment, radios, games, and the like, from which a choice can be made that is suited to the interest, grade level, and ability of the student. Such projects may be in the form of pictures, drawings, or models. They may also be in the form of references to sources, such as books, magazines, and picture files.

The experienced teacher will be well acquainted with all the usual sources of project ideas. For the new teacher, however, the following list of general sources may be of value:

1. Project Books. Some companies give special attention to books of projects covering almost every industrial arts area.

2. Other Teachers. Visits to the laboratories of other teachers of industrial arts often result in valuable "project finds." Visits to a school's "Open House" also may prove rewarding for the teacher seeking project ideas.

3. Magazines. Several professional magazines regularly present projects suitable for industrial arts classes including *Industrial Education* (formerly *Industrial Arts and Vocational Education*) and *School Shop*. In addition, *Popular Science* and *Popular Mechanics* as well as certain hobby and "how to do it" magazines often have excellent ideas for school projects.

4. Project Fairs. Industrial concerns, large department stores, and professional organizations frequently sponsor annual contests with prizes and awards for winning students. Visits to such "Project Fairs" provide excellent ideas for suitable projects of educational value.

5. Conventions. Professional associations at local, county, state, and national levels often feature displays of educational projects at their meetings.

6. Pictures. Illustrations in magazines, books, catalogs, and advertisements frequently furnish ideas that can be developed into worthwhile projects.

7. Museums and Stores. Places like these often provide a rich source of ideas for industrial arts projects.

The alert industrial arts teacher will be constantly watching all of these avenues as well as other sources for new ideas. The more sources that can be made available to students, the easier it will be for them to select projects which they wish to develop and which will elicit the desirable changes in behavior.

PLANNING THE INDUSTRIAL ARTS PROJECT

It was stated in Chapter 2 that from an educational viewpoint it is probably more important that a pupil be able to plan his project correctly than to carry it out skillfully. This means that the instructor must give at least as much attention to helping a student plan his work as to directing his manipulative activities. *Planning is probably the most important phase of the problem-solving method of teaching because it causes students to think.* Figure 8–8 shows a typical industrial arts student planning his project.

Fig. 8–8. From an educational point of view, it is more important for a student to plan his project well than to execute its construction skillfully. (Courtesy, George M. Haney and Alan D. Brown, Earle B. Wood Junior High School, Rockville, Maryland)

Students do not naturally know how to plan; they must be taught, just as they must be taught anything else. For some years progressive teachers in all parts of the country have been using student planning sheets. In these sheets they have found a technique for fostering thinking and planning by their students. Although a few teachers protest that young students do not have the experience background to plan their work, plan sheets have nevertheless proved successful where they have been given a fair and thorough trial.

ADVANTAGES OF STUDENT PLAN SHEETS

There are a number of advantages inherent in the student plan sheet. These become more apparent to the instructor as he becomes adept in supervising their use by students. Among such advantages are the following:

1. Plan Sheets Promote Thinking. Probably the most important single advantage of the plan sheet is that it promotes thinking on the part of the student. No student can go through the planning process for a project (even though he may have lifted the design bodily from some other source) without doing a certain amount of thinking. Often students resist planning because thinking requires effort. For this reason the teacher should continually emphasize the importance of planning and insist on some form of planning for each project. One of the most satisfying experiences that any teacher can have is to watch the development of his students in the ability to think through and solve their own problems.

2. Plan Sheets Take a Minimum of Instructor's Time for Preparation. Some teachers have objected to the use of plan sheets on the basis of the time required to check them. Although this is a time-consuming task, it will become evident that, when one considers the amount of work and time required to prepare teacher-made sheets covering the same number of projects, plan sheets actually save countless hours of work. Once the instructor becomes familiar with the process, checking can be accomplished in a minimum of time.

It must be remembered, too, that planning is one of the most educational benefits accruing in the project method of teaching. The time an instructor spends in teaching students to plan their work and in checking their plans is time well spent—not wasted.

3. Plan Sheets Promote Progress and Change. Since the instructor has no vested interest in the student plan sheets, there will not be the tendency to continue to use the same ideas over and over. Changes in design and the development of new ideas can be encouraged. The course, therefore, remains flexible and can readily be adjusted to meet the needs and abilities of each student in the class.

CHARACTERISTICS OF STUDENT PLAN SHEETS

Student plan sheets differ from other forms of written instruction sheets in several important ways. Most important of these is the fact that they are student-made rather than teacher-made. However, in order to direct and help the student in his planning, some kind of form usually is used. This will consist largely of blank spaces to be filled in by the student. The details of such a form will depend on what the teacher thinks is important in the way of planning, on the subject being taught, and on the grade level at which it is used. There are, however, certain definite characteristics which are common to most plan sheets. These characteristics include the following:

1. Plan Sheets Should Be Simple. The form for a plan sheet should be such that it is easily understood by the student. No work or information should be asked for which is not actually necessary in the planning process. Plan sheets which require more than three pages are seldom successful. If all essential drawings and information can be included on a single sheet, no more should be required. Some teachers prefer to use 5×8-in. cards. Figures 8–9 and 8–10 show the front and back of a suggested plan sheet.

2. Plan Sheets Must Be Easily Checked. Provision should be made to permit the instructor to check the various parts of the sheets at a glance. The sheet should be well organized with specific instructions as to when it should be checked, and with a place provided for initialing by the instructor.

3. Plan Sheets Should Contain Identification Data. Data on a plan sheet will probably include the student's name, grade, shop section, and the name of the project. Some teachers have found it helpful to place this material along the left-hand edge of the page, in order that it will be more easily noticed when the sheet is placed in a folder for filing.

4. Plan Sheets Should Include a Pictorial Sketch of the Project. An important purpose of a plan sheet is to convey to the instructor the idea that is in the mind of the student. This is the first "checking point," and permits the instructor to decide whether the project is suitable and will contribute to the attainment of the course objectives. This checking point also provides an opportunity for the instructor to discuss with the student any necessary changes or desirable modifications in light of the student's ability, available materials, etc. The teacher should not be disturbed if most pictorial sketches seem rather crude, since at this

STUDENT PLAN SHEET

Step 1: In the space below, make a pictoral sketch of your project. If there is not enough room below or if you have several different ideas for the same project, make your sketches on another sheet of paper and staple it to this sheet.

Be sure your sketch is approved by your instructor before going on to Step 2.

Approved by _____

Step 2: Prepare a working drawing of your project on separate sheets of paper and staple to this sheet. Have your working drawing approved by your instructor before going on to Step 3.

Approved by _____

NAME _____ GRADE _____ SECTION _____ NAME OF PROJECT _____ Date Started _____ Date Finished _____ Source of Project Idea _____

Fig. 8–9. Front of a suggested student plan sheet.

The Industrial Arts Project

Step 3: Make a list of materials you will need to complete this project. Have your instructor approve this step before going on to Step 4.

Name of part	Number of pieces	Size T W L	Material	Cost
			Total Cost	

Approved by

Step 4: In the spaces below, list the steps you will follow in making your project and the tools and equipment that will be used. As soon as your instructor has approved this step, you will be ready to begin work on your project.

Procedure	Tools and Equipment
	Approved by

Fig. 8–10. Back of a suggested student plan sheet.

point their purpose is the representation of an idea, rather than artistic perfection.

5. Plan Sheets Should Include a Working Drawing or Sketch. The making of a working drawing or sketch gives the student an opportunity to think through the problems of construction. He must decide just how much material to allow for each piece, how much stock to use, how large to make the joints, how much metal to allow for a seam, and similar considerations. Problems of how a project is to be shaped and how it is to go together must also be faced. This is the second "checking point" for the instructor and permits him to judge whether the student has thought through the details of construction. Figure 8–11 shows an instructor and student conferring about the constructional details of a project.

Fig. 8–11. Student and teacher discuss the details of construction for a project. (Courtesy Penn Manor Senior High School, Millersville, Pennsylvania. Photo by Philip D. Wynn)

6. Plan Sheets Should Include a Stock Bill. The making out of a stock bill provides an extra check on design and construction. The student must review his drawing, not only in light of the number and size of pieces needed but also from the standpoint of how much to allow for squaring, trimming, etc.

7. Plan Sheets Should Give Sequential Listing of Operations. A listing of the sequence of operations is the proof of the planning procedure. Can the student list in their correct order the operations required to make this project? Ability to do this kind of planning differs greatly among students. Some students readily learn to look ahead and to decide on the proper sequence of steps. Others achieve this ability slowly, and at first will be able to list not more than one or two steps at a time. The instructor will need to recognize these individual differences. The listing of these steps gives the instructor a final check on the student's plan. It permits him to help in avoiding mistakes that might otherwise result in a waste of material and serious discouragement.

METHODS FOR PROMOTING THE USE OF STUDENT PLAN SHEETS

The successful use of plan sheets requires special techniques and methods. Unless the teacher makes special plans for their use, the results are likely to be disappointing. Experience has shown that the following methods are essential to satisfactory results.

1. Plan with the Students and Help Them to Learn Planning. Students do not learn to plan unless they are taught. The teacher should plan the first project with the entire class. If it is a beginning group in industrial arts, there will probably be little background for the students to call upon. What little there is, however, should be utilized. Furthermore, it is important for the class to realize that planning is being taught and that each student will be expected to make similar plans for each of his projects in the future.

The planning of this first project is vitally important and may take considerable time. The time spent, however, is more than compensated by the progressive accumulation of educational benefits. An extra class period devoted to planning will pay large dividends later. The following steps are involved.

How to Make a Pictorial Sketch. It will be found that most students have no conception of how to represent an object pictorially. They must, therefore, be given some simple instruction along this line. One of the best methods for achieving results quickly is the oblique projection method. This has several advantages over either perspective or isometric. A few simple directions given during the first lesson, followed by additional instruction as the course progresses, will result in entirely satisfactory sketches.

How to Prepare a Working Drawing. Most of the students will not know how to make a working drawing. This must also be taught. Only the simplest elements necessary for the first job should be given during the initial lesson. Further instruction can be given as the need arises. In some cases the pictorial sketch can also serve as a working drawing.

Where this can be done satisfactorily, the working drawing may be omitted.

How to Prepare a Bill of Materials. It is desirable to tell students why a bill of materials is necessary. Following this a sample drawing can be used to illustrate how to determine the number and sizes of pieces needed. Students must be cautioned to make allowances for squaring and trimming of stock.

Outlining the Procedure. Plan with the class the procedure for making the project. If the project is somewhat involved and more than one demonstration is required, it may be preferable to plan only those steps to be covered in the first lesson. The remainder may be completed in a succeeding lesson.

The planning of the first project should set the pattern for the other plan sheets which will follow. This does not mean, however, that the teacher will not need to continue to teach planning. Especially at first he will need to work continually with individuals and groups to improve their ability to think through and solve their problems. Additional lessons will also need to be given to the whole group on various phases of planning such as pictorial drawings, and working drawings.

2. Permit No Work without a Plan Sheet. Thinking is one of the hardest tasks that can be set for any student. For this reason students may not take readily, at first, to the making of plan sheets. However, persistence on the part of the instructor and constant checking during the first few weeks of the course will go far in building up the idea that plans are necessary. When this concept is thoroughly engrained, the most difficult part of teaching planning is accomplished.

3. Post Plan Sheets That Are Especially Well Done. Just as teachers are accustomed to exhibiting well-made projects, so one may post well-written plan sheets. *Care should be taken not to confuse neatness and artistic ability with excellence in planning,* although these qualities frequently are found together. The essential features of a good plan sheet, however, are the clearness with which the idea is expressed and the problems involved in its construction. These points should be kept in mind when choosing sheets for display.

4. Check Plans Carefully. Students sense immediately whether the teacher feels that plan sheets are important. The amount of effort which they will put into their preparation varies accordingly. One way in which the teacher can most effectively show his feelings in this matter is to check the sheets carefully. Natural checking points are: (1) after the pictorial sketch has been made, (2) after the working sketch has been prepared, (3) when the stock bill is completed, and (4) when the procedure has been listed. After a brief period of practice, the teacher can check at each of these points without using an undue amount of time.

MISUSES OF THE PROJECT AND THE PROJECT METHOD

Some teachers misuse the project method simply because they overwork it. It should be obvious that all of the objectives of industrial arts cannot be achieved with a single method of instruction. Yet too many teachers seem to rely on the project method alone to the exclusion of all other methods. It must be remembered that the project method of teaching is a valuable one which should be in the "teaching kit" of all teachers, but it should be used with discrimination and understanding of its place and contribution to the total learning process. A project cannot be all things to all persons.

Then too, some teachers confuse pupil activity with the project method of teaching. They seem to feel that if everyone is busy, learning is taking place. At the risk of repetition, it must be pointed out again that not all objects made in the industrial arts laboratory are *true* projects in the sense that the project method of teaching has been employed. Using tools and materials in accordance with a ready-made plan may result in what is termed a project, but this product is not an educational project in the true sense of the word nor has the project method of teaching been used. Teachers rob students of vital educational experiences by requiring them to make objects from teacher-selected plans. To reemphasize a point previously made: *Construction alone does not make an object a project.* The project method involves pupil planning based on needs and interests with construction, experimentation, or manipulation followed by student evaluation of results.

Teachers often ignore time as a factor in the project method of teaching. It must be remembered that timewise, the project method is expensive. The teacher should ask himself if the time required to complete the project is commensurate with the expected outcomes, that is, the desirable behavior changes.

Finally, some teachers utilize required projects exclusively so that in only a short time their courses become stereotyped and static. Of course, it is easier to teach this way, but the course will soon become dull for most students.

PROJECT EVALUATION

The literature of the field is replete with devices for evaluating student products, such as score cards, check lists, and rating scales. Unfortunately, these devices tend to focus attention on the project itself, especially to an evaluation of the manipulative skill with which the project was completed. Points relative to purpose and planning are generally ignored. Furthermore, these devices are usually teacher-centered, that is, the teacher makes the evaluation. This latter point is not in complete harmony with the philosophy of the project method of teaching. In short, scoring devices direct attention to what the boy did to the project; instead, the teacher should be more concerned with what the project did to the boy.

It will be recalled that the prime purpose of all subject matter is to help students reach educational goals. The selection of learning activities, including projects, should always be in terms of desired behavior changes. It follows, then, that evaluation should also be in terms of student behavior changes.

It was not the intent here to present a scheme or plan of project evaluation, but rather to defer this important topic to a later chapter, "Evaluation in Industrial Arts," to which the reader is referred.

SUMMARY

A project is a worthwhile educational activity planned by both students and teacher to reach a given educational objective. The project is not an end in itself, but is only a means to an end, that is, it is a way to achieve desirable behavior changes.

A *true* project is a high form of problem solving and consists essentially of (1) feeling a need or desire, (2) making a plan, (3) creating or constructing, and (4) evaluating the results.

A project that is educationally acceptable will meet most of these criteria: Does the project *promote desirable behavior changes* and *simulate industrial methods?* Does it *possess real "boy interest"* which is *within his ability?* Is it *well designed* and does it *present a challenge?* Is it *economical of materials and student time?* Does it allow students to *work together on a common problem?*

Three methods for using projects include: (1) all projects assigned, (2) choice within groups, and (3) free choice. A workable system includes combinations of these depending upon the needs, interests, and competencies of the students.

Most projects are individual, but many teachers utilize class projects involving line production activities and groups working on a single project.

Planning is the most important phase of the problem-solving method of teaching. It is more important for a student to plan his project well than to execute it skillfully. This calls for instruction in planning. Students do not naturally know how to plan; they must be taught. Student plan sheets are useful in teaching students how to plan. Three advantages of student plan sheets are: (1) they promote student thinking, (2) they take little instructor time in preparation, and (3) they foster flexibility and change in a course.

Some desirable characteristics of student plan sheets are: identification data, simplicity, ease of checking, space for pictorial and working sketches, bill of materials, and steps of procedure.

After showing students how to plan, the teacher should (1) check plan sheets carefully before permitting students to begin construction, (2) permit no work without a plan sheet, and (3) post on the bulletin board plan sheets that are well done.

DISCUSSION TOPICS AND ASSIGNMENTS

1. Discuss the historical development of the project as an industrial arts vehicle.

2. Differentiate clearly, using references, between the meaning of the term *project* as used in industrial arts and its meaning as used in relation to the elementary school curriculum.

3. Select a given project and show how its construction would contribute to achieving the objectives of industrial arts.

4. Search the literature of the field for examples of project work in which the author describes the educational benefits that will accrue to those who complete the project.

5. Make a list of ten "boy interest" projects which you feel might be suitable for a seventh-grade or an eighth-grade class in industrial arts.

6. Interview selected pupils from a seventh-grade or an eighth-grade class concerning projects of interest to them. Compare this list with (5) above.

7. Discuss the relative merits of (a) assigned projects, (b) choice within groups, and (c) free choice.

8. Design a student plan sheet which you feel better suits your needs than is illustrated in this chapter.

9. Research the four types of projects identified by William Heard Kilpatrick in 1925. Can industrial arts projects be classified under any of these? Explain.

10. Prepare a lesson plan designed to teach students in a seventh-grade class how to plan (see Chapter 7).

11. Survey a class of boys in (1) the middle school, (2) the junior high school, and (3) the senior high school and compare their interests in projects.

12. Explain this statement: The project is a means to an end and is not an end in itself.

13. Do all projects utilize the project method of teaching? Explain your answer.

14. Using references from the field of industrial engineering, compare the terms mass *production* and *line production*.

SELECTED REFERENCES

Anderson, Agner and Marlyn Tibbetts. "Sport Biplane," *Industrial Arts and Vocational Education*, vol. 55, no. 9 (November 1966), pp. 46–47.

Bell, Lawrence L. "Student-Centered Instruction in Manufacturing Technology," *School Shop*, vol. XXVII, no. 5 (January 1968), pp. 38–39.

Bensen, M. James. "The 'Project' in a New Perspective," *Industrial Arts and Vocational Education*, vol. 58, no. 9 (November 1969), pp. 34–35.

Bishop, James H. "Team-Producing a Metals Band Saw," *School Shop*, vol. XXIX, no. 5 (January 1970), pp. 40–41.

Dutton, Bernard. "Mass Production—Principles, Applications, and Operations," *School Shop*, vol. XXVI, no. 1 (September 1966), pp. 44–46.

Feirer, John L. "Is the Project Dead?" *Industrial Arts and Vocational Education*, vol. 59, no. 6 (September 1970), p. 21.

Figurski, Arthur J. "Manufacturing Products for Christmas," *School Shop*, vol. XXX, no. 1 (September 1970), pp. 58–59.

Glazener, E. R., Ivan Hostetler, John Mitchell, Howard Reed, and Welcome E. Wright. "The Future Role of the Industrial Arts Project," *Industrial Arts and Vocational Education*, vol. 49, no. 8 (October 1960), pp. 24–26, 60, 62, 64.

Hunter, E. Max. "Another Project File System," *Industrial Arts and Vocational Education*, vol. 56, no. 3 (March 1967), p. 88.

Israel, Everett. "Planning a Line-Production Experience," *Industrial Arts and Vocational Education*, vol. 55, no. 5 (May 1966), pp. 39–41.

Johnson, Ira H., Larry E. Hoffer, Harold V. Johnson, and John G. Miller. "Are Metalworking Projects Providing Exploratory Experience in Industrial Arts?" *Industrial Arts and Vocational Education*, vol. 54, no. 8 (October 1965), pp. 26–29.

Kauffman, Henry J. "Restating the Concept of Projects," *School Shop*, vol. XVIII, no. 3 (November 1958), p. 34.

———. "Projects and Problems," *The Industrial Arts Teacher*, vol. XXIII, no. 2 (November-December 1963), p. 12.

Klammer, Waldemar E. "Project Information Index Cards," *Industrial Arts and Vocational Education*, vol. 55, no. 5 (May 1966), p. 41.

Kleinbach, M. H., George R. Keane, and Ethan A. Svendsen. "Should the Nature of Manipulative Work in Industrial Arts Change?" *Industrial Arts and Vocational Education*, vol. 51, no. 2 (February 1962), pp. 22–24.

Nelson, Hilding E. "The Production Project as a Unifying Experience," *School Shop*, vol. XXVII, no. 3 (November 1967), pp. 50–51.

Pendered, Norman C. "Once Upon a Time," *School Shop*, vol. XXIV, no. 6 (February 1965), p. 27.
Pfister, Herbert R. "New Cars in the Old Time Style," *Popular Science*, vol. 174, no. 6 (June 1959), pp. 125–131.
Slepicka, Frank V. "Learning Manufacturing Processes," *School Shop*, vol. XXVI, no. 2 (October 1966), pp. 58–59, 62.
Teel, Dean. "Unit in Manufacturing Culminates Beginning E-E Course," *Industrial Arts and Vocational Education*, vol. 56, no. 2 (February 1967), pp. 50–51.
Magowan, Robert E. "Operation Process Chart—Tool for Mass Production," *Industrial Arts and Vocational Education*, vol. 56, no. 5 (May 1967), pp. 62–63.
"The Project: Out or Redirected?" *Industrial Arts and Vocational Education*, vol. 59, no. 6 (September 1970), p. 22.
Tinkham, Robert A., Lewis H. Hodges, and John R. Lindbeck. "Individualizing the Project," *Industrial Arts and Vocational Education*, vol. 50, no. 9 (November 1961), pp. 40–46, 59, 60.
Whaley, Don. "Simulate Industry in a Laboratory 'Factory,' " *Industrial Arts and Vocational Education*, vol. 59, no. 7 (October 1970), pp. 39–42.
Wynn, Philip D. "Do Your Students' Projects Have Blue Collars?" *Industrial Arts and Vocational Education*, vol. 56, no. 3 (March 1967), pp. 72–73.
Young, Talmadge B. "The Group Project as a Beginning I-A Experience," *Industrial Arts and Vocational Education*, vol. 49, no. 8 (October 1966), pp. 27–29.

Chapter 9

Written Instructional Materials

One of the most important forms of written instructional materials is the student plan sheet which was discussed at length in the preceding chapter. It will be recalled that to use this sheet the student is required to do all the thinking and planning and then record his plans in both pictorial and written form. In the present chapter, however, the term *written instructional sheet* refers to the use of teacher-prepared materials designed largely for self-instruction of pupils.

While there is evidence that various kinds of written instructions were used in early manual-training classes, they did not come into common use until after World War I. During the decade from 1920 to 1930, however, one type or other of written instructional sheets became accepted in the teaching of industrial arts classes. While the general use of old-type instructional sheets has been discontinued in industrial arts, some kinds are still in use. One new form is still in developmental stages so far as its application to industrial arts is concerned. For this reason a discussion of the various types and how they came to be developed as well as their uses and limitations seems pertinent at this point. A major portion of this chapter is devoted to the industrial arts adaptation of written instructional material based on programed learning.

ORIGIN OF USE OF WRITTEN INSTRUCTIONAL MATTER IN INDUSTRIAL EDUCATION

During World War I, as was later true in World War II, the United States found itself critically short of skilled workers needed in vital war industries. A method was needed for training large numbers of men quickly and effectively. In this crisis Charles R. Allen, a leader in vocational education, was asked to study the situation and to develop a method for the training of shipyard workers. The study culminated in an analysis of all the "jobs" performed by shipyard workers and the

Written Instructional Materials

preparation of carefully planned "job sheets" which explained exactly how each should be performed. A "job" was defined as "anything for the doing of which a man is paid." In his book, *The Instructor, the Man and the Job,* Allen[1] described his method of analysis, the preparation of "job sheets" and how to use them. This book was widely used as a text.

However, certain other leaders in the field of vocational education disagreed with Allen's method in analyzing a trade. Foremost among these was Robert Selvidge. He argued that the "operation," not the "job," was the basic element which should be considered. In his two books, *How to Teach a Trade*[2] and *Individual Instruction Sheets,*[3] he presented his method of analysis. During recent years the techniques of trade analysis for vocational education have been greatly improved and refined. Texts by Fryklund[4] and by Bollinger and Weaver[5] are examples of modern practice.

Vocational educators were divided in their support of these two methods of analysis, but in one thing they were practically unanimous —some type of written instruction sheet should become the basis for instruction in vocational classes. This movement was given still greater impetus by the development of the Continuation School, where, because of the individual nature of most of the instruction, some such device as the written instruction sheet became a necessity. By the middle 1920's the unit instruction sheet had become essential in vocational education.

It should be noted that originally the technique of using written instructional material was developed solely for the purpose of teaching trade skills. However, many industrial arts teachers began experimenting with this newly developed method as a means for teaching their classes. Almost every type of written instruction sheet has been tried at one time or another with industrial arts classes. Some types of these sheets have probably contributed materially to standardizing courses of study and to the development of tool skills. Since no serious attempt has been made to relate these sheets to other than the skill and information objectives of industrial arts, it is likely that their effect has been to place still greater emphasis upon these two objectives, and to detract from the attainment of the others.

[1]Charles R. Allen, *The Instructor, the Man and the Job* (Philadelphia, Pa.: J. B. Lippincott Company, 1919).
[2]Robert W. Selvidge, *How to Teach a Trade* (Peoria, Ill.: Manual Arts Press, 1923).
[3]Robert W. Selvidge, *Individual Instruction Sheets* (Peoria, Ill.: Manual Arts Press, 1926).
[4]Verne C. Fryklund, *Analysis Technique for Instructors* (Milwaukee, Wis.: Bruce Publishing Company, 1965).
[5]Elroy W. Bollinger and Gilbert C. Weaver, *Trade Analysis and Course Construction for Shop Teachers* (New York, N.Y.: Pitman Publishing Corporation, 1955).

PURPOSES OF WRITTEN INSTRUCTIONAL MATERIALS

The use of the older forms of written instructional materials for purposes of industrial arts instruction is usually a compromise. They are used either because classes are so large that the teacher cannot give adequate attention to individual instruction, or because the teacher is willing to emphasize the development of skill and information in place of an all-around attainment of industrial arts objectives.

The purposes for which written instructional material is commonly used by industrial arts teachers are:

1. To Provide for Self-paced Instruction. In some cases where classes are large, it becomes almost impossible for the teacher to give the necessary individual instruction to each student precisely at the time when he needs it. Individual instruction sheets are frequently used to permit students to progress at their own rate of learning. The assignment of new projects involving processes which have previously been demonstrated is possible when this technique is used. Some of the pressure on the instructor is thus reduced, because full directions for proceeding with the lesson or project are contained on the sheet. The teacher thereby gains more time to work with students having major difficulties.

2. To Help in Demonstration Recall. In the general shop type of organization, it becomes impossible to give each demonstration at exactly the time when it is most needed by each student. During the time which elapses between a class demonstration and its application by individual students, many of the operations may be partially or entirely forgotten. An operation sheet which describes the step-by-step procedure may, therefore, be useful in the hands of the student to aid him in recalling former instruction. Many individual demonstrations may be eliminated by this technique.

3. To Provide Supplementary Information. Sometimes it is desirable to provide students with supplementary information concerning certain materials or occupations which is not readily available in standard sources. In such cases it is possible to prepare an information sheet containing this material and make it available to the students. This method of presenting material appears to be justified only when it does not appear in class textbooks or is not available in the library.

4. To Provide Homework Assignments. Laboratory time is valuable time and every moment available should be devoted to activities designed to achieve the objectives of industrial arts. Since the informative aspects are as important as the manipulative, laboratory time may, perhaps, be best used for experiences with tools, materials, and processes of industry while some of the informative experiences can be covered by self-study during out-of-shop hours. Whenever appropriate, homework assignments should be made to save valuable laboratory time. Assignment sheets, information sheets, and especially programed instruction sheets lend themselves well to out-of-shop assignments.

LIMITATIONS OF WRITTEN INSTRUCTIONAL MATERIAL

The extensive use of teacher-prepared instructional material for industrial arts classes has a number of rather serious limitations. These are:

1. Some Written Instruction Sheets Tend to Provide a Substitute for Thinking. It has been pointed out that an important objective of industrial arts (and every other subject in the curriculum also) is to promote critical thinking on the part of all students. Surely an effective way to accomplish this is to face students with problem situations. Much of the opportunity to provide such problem situations in industrial arts lies in planning how to construct projects or accomplish jobs. In written instructional sheets, other than the assignment sheet, the student plan sheet, and programed instruction sheets, the teacher selects and analyzes the material or the problem and in so doing removes most of the problem-solving experiences; this leaves the student only a cut-and-dried set of detailed instructions to be followed. Thus when the instructor hands the student a sheet which has all the answers, there is no longer any necessity for student thinking and planning. This is undoubtedly the most serious limitation to the continued use of written instructional material in industrial arts classes.

2. It Is Difficult and Time-consuming to Prepare Good Written Instruction sheets. To be effective, instruction sheets not only must be written explicitly, but they should be well illustrated. A sheet which does not contain well-made drawings, pictures, or sketches is often wasted.

3. Instruction Sheets Make No Allowance for Individual Differences. Although instruction sheets provide self-paced instruction, herein lies another serious limitation. Written instruction sheets provide uniform instruction at the same level for one and all. There is no allowance made usually for differences in student abilities, especially in reading.

4. Instruction Sheets Tend to Make Learning Impersonal. By its very nature, an instruction sheet as well as the way it must be handled in the shop, tends to promote an impersonal relationship between the student and the teacher. Personal contacts become less frequent. Since individuality is stressed, there will be fewer reasons for treating the class as a whole.

Written instruction sheets tend to discourage group enterprises such as group projects or cooperative activities. This in itself is a serious disadvantage of instruction sheets because leaders today are recognizing more and more the unique value of group effort in industrial arts. (see Chapter 8).

5. Instruction Sheets Penalize the Poor Reader. For those students who are not able to read well, the effectiveness of written instruction sheets is completely lost. Some students will be unable to perform the required manipulative activities simply because they are unable to read, understand, and follow the printed word. While it behooves the teacher to assist the learning process in any way possible, there is precious little time in industrial arts to teach remedial reading.

6. Students Are Reluctant to Read Instruction Sheets. Another problem which faces the teacher in the use of written instructional materials is to induce the students to read the sheets carefully. Students seem to have a natural tendency to go ahead immediately with the construction of a project or job without recourse to written instructions. The number of students who will read instructions carefully and try to interpret their meaning seems to be in the minority.

7. Written Instruction Sheets Tend to "Set" the Course in a Definite Pattern Which Discourages Change. As has been noted, it is both difficult and time-consuming to prepare good instruction sheets. Once a set of sheets is completed there is a natural reaction for the teacher to use them over and over, rather than to make new ones. The course thus tends to become stereotyped and the same projects are made year after year. This leads to the situation where a given family may have a half-dozen taborets and an equal number of foot scrapers of identical design, representing the progress of as many sons through the industrial arts shop. The obvious answer to this problem is for the teacher to make a few new instructional sheets each year and to discard some of the older ones, but rarely do teachers seem to find time to do so.

TYPES OF WRITTEN INSTRUCTION SHEETS

The seven types of written instruction sheets include: (1) job sheet, (2) project sheet, (3) student plan sheet, (4) operation sheet, (5) assignment sheet, (6) information sheet, and (7) programed instruction sheet.

Instruction sheets came into existence because in times past there was a dearth of instructional material available commercially. Even with the increasing amount of material which is available today, instruction sheets have the added advantage of allowing the teacher to pick and choose selected materials for presentation in short unit form.

Each of the several types of instruction sheets is designed for a specific purpose and will be described in the following sections.

1. Job Sheets. Long ago Selvidge[6] defined job sheets as "instruction sheets that tell how to do complete jobs which may involve a number of operations . . ." A job may be the construction of a useful object or it may involve installation, repair, or maintenance. Examples of jobs include the following: making a center punch, installing a duplex outlet, or repairing a cord for an electric iron. In this latter instance the operations involved are removing the defective cord, testing for a broken wire, splicing the wire, and taping the wire.

Job sheets do not teach *how* to do something because it is assumed that the user already knows how to use the necessary tools or equipment. Instead, the job sheet lists in correct order the steps involved in completing the job. Job sheets tend to become a mere listing of operations for if they actually gave complete instructions they would become long and cumbersome—difficult to read and follow.

Unfortunately for educational purposes the emphasis in a job sheet is focused on the job rather than on student needs or interests. In times past the job sheet was a mainstay in programs of vocational education. Today these sheets are still used in vocational education, but somewhat less extensively, especially in those programs that emphasize student planning. In industrial arts, job sheets have long been superseded by project sheets and student planning sheets.

2. Project Sheets. A project sheet is similar to the job sheet in that it is primarily a listing of operations to be performed in the making of some industrial arts project. It differs, however, in that the "end-in-view" is the completed project, rather than the mastery of any specific trade skill as in the case of the job sheet. Often "how-to" projects appearing in home workshop and hobby magazines are actually project sheets. Likewise, the plans accompanying model boats, cars, or planes are project sheets. Such sheets will usually contain a working drawing of the proposed project, a bill of materials, and a step-by-step procedure to be followed.

Project sheets are frequently used with large classes in industrial arts to conserve the instructor's time or in classes where students are working on an individual basis. An inherent danger is, of course, that the program will become stereotyped unless new project sheets are continuously developed.

[6]Selvidge, *Individual Instruction Sheets*, p. 9.

3. Student Plan Sheets. These are the planning sheets to which detailed reference was made in the previous chapter. These sheets are without doubt one of the most valuable written instructional sheets presently used in industrial arts. See Chapter 8 for details and examples of plan sheets.

4. Operation Sheets. Operation sheets, or process sheets as they are sometimes called, are designed to show a student how to do a specific operation. Often these sheets are illustrated with pictures or diagrams. It is first necessary for the teacher to analyze each project or job in terms of the operations involved. He must then prepare an operation sheet for each operation that he is interested in teaching.

Thus, in the example previously cited, "Repairing a Cord for an Electric Iron," the student would be directed (by a separate sheet or otherwise) to the initial operation sheet, as well as to others, as the work progressed. To continue the example, a student would first study carefully the operation sheet, "How to Remove a Defective Cord." With the necessary information at hand, he would proceed to perform this operation. He would then be directed to the next operation sheet entitled, "How to Test for a Broken Wire," and would perform that operation, and so on.

Theoretically, this method appears excellent, but in actual practice, especially with younger pupils, it has been found almost impossible to persuade the students to read the necessary directions before starting their work. Students like to get to work immediately with little, if any, prerequisite reading. So often the adage applies to students as much as to home handymen: When all else fails, read the directions.

Like the job sheet, the operation sheet is not pupil-centered. Unfortunately, most if not all of the problem solving has already been done, leaving only a casehardened step-by-step procedure for the student to follow. Perhaps the best use of operation sheets in industrial arts classes is to place in the student's hands a visual summary of a teacher demonstration of an operation or process.

5. Assignment Sheets. Written instruction sheets which are designed primarily to direct the reading, study, investigation, or experimentation of a student are called assignment sheets. They usually consist of a series of questions, the answering of which requires careful study of the references indicated on the sheet. Since these sheets guide the independent outside reading of a student, they are often called student study guides. For some years study guides have been the principal method of correspondence instruction. In industrial arts the device is extremely useful in homework assignments. Whether used in or out-of-class, assignment sheets can be educationally beneficial especially if

Written Instructional Materials

used in conjunction with models (see Chapter 14) or with flat pictures (see Chapter 15). Figure 9–1 illustrates a typical assignment sheet.

SAMPLE ASSIGNMENT SHEET

Course Title: Mechanical Drafting — Assignment Sheet #9

Unit Title: Fasteners

Purpose:
1. To become acquainted with the various means used to fasten parts together in mechanical assemblies.
2. To understand the types of fastening devices and their correct callout on the assembly drawing.
3. To gain an understanding of the different fastening processes.

Text Reference: Giachino and Beukema; *Engineering-Technical Drafting and Graphics*; pp. 191-227.

Assignment: Study carefully the above reference and write the answers to the following questions.

1. If nails and rivets are considered permanent fasteners, how would you classify cap screws? _____

_____

2. What type of threaded fastener is generally the cheapest and most convenient to use in the assembly of sheet metal items? _____

_____

_____

3. Give the correct specifications (as on a drawing) for a self-tapping screw. _____

_____

4. List the three uses for screw threads.

(1) _____

(2) _____

(3) _____

5. Be prepared for class discussion to identify and/or define the terms relating to screw threads as found on pages 195 to 199.

Fig. 9–1. An example of an assignment sheet. (Courtesy, William A. Williams, The Pennsylvania State University)

6. Information Sheets. Written instruction sheets whose objective is the presentation of information are called information sheets. Their purpose is not to teach how to do an operation or a process, but rather to impart certain pertinent information about some subject or topic. The information may be of a technical nature, guidance or occupational material, or any other information which the teacher deems important.

Typical examples of information sheets might be: "Kinds of Solder," "Job Opportunities in Graphic Arts," "The History of Interchangeable Parts in Mass Production." Information sheets are really supplemental textual material; they often become an important part of the student's notebook. Figure 9-2 is an illustration of an information sheet.

SAMPLE INFORMATION SHEET

Title: History of Numerical Control   Information Sheet #1

Course Title: Machine Shop Practice

   Numerical control, as applied to modern machine tools is relatively new but the theory behind the operation is several hundred years old. In 1784, a Philadelphia miller by the name of Oliver Evans built a completely automatic flour mill. No human labor was needed from raw grain to finished flour. In 1801 Parisian Joseph Marie Jacquard produced an automatic loom controlled by punched paper cards. Eleven years later, when Napoleon was hammering at the gates of Moscow, there were 11,000 automatic Jacquard looms being used in France alone.
   Automatic control of machines has a long and honorable history. The type of automatic control so necessary for modern machine tools, however, incorporates a feature absent in most of the earlier systems — <u>feedback</u> — a shortened word for "self-correcting". To be completely successful, feedback must make allowances or degrees of correction that are usually obtained through mathematical formula or geometric elements. Inevitably, it became necessary to provide some type of mechanism to control feedback.
   An Englishman by the name of Charles Babbage in 1822 conceived an idea and called it the "difference engine". It was the forerunner of today's calculators or computers. Because of low standards of technical progress and mechanical competence, Babbage used up his personal fortune and a government grant in a fruitless attempt to build a nineteenth century computer. It would have been difficult to build a simple adding machine and much harder to build a calculator which is actually a second-stage adding machine. To build a computer — a second-stage calculator carried to the third stage — was an impossible feat. It was not until a hundred years later that a machine, a computer, was built which was capable of solving differential equations.
   Numerical control, as a concept, was partially developed by Parsons Corporation, Traverse City, Michigan for the Air Material Command, United States Air Force. The development no doubt originally grew from the necessity during World War II of more production in less time with fewer workers.
   Working with 20th century machine tools at the Massachusetts Institute of Technology, Vannevar Bush designed and built the first differential analyzer. This computer worked entirely mechanically, using an electric motor drive. The machine, a milling machine, was controlled magnetically. Its various motions or movements were controlled by sets of numerical commands presented in sequence from punched paper tape. The second step had been made: self-controlled machine tools were a reality.

Fig. 9-2. An example of an information sheet. (Courtesy, William A. Williams, The Pennsylvania State University)

It has been emphasized throughout this text that the informative aspects of industrial arts are fully as important (if not more so) than the tool processes. Information sheets are useful in supplementing informative lessons. Since most students are visually rather than acoustically oriented, information sheets reinforce the usual informative presentation (class lecture) and also provide a means to facilitate student recall.

If the information contained in information sheets is really important, and if some means is provided to make sure that it is read (such as questions to be answered), these sheets may prove a valuable aid in helping to attain some of the objectives of the course.

7. Programed Instruction Sheets. These refer to teacher-made sheets based on the principles of programed learning. In scope they may cover a lesson or a unit of instruction. Programed instruction sheets are well suited to teaching the informative aspects of industrial arts; in fact, their potential is tremendous. For this reason the remaining sections of this chapter are devoted to the preparation of teacher-made programed instruction sheets. Before considering the actual construction of such sheets, however, it is desirable first to survey some of the general aspects of programed learning.

PROGRAMED INSTRUCTION

The newest form of written instructional material in education today is called programed instruction. It holds great promise of becoming one of the most effective methods of written instruction yet devised. Generally, programed instruction is a self-study technique in which the student reads one or more short statements and then responds to a question. Immediately after responding, the student is shown the correct answer. After comparing both answers, he then moves to the next "frame" which is another informative statement and a question to be answered. The small bits of subject matter are arranged on sequential frames to carry the learner from the simple to the complex in short, easy steps. Each instructional frame is dependent upon previous frames. Ideally, the program is designed so that the average student will select the correct answer every time.

The foundations for modern programed instruction were laid by S. L. Pressey at Ohio State University in the 1920's, but his work received little recognition or support. In 1954 the work of B. F. Skinner of Harvard aroused new interest in programed instruction. Since that time numerous studies and investigations have provided convincing support for programed instruction. New advances are continuously being made. Some of America's most eminent scholars, abetted by generous research funds, are refining old practices and exploring new concepts of programing. Education, industry, and the military services are vitally concerned with programed instruction. Whether or not it will ever become as widely used or as highly effective in everyday teaching as its proponents have predicted remains for the future to decide.

FEATURES OF PROGRAMED INSTRUCTION

The tenets of programed instruction lie deeply rooted in the theories and practices of educational and experimental psychology. From a psychological viewpoint, the sequence of programed learning is basically as follows:

1. *Exposure*—a small bit of information to be learned is shown to the learner.

2. *Stimulus*—a question challenges the learner to interpret or apply what has just been learned.

3. *Response*—the learner responds to the stimulus by writing or selecting an answer.

4. *Reinforcement*—immediate confirmation of the answer (feedback) is given to the learner.

The principal psychological basis on which programed instruction rests is that of reinforcement which is rewarding desired behavior so that it will tend to reoccur. If a learner knows immediately that his answer is correct, his learning is said to have been reinforced. Psychologists have found that learning is most efficient and retention is enhanced when reinforcement is present.

It must be remembered that programs are structured purposely so that the student will usually get the correct answer. When an incorrect answer does occur, the student is informed immediately of his error. By correcting his response at once, he receives immediate reinforcement. At the same time the learner's self-awareness of his success breeds a growing self-confidence which is inherently rewarding thus providing another source of reinforcement.

Good programed instruction will include the following characteristic features:

1. Statements of objectives in terms of expected behavior or performance outcomes.

2. Logical organization of instructional material to achieve the stated objectives.

3. Presentation of information to be learned in short, easily comprehended steps built one upon another.

4. Active participation of the learner through reading the material with covert or overt responses to the questions.

5. Immediate feedback of correct answers.
6. Provision for students to proceed at individual rates of learning.

TYPES OF PROGRAMED INSTRUCTION

Programed instruction appears in several forms such as (1) programed textbooks, (2) teaching machines, and (3) programed lessons or units of instruction. It is with programed lessons that subsequent sections of this chapter are primarily concerned.

1. Programed Textbooks. Industrial arts textbooks especially programed for instruction are conspicuous by their absence. The pioneer work of Coover and Helsel[7] called *Programed Blueprint Reading* is exemplary. There are other programed texts which have applicability to industrial arts, but most of these are in the area of electricity-electronics. Typical examples include *Programed Basic Electricity Course*[8] and *Capacitance and Capacitors*, Part I[9] and Part II.[10] In addition, the New York Institute of Technology has programed three textbooks called *Basic Electricity*,[11] *Basic Electronics*,[12] and *Basic Transistors*.[13]

2. Teaching Machines. A variety of teaching machines is available with programs in academic disciplines. Some of these, particularly in science education, may have limited usefulness in industrial arts. Unfortunately, however, programs are unique to their respective machines, that is, programs cannot be transferred from one type of machine to another.

It must be remembered that the teaching machine is merely a device with three basic purposes: (1) to present the material to the learner, (2) to provide a means of student response, and (3) to provide feedback. The heart of the teaching machine is the program, not the mechanical device itself.

---

[7]Shriver L. Coover and Jay D. Helsel, *Programed Blueprint Reading* (New York, N.Y.: McGraw-Hill Book Company, 1963).
[8]Milton Rosenberg, *Programed Basic Electricity Course* (Indianapolis, Ind.: Howard W. Sams and Company, Inc., 1964).
[9]Robert H. Kantor, *Capacitance and Capacitors*, Part I (Palo Alto, Calif.: Fearon Publishers, Inc., 1963).
[10]Robert H. Kantor, *Capacitance and Capacitors*, Part II (Palo Alto, Calif.: Fearon Publishers, Inc., 1963).
[11]New York Institute of Technology, Staff of Electrical Technology Department, *A Programmed Course in Basic Electricity* (New York, N.Y.: McGraw-Hill Book Company, 1963).
[12]New York Institute of Technology, Staff of Electrical Technology Department, *A Programmed Course in Basic Electronics* (New York, N.Y.: McGraw-Hill Book Company, 1964).
[13]New York Institute of Technology, Staff of Electrical Technology Department, *A Programmed Course in Basic Transistors* (New York, N.Y.: McGraw-Hill Book Company, 1964).

The latest and most sophisticated form of teaching machine is one which is computer-assisted. The student works at a typewriter terminal and receives instruction via typed messages, colored slides, and tape recorder all controlled by the computer. Figure 9-3 illustrates a student terminal in a computer-assisted program. In some computer-controlled programs, student capabilities are evaluated and incorporated into a control program which determines the level of material to be presented, its sequence, and the mode of instruction. Objective-type questions are employed, but in addition, some programs require the student to word his own response. The computer then compares this response with those stored in its memory. A correct response will instantly forward the next bit of information to be mastered. Key responses may speed the learner ahead bypassing material which he understands. An incorrect response may bring a diagnostic comment and branch the student to remedial instruction. Records of student progress are kept and the computer can instantly present the right portion of the program regardless of time lapse since the student last worked on the program. Refined systems are in use which provide this highly individualized self-instruction to as many as 32 learners simultaneously. Newly developed experimental systems can now accommodate as many as a few hundred learners simultaneously, depending upon the size of the computer utilized.

3. Programed Lessons or Units of Instruction. Programed materials in the form of instruction sheets are not available commercially for use in industrial arts classes. However, enterprising industrial arts teachers have been experimenting with this new media of instruction. Descriptions of their experiences and samples of their programs may be found in the current literature of the field. Some recent examples include the following: "Programed Unit—The Board Foot,"[14] "Teacher-Made Program—Electricity,"[15] "How to Adjust the Ion Trap,"[16] "Programed Instruction—Work and Energy,"[17] and "Learning How to Read a Rule and Measure Accurately."[18] This latter program is unique because it is used in conjunction with geometric models which the student is to measure at the end of the program. This is the acid test because if a

---

[14]Richard C. Quesenberry, "Programed Unit—The Board Foot," *Industrial Arts and Vocational Education*, vol. 53, no. 8 (October 1964), pp. 53–56.

[15]Norman F. Plezia, "Teacher-Made Program—Electricity," *Industrial Arts and Vocational Education*, vol. 53, no. 8 (October 1964), p. 48.

[16]William W. Jerold, "Programmed Learning in Electronics," *Industrial Arts and Vocational Education*, vol. 52, no. 9 (November 1963), pp. 22–23.

[17]Terence J. Trudeau, "Programed Instruction—Work and Energy," *Industrial Arts and Vocational Education*, vol. 54, no. 4 (April 1965), pp. 48–49.

[18]Clarence H. Preitz, "Writing Programed Instruction in Industrial Education," *Industrial Arts and Vocational Education*, vol. 55, no. 4 (April 1966), pp. 45–48.

Written Instructional Materials

Fig. 9-3. A terminal of a computer-assisted teaching system. (Courtesy, Scott Kostenbauder, The Pennsylvania State University)

student can measure the models successfully, learning obviously has taken place. An experimental study in which programed materials were utilized to teach manipulative skill is described in Chapter 10.

The preparation of programed lessons is well within the capability of the progressive industrial arts teacher. In fact, the instructor who can prepare well-written instructional materials such as job and operation sheets as well as project and information sheets will enjoy an immediate and enviable advantage. His previous experience in analyzing subject matter into small steps and organizing these into a logical step-by-step procedure will stand him in good stead. For these instructors the transition from preparing conventional instruction sheets to preparing programed sheets is not difficult.

PROGRAM FORMAT

Before a teacher can begin to prepare a programed lesson or unit, he should first decide whether the format of his program will be linear or branching.

Linear programming is a straight-line program in which the steps are in a fixed order of succession. The frames are presented in the same sequence to each student whether or not he gives the correct answer. Primary credit for linear programing is usually given to B. F. Skinner.

The frames of linear programs which are presented by means of printed materials (nonmechanical presentations) may be arranged vertically on the page so that the learner reads downward much as one reads a book (see Fig. 9–4 and Appendix 2). Or the frames may be arranged horizontally across successive pages as shown in Fig. 9–5.

HOW TO PLANE A SMOOTH SURFACE

Name _____

1. The <u>plane</u> is a special tool with a blade for <u>smoothing</u> and <u>removing wood</u> as shavings. The modern plane developed from the <u>chisel</u>.

   The <u>p   </u> removes wood as <u>s   </u> and creates a <u>s   </u> surface. The plane developed from the <u>c   </u>.

   > plane, shavings,
   > smooth, chisel,

2. The <u>first step</u> in the use of the plane is to <u>check</u> the <u>double plane iron</u> for proper fit.

   The <u>d   p   </u> iron should be checked for proper fit before other adjustments are made.

   > double plane,

3. Sight along bottom of plane to see if the <u>cutting edge</u> is <u>parallel</u> to the bottom of the plane.

   If the <u>c   e   </u> is not <u>p   </u> to the plane bottom, adjustments are made with the <u>lateral adjusting</u> lever.

   > cutting edge,
   > parallel,

4. The cutting edge should be adjusted with the <u>depth adjusting nut</u> to allow it to stick out about the <u>thickness of a hair</u>.

   The cutting edge should stick out about the <u>t   </u> of a <u>h </u>. This adjustment is accomplished by turning the <u>d   a   n   </u>.

   > thickness of a hair,
   > depth adjusting nut,

Note: Always place the plane on its <u>side</u> when not in use. By placing the plane on its <u>s   </u> no damage will occur to the cutting edge.

   > side,

5. Before it is planed, the stock must be <u>fastened</u> in the <u>vise</u> or with a clamp.

   The <u>v   </u> or <u>c   </u> secures the stock to aid in the planing operation.

   > vise, clamp,

6. All <u>surface</u> and <u>edge</u> planing should be done in the <u>direction</u> of the <u>grain</u>.

   Planing in the <u>d   </u> of the grain for all <u>s   </u> and <u>e   </u> produces a smooth finished surface.

   > direction, surfaces,
   > edges,

Fig. 9–4. *Above and facing page:* An example of vertical linear programing. (Courtesy, M. James Bensen, University of Wisconsin—Stout)

7. Apply pressure on the knob at the beginning of the cutting stroke. This pressure on the k____ at the b_____ of the stroke permits a good start to the cut.	
	knob, beginning,
8. When the whole plane surface is in contact with the work surface, then pressure should be applied to both the knob and handle. This even pressure to both the k____ and h_____ permits a smooth even cut.	
	knob, handle,
9. As the toe of the plane reaches the end of the stroke, release knob pressure, but retain handle pressure to complete the stroke. The releasing of the _____ pressure while retaining h_____ pressure, insures a straight edge.	
	knob, handle,
10. Planing End Grain: (A) The pressure principles are the same for end grain as for the surface or edge. Special precautions have to be observed, however, to avoid splitting the end grain. (B) One precaution for planing end grain is to plane halfway across the grain from the two opposite edges. By planing h_____ across the end from opposite edges, no splits are likely to occur.	
	halfway,
(C) Another precaution is to place a piece of waste stock at the edge and plane all the way across the end from one direction. All splits will occur in the w____ stock as it protects the edge of the workpiece.	
	waste,

After a question has been answered, the page must be turned to reveal the correct answer. In linear programs the student provides a constructed response; in other words, he is usually called upon to write the answer in the space provided.

Ordinarily in a linear program the student works his way through frames in consecutive order. However, in certain types of linear programs it is possible to utilize diagnostic frames for forward-branching to material best suited to the learner.

Branching or intrinsic programing was devised by Norman A. Crowder. In this format the frames present more information than do frames in linear programing. In some cases several paragraphs of material are presented. In branching programs the response consists of a multiple-choice item with a page number following each response. Depending upon the answer selected, the student is directed to a certain page for further instruction. If the correct answer has been selected, the student is sent to the page with the next frame in sequence. However, if a wrong answer was selected the student is referred to a remedial frame on another page and later guided back to the main program.

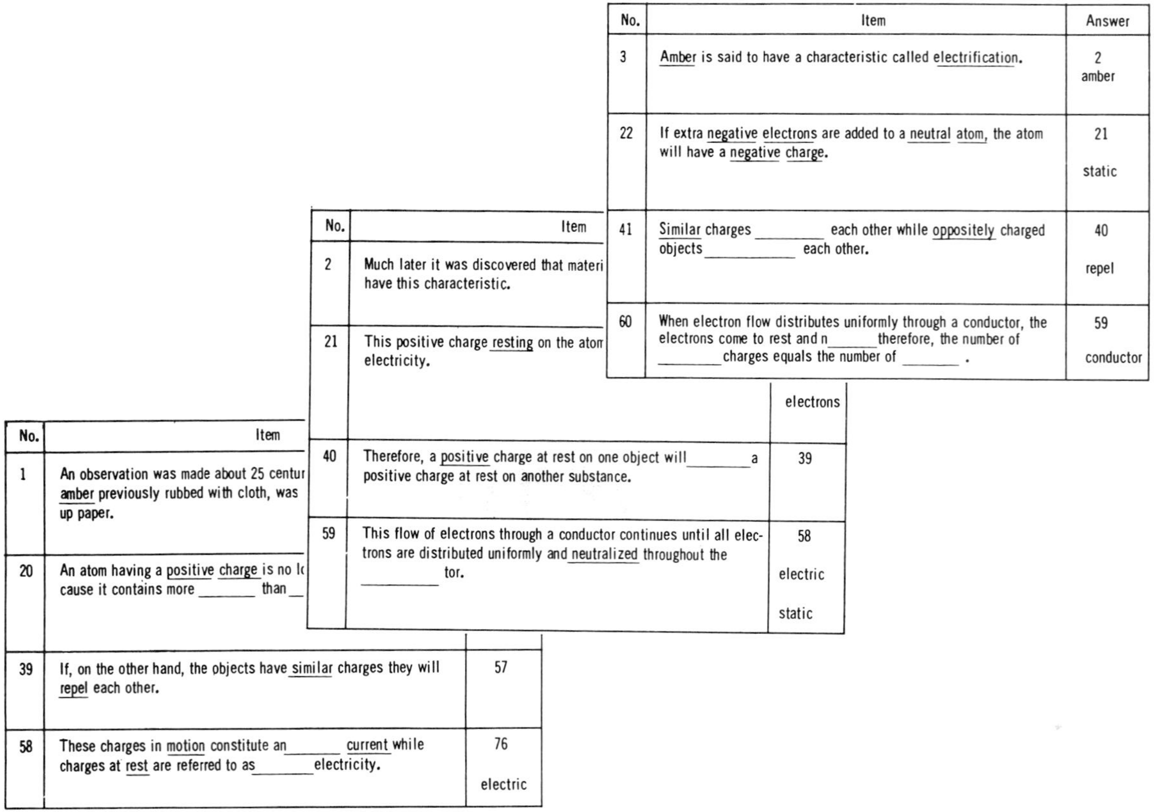

No.	Item	Answer
3	Amber is said to have a characteristic called electrification.	2 amber
22	If extra negative electrons are added to a neutral atom, the atom will have a negative charge.	21 static
41	Similar charges _____ each other while oppositely charged objects _____ each other.	40 repel
60	When electron flow distributes uniformly through a conductor, the electrons come to rest and n _____ therefore, the number of _____ charges equals the number of _____.	59 conductor

No.	Item	
2	Much later it was discovered that materi[al] have this characteristic.	electrons
21	This positive charge resting on the atom electricity.	39
40	Therefore, a positive charge at rest on one object will _____ a positive charge at rest on another substance.	58 electric static
59	This flow of electrons through a conductor continues until all electrons are distributed uniformly and neutralized throughout the _____tor.	57

No.	Item	
1	An observation was made about 25 centur[ies ago that when] amber previously rubbed with cloth, was [able to pick] up paper.	
20	An atom having a positive charge is no l[onger neutral be]cause it contains more _____ than _____	
39	If, on the other hand, the objects have similar charges they will repel each other.	76 electric
58	These charges in motion constitute an _____ current while charges at rest are referred to as _____ electricity.	

Fig. 9–5. An example of horizontal linear programing. (Courtesy, Norman F. Plezia, Central Schools, Williamsville New York)

208

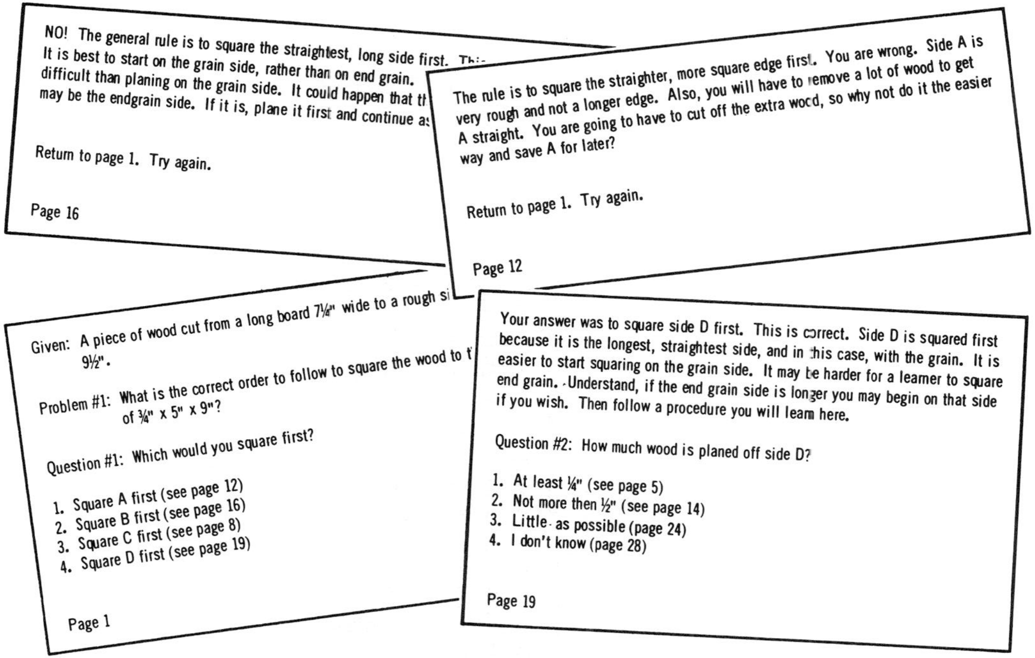

NO! The general rule is to square the straightest, long side first. It is best to start on the grain side, rather than on end grain. This may be the endgrain side. If it is, plane it first and continue as difficult than planing on the grain side. It could happen that the

Return to page 1. Try again.

Page 16

Given: A piece of wood cut from a long board 7¼" wide to a rough size of ¾" x 5" x 9".

Problem #1: What is the correct order to follow to square the wood to 9½"?

Question #1: Which would you square first?

1. Square A first (see page 12)
2. Square B first (see page 16)
3. Square C first (see page 8)
4. Square D first (see page 19)

Page 1

NO! The general rule is to square the straightest, long side first. This Side A is It is best to start on the grain side, rather than on end grain. You are wrong. Side A is difficult than planing on the grain side, more square edge first. Also, you will have to remove a lot of wood to get may be the endgrain side. If it is, plane it first and continue as very rough and not a longer edge. You are going to have to cut off the extra wood, so why not do it the easier A straight. You are going to have to cut off the extra wood for later?
way and save A for later?

Return to page 1. Try again.

Page 12

Your answer was to square side D first. This is correct. Side D is squared first because it is the longest, straightest side, and in this case, with the grain. It is easier to start squaring on the grain side. It may be harder for a learner to square if you wish. Understand, if the end grain side is longer you may begin on that side if you wish. Then follow a procedure you will learn here.

Question #2: How much wood is planed off side D?

1. At least ¼" (see page 5)
2. Not more then ½" (see page 14)
3. Little as possible (page 24)
4. I don't know (page 28)

Page 19

Fig. 9–6. Pages from a teacher-made scrambled booklet, "Correct Procedure for Squaring Stock to Size." (Courtesy, William L. Kadoich, Lincoln School, Bethlehem, Pennsylvania)

Some branching programs utilize diagnostic frames whose correct answers will branch the learner on a fast track to new material ahead. Thus branching can lead a student into remedial frames, into optional enrichment frames, or into normal-sequence frames. The primary purpose of branching is to accommodate individual differences in students.

Two examples of branching format are Kantor's programed texts, *Capacitance and Capacitors,* Parts I and II.[19] The "scrambled textbook" is a form of branching or intrinsic programing. Figure 9–6 is an example of a teacher-made scrambled booklet in industrial arts.

On the surface it may appear that branching programs possess more advantages and are far superior to linear programs. This is not the case, however, because there is no research evidence to support such claims. On the other hand, psychologists have been somewhat critical of branching programs for a number of reasons, mostly pedagogical. The issue has not been settled and will probably occupy those in the field of psychology for some time to come. In the meantime, the industrial arts teacher probably should begin with linear programing which, after all, is less complex and has yielded excellent results.

STEPS IN DEVELOPING PROGRAMED INSTRUCTION SHEETS

1. Identifying Terminal Behavior

The first step in programing is to identify the terminal behavior of the student—that is, the expected behavior *after* he has completed the program. This is accomplished in the same manner as the behavior changes were identified for each of the objectives of industrial arts (see Chapters 5 and 6).

In programed instruction it is especially important to state the objective in terms of specific behavior or performance. Mager describes it this way:

> The way to write an objective . . . is to write a statement describing one of your educational intents and then modify it until it answers the question, "What is the learner DOING when he is demonstrating that he has achieved the objective?"[20]

If the instructor wishes to develop a programed lesson on, say, board measure he must first state the objective of the lesson in performance

---

[19] Kantor, *op. cit.*
[20] Robert F. Mager, *Preparing Instructional Objectives* (Palo Alto, Calif.: Fearon Publishers, Inc., 1962), p. 14.

terms. Certain words such as *know, understand,* and *appreciate* are subject to misunderstanding and should be avoided in writing objectives for programed instruction. For example, the lesson objective *to know board measure,* is both weak and ambiguous because it is not clear what kind of knowledge is being specified. What is the student expected to be able to do after he completes the lesson? Should he know the formula for computing board measure? Should he know why board measure is used? Should he be able to recite board measure charts? The instructor must first limit or narrow his objective so that it focuses on a single behavior outcome. If the instructor expects his students to solve problems in board measure, then it would be better to state the objective in this way: *Upon completing this lesson, the student is expected to be able to solve board measure problems.* Obviously this answers the question: What will the learner be doing when he is demonstrating that he has achieved the objective?

Mager recommends several ways to tighten up an objective by specifying the conditions under which the student must operate in demonstrating that he has achieved the desired behavior.[21] These involve specifying certain limiting conditions, acceptable performance levels, time, *et al.* To illustrate, the above objective would read: "Upon completing this lesson, the student will be able to solve board measure problems *without reference to any printed material.*"

Another way is to specify acceptable levels of performance, that is, specify that a student must get so many correct responses or attain a given percentage right, and so on. The objective referred to above would now read: "Upon completion of this lesson, the student will be able to solve *four out of five board measure problems* without reference to any printed materials."

Still another way involves time, that is, the teacher may require that the student demonstrate his achievement of the desired outcome within a given interval of time. Applying this to the board measure lesson, the objective would read: "Upon completion of this lesson, the student will be able to solve four out of five board measure problems *in fifteen minutes* without reference to any printed materials."

In this instance it may be that a time limit is of no importance to the teacher. He may be more concerned with successful completion of the criterion test without consideration of time. On the other hand, time may well be an important factor in student performance on this, or other, objectives. The point is that the teacher must specify exactly what the student is expected to demonstrate and under what conditions the performance is to be carried on. The teacher not only specifies the

[21] *Ibid.,* pp. 25–44.

terminal pupil behavior but also determines the standards of acceptable performance. The greater the specificity of the objective, the easier it will be later to prepare the instructional frames and the criterion test to measure final pupil outcomes.

Of course, a programed lesson may have more than one objective, but each should be stated succinctly and specifically in terms of expected behavior outcomes. Usually two or three objectives are sufficient for a programed lesson. If more objectives are required, it may be better to plan two short programs rather than one long one.

2. Criterion Test

As soon as the terminal behavior has been identified and stated specifically in measurable terms, it is a good plan to prepare the criterion test. This is usually a simple evaluative measure to provide objective evidence whether or not the student has achieved the desired outcome. In the case of the programed lesson on board measure, the criterion test would present the student with one or more problems in board measure. When the student gets the right answers (and he is expected to), he has demonstrated his ability to solve board measure problems and has, in fact, achieved the objective of the lesson.

Criterion tests can be prepared easily by examination of the lesson objectives as expressed in specific terminal behavior. For the board measure example, the criterion test would be to present to the pupil five problems on board measure. If he correctly answers four of these five problems without the aid of printed materials and within the time limit of 15 minutes, it is presumed that he has achieved the stated behavior outcome. In other words, the criterion test score stands as objective evidence that the objective of the lesson has been reached.

3. Constructing the Instructional Frames

After the terminal behavior has been specified, the actual construction of the program can begin. The core of programed instruction is the instructional frame which contains the material to be presented, a question (stimulus), and the answer to the preceding frame (reinforcement). It is necessary, of course, that the lesson content be first analyzed by means of a detailed outline. This is sometimes called a *task analysis*. The effort exerted in detailed outlining will pay dividends later. Based on this outline, statements should be prepared which cover all points in the outline. This is the laborious part of programing and it will take considerable effort to get the statements concise and succintly worded.

Now the statements may be arranged in steps. Each step should consist of one or two statements. Avoid the temptation to include too much information in a step. The estimated difficulty between any two steps should be approximately the same.

At this point the stimulus (question) should be prepared for each step. The answer should be prepared, too, and readied for placement on the appropriate step. Each step or frame should contain only enough information to answer the question. Superfluous material should be avoided at all costs.

It is during this phase of programing that the industrial arts teacher who is already skillful in writing instructional sheets comes into his own. Previous experience will be most helpful in analyzing subject matter and arranging it in step-by-step procedures as is required in the preparation of instruction sheets. Existing job, operation, and information sheets are sometimes useful in providing valuable programing material. The subject matter contained on these sheets has usually been analyzed and arranged in an orderly, logical, step-by-step arrangement.

4. Cueing and Fading

Depending on the nature of the subject matter and the age level of the learner, cueing may be employed. A cue is a hint, prompt, or aid to help the student get the correct answer. One common method of cueing is to give one or more letters in the answer space. An example of this kind of cueing is illustrated in Fig. 9–7.

Cues may be continued throughout the program. Usually, however, they are gradually withdrawn (fading) over a series of frames. The skill with which fading is accomplished is a mark of good program design.

Cueing is an optional feature which may be included at the discretion of the programer. Many outstanding programs do not utilize cueing. One of the inherent dangers is overcueing, that is, making the responses too easy.

5. Preparing Review Frames

Even though the material to be learned is presented in small steps, it is desirable to review at certain intervals. This review should not take the form of a mid-program test, but rather should be worked into the frames throughout the program. It is good teaching to teach and test, reteach and retest in an endless cycle; the same applies to programing. Sometimes reviewing can be accomplished simply with a frame headed "Review" followed by the review question.

EXAMPLES OF CUEING	
5. Files should always have handles on the "tang". To use a file without a handle is UNSAFE! A handle should always be placed on the t_____ of the file.	Answer 4 SHAPING

6. There are FOUR ways of identifying or classifying files. They are by: (1) size, (2) shape, (3) type of cut, and (4) coarseness.
Name four ways of identifying files:
  (1) s_____   (3) t_____ of c_____
  (2) sh_____   (4) coa_____
	Answer 5
	
	TANG

7. The size (length) of a file is measured from the point to the heel. The distance from the point to the heel of a file (length) is called its s_____.	Answer 6 (1) SIZE (2) SHAPE (3) TYPE OF CUT (4) COARSENESS

Fig. 9–7. Examples of cueing. (Courtesy, Franklin L. Busch, Dover High School, Dover, Pennsylvania)

6. Use of Illustrations

Illustrations have been used to some extent in programed instruction but their effectiveness, if any, has not yet been confirmed by research although some progress is being made in this direction. One study investigated the effects of illustrated and nonillustrated programed instruction sheets in the teaching of a manipulative operation; see the *Bensen Study* reported toward the close of this chapter. An example of the use of illustrations in a programed unit of instruction is shown in Fig. 9–8. Compare this program with that shown in Fig. 9–4.

7. Testing and Revision

It is not necessary to wait until a program is completely finished before trying it out. In fact, it is highly desirable to begin tryouts as soon as the first frames are completed. This is called *developmental testing*. Trial readings with students will soon reveal errors and weaknesses. Frames may be ambiguously worded; they may be too long; they may contain extraneous material; or the cues (if used) may be giveaways. Improvements, refinements, and revisions in the frames should be made continuously during this developmental testing period.

When all the frames are ready, the second phase of testing called

Written Instructional Materials

	HOW TO PLANE A SMOOTH SURFACE	
	Name _____	
SHAVING / PLANE / CHISEL	1. The plane is a special tool with a blade for smoothing and removing wood as shavings. The modern plane developed from the chisel. The p____ removes wood as s____ and creates a s____ surface. The plane developed from the c____ .	plane, shavings, smooth, chisel,
POOR FIT / PROPER FIT	2. The first step in the use of the plane is to check the double plane iron for proper fit. The d____ p____ iron should be checked for proper fit before other adjustments are made.	double plane,
LATERAL ADJUSTING LEVER UNDER PLANE IRON	3. Sight along bottom of plane to see if the cutting edge is parallel to the bottom of the plane. If the c____ e____ is not p____ to the plane bottom, adjustments are made with the lateral adjusting lever.	cutting edge, parallel,
DEPTH ADJUSTING NUT	4. The cutting edge should be adjusted with the depth adjusting nut to allow it to stick out about the thickness of a hair. The cutting edge should stick out about the t____ of a h____ . This adjustment is accomplished by turning the d____ a____ n____ .	thickness of a hair, depth adjusting nut,
NOT IN USE	NOTE: Always place the plane on its side when not in use. By placing the plane on its s____ no damage will occur to the cutting edge.	side,
VISE CLAMP	5. Before it is planed, the stock must be fastened in the vise or with a clamp. The v____ or c____ secures the stock to aid in the planing operation.	vise, clamp,
WITH DIRECTION OF GRAIN	6. All surface & edge planing should be done in the direction of the grain. Planing in the d____ of the grain for all s____ and e____ produces a smooth finished surface.	direction, surfaces, edges,

Fig. 9–8. An example of the use of illustrations in programed material. (Courtesy, M. James Bensen, University of Wisconsin—Stout)

field testing, may begin. Here the completed program is tried out under actual teaching conditions. More revisions will probably have to be made in some of the frames based on the field tests. If a number of students run into difficulty on the same frame, it may be indicative of weakness not necessarily in that frame, but in the frames preceding it. While a low student-error rate on each frame is desirable, this in itself does not ensure a good program. The program must teach the material so that the student will achieve the desired terminal behavior as measured by the criterion test.

It is not to be expected that the industrial arts teacher will produce excellent programed instruction sheets on his first attempt. Programing is time-consuming and demands a high order of knowledge and skill not only in industrial arts, but in the principles and practices of programing as well. However, improved programs will doubtless ensue as the teacher gains experience in this new media. Industrial arts instructors who are seriously interested in programed instruction will consult the many fine references available on programing and will study carefully both teacher-made and commercial programs.

EXAMPLE OF A PROGRAMED INSTRUCTION SHEET

Appendix 2 contains an example of a teacher-made programed lesson on the history of mass production. It is complete with terminal objectives, instructional frames, and criterion test.

RECENT RESEARCH FINDINGS

Like the cambium layer of a tree, research is the growing edge of the teaching profession. Without research our profession would soon take on the characteristics of a trade. With research new growth and development lie within our grasp. This section deals with selected research findings related to written instructional materials. The first two studies examine the important question: Does the use of written instructional material increase the effectiveness of the class demonstration?

1. The Bensen Study[22]

The primary purpose of this doctoral study was to compare the relative effectiveness of programed operation sheets as a supplement to

---

[22]M. James Bensen, "An Investigation in the Use of Programed Operation Sheets as a Supplement to the Group Demonstration in Teaching Manipulative Operations" (unpublished doctoral dissertation, The Pennsylvania State University, 1967).

the class demonstration in the teaching of manipulative operations. A sub-problem was to study the effects, if any, of a technical line drawing as an illustration in each frame of the programed instruction sheet.

Two experimental groups (one using a conventional programed operation sheet and the other using an illustrated program operation sheet) were compared with a control group which utilized a conventional operation sheet. These self-instructional materials were used to supplement class demonstrations in seventh-grade woodworking classes. The three groups were studied in relation to the following: (1) amount of technical knowledge acquired, (2) amount of construction time required to complete the job, (3) amount of teacher assistance required by the learner in completing the job, and (4) quality of the finished job.

Analysis of the data revealed that both experimental groups showed a significant difference (.01 level) over the control group in the variables, "technical knowledge" and "teacher assistance required by the learner." This indicated that pupils who used programed instruction sheets (either illustrated or nonillustrated) acquired more technical knowledge and required less teacher assistance in completing their jobs than did pupils who used the conventional operation sheet. However, there was no difference noted between the two experimental groups on either of these two variables.

The data showed no significant differences among the groups on either (1) "time required to complete the job," or (2) "quality of the completed job." Thus, illustrated, nonillustrated, and conventional operation sheets produced equal results on these two variables.

No significant differences were found in this study between the two experimental groups using the illustrated and nonillustrated versions of the operation sheets.

In discussing his findings Bensen noted the following:

> No significant difference was found in the amount of time it took the students to complete the job. Previous research indicated that the use of programed materials required significantly more time to complete specific operations. The class demonstration, which preceded the construction phase, apparently provided the students with an orientation to the overall performance of the operation and they only needed the programed material as a guide.
>
> Previous research also indicated that illustrated programs are superior to nonillustrated programs in studies involving the learning of skills. This study did not substantiate these previous findings. When programed materials are used as a supplement to the class demonstration, students tend to have a complete orientation to the operation. Hence,

any effect the illustrations have on understanding the concept and aiding in the performance of the operation are cancelled out by instruction received through the demonstration.[23]

2. The Repp Study[24]

This research study compared two different ways of using the class demonstration to teach engine lathe operations. The conventional teacher demonstration was compared with the same demonstration but supplemented with a printed demonstration-performance guide. These guides were unique. They resembled an illustrated operation sheet, but contained blank spaces to be filled in by students as the lesson-demonstration progressed. The study involved six lessons presented to intact high school classes (109 students) for two consecutive semesters. The experimental group (using the demonstration-performance guides) was compared to the control group (using the conventional teacher demonstrations) on the following factors: (1) amount of technical knowledge acquired by the learner, (2) degree of manipulative skill developed, (3) speed of performance, (4) teacher time required to present the class demonstrations, (5) quantity of material consumed in the learner's performance of the demonstrated tasks, (6) number of follow-up individual demonstrations required, and (7) number of follow-up small group demonstrations required.

Analysis of the data revealed that significant differences between the experimental and control groups did not appear for any of the seven variables studied. This means simply that in this study equal results were attained on the variables mentioned regardless of which teaching technique was employed. In other words, the use of the demonstration-performance guides did not increase the effectiveness of the class demonstrations in this investigation.

3. The Warner Study[25]

In 1969 Warner studied programed instruction in junior high school metalworking. The primary purpose of his study was to compare results of the Min-Max III teaching machine with the lecture-discussion method. Programed materials for eighth- and ninth-grade students

[23] *Ibid.*
[24] Victor E. Repp, "The Relative Effectiveness of Two Demonstration Techniques in Teaching Industrial Arts" (unpublished doctoral dissertation, The Pennsylvania State University, 1970).
[25] Richard A. Warner, "The Effect of Programed Instruction in Industrial Arts at the Junior High School Level" (unpublished doctoral dissertation, Texas A&M University, 1969).

were prepared for five units: (1) iron making, (2) steel making, (3) non-ferrous metals, (4) properties of metals, and (5) hack saws.

Analysis of the data indicated that the lecture-discussion method of teaching produced results superior to those of programed instruction utilizing the Min-Max III teaching machine. But it was found that programed materials saved preparation and lecture time for teachers thus providing them with more time for individual instruction. In addition, programed materials provided an opportunity to presented related material to the classes which could not have profited by this experience otherwise. Students and teachers (with some reservations) favored the use of the teaching machine. The researcher noted, however, that many of the programed lessons were not given the proper time or attention by the students, which may have affected the results of the study.

4. The Traum Study[26]

A study of the relative effectiveness of programed learning and videotape recording was conducted by Traum in 1970. High school students (119) were assigned to one of two reading ability levels in the remedial teaching of reading a rule and performing field measurements. Initial learning of subject matter by these media was measured as well as its retention four weeks after instruction.

The findings revealed there were no significant differences among the means of the high or low reading ability subjects in either the programed learning or the closed-circuit television groups. A statistically significant difference was found between the means of the high and low reading groups in the programed learning classes. Students at both the high and low reading ability levels of the programed instruction group scored higher than subjects in corresponding levels of the videotape group.

Based on these data Traum concluded that either programed learning or videotape recording used singly can be as effective as regular instruction in teaching similar performance tasks and related cognitive information for students at both high and low reading ability levels. He noted that videotaped instruction is adaptable to a group-paced review lesson with self-paced programed lessons providing remedial instruction for the low reading ability level readers.

It is interesting to note the analysis of student attitudes disclosed

[26]Emil F.Traum, "An Experimental Comparison of Closed-Circuit Television and Programed Learning Used as Instructional Media for the Performance of Measurements in Industrial Education" (unpublished doctoral dissertation, The University of Michigan, 1970).

that students in both experimental groups felt that teachers can teach much better than either of the newer media investigated. A significant proportion of the students also believed that using either medium is an excellent way of teaching the learning task to those with low reading ability.

5. The Moegenburg Study[27]

Moegenburg made a three-way comparison among two ways to administer programed instruction materials (with and without the teacher present) and a videotape presentation. His study was conducted in the area of orthographic projection at the college level.

Moegenburg found that a significant difference in learning did exist between videotape television presentations and programed instruction. He also found that it makes no difference in initial learning whether students study programed instruction materials with or without the teacher being present. Students in the study expressed a preference for programed instruction over videotape presentations.

SUMMARY

Seven types of written instructional material include: (1) job sheets, (2) operation sheets, (3) project sheets, (4) information sheets, (5) assignment sheets, (6) student plan sheets, and (7) programed instruction sheets. Of these types, the most important ones for industrial arts purposes are those in which the student participates by thinking and writing, namely, the assignment sheet, the student plan sheet, and programed instruction sheets. See Appendix 2 for an example of a programed instruction sheet.

Some of the purposes for which instruction sheets are used in industrial arts are: (1) to provide self-paced instruction, (2) to help in demonstration recall, (3) to provide supplementary information, and (4) to provide homework assignments.

With the exception of the student plan sheet, there are some disadvantages to the extensive use and reliance upon written instructional sheets.

Programed instruction adapted to industrial arts appears to be emerging slowly. Progressive industrial arts teachers are experimenting with this media in the form of programed instruction sheets. These

---

[27]Louis Arthur Moegenburg, "An Experimental Comparison of Programed Instruction Versus Video-Tape Television in Teaching Selected Orthographic Projection Concepts" (unpublished doctoral dissertation, Texas A&M University, 1969).

sheets, together with student plan sheets and assignment sheets, have great potential for achieving certain of the objectives of industrial arts.

Although research evidence is accumulating, additional experimental studies are needed to refute or support present practices with written instructional materials.

DISCUSSION TOPICS AND ASSIGNMENTS

1. Discuss the historical development of written instructional material in vocational education.
2. How does a job sheet mainly differ from an operation sheet?
3. In parallel columns list the essential differences between a project sheet and a student plan sheet.
4. Prepare an information sheet on one of the units in metalworking for an eighth-grade class.
5. Prepare an assignment sheet in the area of visual communications for a ninth-grade class.
6. Why are job sheets and operation sheets of more importance to vocational education than to industrial arts?
7. What is the psychological basis on which programed instruction rests?
8. Briefly explain the sequence of programed learning from a psychological viewpoint.
9. Identify terminal behavior for a programed lesson in an area of your choice.
10. Outline a plan for a programed lesson for content of your choice.
11. Find additional research studies on written instructional materials and summarize the principal findings.
12. Locate as many research studies as you can on the subject of programed learning. On the basis of this evidence, summarize the advantages and disadvantages of programed learning.

SELECTED REFERENCES

Baker, G. E., Richard J. Fowler, and Charles A. Schuler. "Research in the Area of Electricity-Electronics," *Industrial Arts and Vocational Education*, vol. 56, no. 10 (December 1967), pp. 54–56, 72.

Baron, Denis. "Programed Self-Instruction," *Industrial Arts and Vocational Education*, vol. 54, no. 4 (March 1965), p. 6.

Becker, James L. *A Programed Guide to Writing Auto-Instructional Programs.* Camden, N.J.: RCA Service Company, 1963.

Bensen, M. James. "An Investigation in the Use of Programed Operation Sheets as a Supplement to the Group Demonstration in Teaching Manipulative Operations" (unpublished doctoral dissertation), The Pennsylvania State University, University Park, Pennsylvania, 1967.

Coover, Shriver L. "Programed Blueprint Reading," *Industrial Arts and Vocational Education*, vol. 53, no. 8 (October 1964), pp. 52–53.

Dalinsky, Isadore. "Programed Learning in Mechanical Drawing," *School Shop*, vol. XXIV, no. 3 (November 1964), pp. 24–25.

Drozdoff, Gene. "Teacher-Prepared Programed Units for Industrial Subjects," *Industrial Arts and Vocational Education*, vol. 53, no. 8 (October 1964), p. 43.

Feirer, John L. "Programed Learning in Our Instructional Tool Kit," *Industrial Arts and Vocational Education*, vol. 53, no. 8 (October 1964), p. 25.

Fryklund, Verne C. *Analysis Technique for Instructors*. Milwaukee, Wis.: Bruce Publishing Company, 1965.

Goldstein, David K. "Programed Instruction for Teaching the Point System of Measurement," *Industrial Arts and Vocational Education*, vol. 57, no. 2 (February 1968), pp. 42–43.

Hofer, Armand G. "Teaching Manipulative Operations with Programed Materials," *Industrial Arts and Vocational Education*, vol. 53, no. 8 (October 1964), pp. 49–51.

Mager, Robert F. *Preparing Instructional Objectives*. Palo Alto, Calif.: Fearon Publishers, 1962.

Moegenburg, Louis Arthur. "An Experimental Comparison of Programed Instruction Versus Video-Tape Television in Teaching Selected Orthographic Projection Concepts" (unpublished doctoral dissertation), Texas A&M University, College Station, Texas, 1969.

Plezia, Norman F. "Teacher-Made Program—Electricity," *Industrial Arts and Vocational Education*, vol. 53, no. 8 (October, 1964), p. 48.

Pratzner, Frank C. "Writing Programs of Instruction," *Journal of Industrial Arts Education*, vol. XXIV, no. 3 (January-February 1965), pp. 31–33.

Preitz, Clarence H. "Writing Programed Instruction in Industrial Education," *Industrial Arts and Vocational Education*, vol. 55, no. 4 (April 1966), pp. 45–48.

Quesenberry, Richard C. "Programed Unit—The Board Foot," *Industrial Arts and Vocational Education*, vol. 53, no. 8 (October 1964), pp. 53–56.

Repp, Victor E. "The Relative Effectiveness of Two Demonstration Techniques in Teaching Industrial Arts" (unpublished doctoral dissertation), The Pennsylvania State University, University Park, Pennsylvania, 1970.

Roberts, Arthur D. and Perry A. Zirkel, "Computer Applications to Instruction," *Journal of Secondary Education*, vol. 46, no. 3 (March 1971), pp. 99–105.

Silvius, G. Harold and Estell H. Curry. *Teaching Successfully in Industrial Education*. Bloomington, Ill.: McKnight and McKnight Publishing Company, 1967.

_____. *Managing Multiple Activities in Industrial Education*. Bloomington, Ill.: McKnight and McKnight Publishing Company, 1971.

Spence, William. "Research and Programed Instruction," *Industrial Arts and Vocational Education*, vol. 53, no. 8 (October 1964), p. 57.

Traum, Emil F. "An Experimental Comparison of Closed-Circuit Television and Programed Learning Used as Instructional Media for the Performance of Measurements in Industrial Education" (unpublished doctoral dissertation), The University of Michigan, Ann Arbor, Michigan, 1970.

Trudeau, Terence J. "Programed Instruction—Work and Energy," *Industrial Arts and Vocational Education,* vol. 54, no. 4 (April 1965), pp. 48–49.

Warner, Richard A. "The Effect of Programed Instruction in Industrial Arts at the Junior High School Level" (unpublished doctoral dissertation), Texas A&M University, College Station, Texas, 1969.

Chapter 10

The Demonstration

From time immemorial man has learned how to do things by a process called conscious imitation. He watched someone perform and then, following his natural instinct to imitate, tried to do it himself. By this means skills have been handed down from generation to generation. Today conscious imitation is aided by the teacher demonstration. In simple terms a demonstration is nothing more than a formal way to encourage conscious imitation. Whenever the teaching of a skill is involved, the teacher demonstration is nearly indispensable. Although the technique of demonstrating has been improved over the years, still the underlying principle of conscious imitation remains unchanged.

In the schools a demonstration is a learning experience for students in which the teacher shows and tells how to perform a manipulative operation, process, or experiment. As defined in the *Dictionary of Education*, the term means

> The procedure of doing something in the presence of others either as a means of showing them how to do it themselves or in order to illustrate a principle, for example, showing a group of students how to set the tilting table on a circular saw or how to prepare a certain food product or performing an experiment in front of a class to show the expansion of metals under heat.[1]

At times during a simple demonstration the student is an observer who is required only to listen and to observe the teacher's performance. At other times he may be involved to the extent of answering questions during the demonstration; afterwards he may perform the operation or process himself under the supervision of the teacher. In this case the pupil is both observing and doing and learns through guided performance. The objectives of the lesson will determine the degree of pupil involvement.

[1] Carter V. Good, *Dictionary of Education* (New York, N.Y.: McGraw-Hill Book Company, Inc., 1959), p. 161.

The Demonstration

The demonstration has been developed to a high state in industrial arts and has been used so frequently that it has become a mainstay in many programs. Certainly it is widely accepted as one of the most effective techniques employed in industrial arts teaching. Much of the students' success in project work is dependent upon successful teacher demonstrations of manipulative operations. Figure 10–1 shows a typical class demonstration in an industrial arts laboratory.

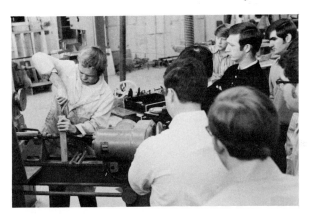

Fig. 10–1. The class demonstration is one of the most important methods of teaching in industrial arts. (Courtesy, Frank Wilson, East Carolina University)

The demonstration is effective because it is based upon established principles of learning. To be specific, a laboratory demonstration is educationally beneficial for the following reasons:

1. A demonstration is given to satisfy a pupil need or interest. For example, when a student is ready to perform a manipulative operation on his project, he has immediate need for information and a demonstration. He will learn from the demonstration because it is given at a time when he is ready and wants to learn. Student interest is easily held, for the demonstration involves action—action with tools and materials which naturally intrigue and interest youth. The demonstration recognizes pupil interest and motivation as a strong factor in the learning process and takes full advantage of it.

2. A demonstration utilizes at least two of the sensory avenues of learning. Typically a demonstration engages both the sense of sight and the sense of hearing. A student who observes and listens can understand symbolically and literally what the teacher is demonstrating. For example, if the teacher is demonstrating knurling on a lathe, his words come to life and become meaningful to the student as the action (demonstra-

tion) progresses. The concept of knurling becomes clear to the student in a manner far better than words alone. In other words, visual and verbal faculties unit to produce vivid mental impressions which facilitate learning and help make it long lasting.

In the hands of a skilled teacher the demonstration may be made to appeal to several of the senses simultaneously. For example, not only does the student actually see what is taking place, in addition to hearing the teacher's words, he also hears the sound of saw, hammer, or machine; he smells the odor of freshly cut shavings or of cutting oil; and he may even feel the texture of a newly-sanded surface of a well-made joint. In rare instances the sense of taste may be involved in some industrial arts demonstrations.

3. A demonstration begins where the pupil is (not where the teacher is) and takes him to new learning. It is built upon knowledge and skills which the pupil already possesses and carries him onward to additional or more complex skills and areas of knowledge.

4. From a psychological point of view, the demonstration consists of an exposure, a stimulus, a response, and reinforcement. A demonstration provides a sound basis for the student to apply his learning and to develop his skills under the guidance of the teacher where prompt reinforcement is possible. In this manner the demonstration takes full advantage of learning by doing.

USES OF THE SHOP DEMONSTRATION

Demonstrations can be given at any grade level and are adaptable to any materials area in industrial arts. Although useful in a wide variety of teaching situations, the most common uses for the demonstration probably include the following:

1. To transmit basic manipulative skills involving the use of hand tools, machines, and equipment as well as to provide opportunities for students to practice and perfect their skills.

2. To establish levels of craftsmanship desired by the teacher for a particular course or class of students.

3. To enrich informative lessons, lectures, and class discussions by demonstrating industrial processes and operations, experiments, and the application of scientific principles. By such means the instruction becomes "real" and more meaningful to the learners.

4. To start classes at the beginning of a term. Both the single and the multiple demonstration are useful in getting a course underway (see Chapter 11).

TYPES OF DEMONSTRATIONS

There are three types of demonstrations commonly given by the industrial arts teacher: (1) class demonstration, (2) small group demonstration, and (3) individual demonstration.

1. Class Demonstration

This type is usually considered to be the most important because it is designed to do the most good for the most pupils at one time. It is the only type of demonstration for which advance preparation can be made. Since it is given to the entire class, this demonstration is usually reserved for core material for which all students are held responsible. Class demonstrations are particularly effective in opening up the possibilities and opportunities within a given course.

2. Group Demonstration

The group demonstration is given as the need arises, that is, when two or more students are ready for certain new material or are having difficulty with some phase of the work previously demonstrated. In such cases the group is called together and the demonstration is given while the remainder of the class continues with its work.

3. Individual Demonstration

The individual demonstration (see Fig. 10–2) is the one most frequently given. Regardless of how well the class and the group demonstrations have been presented, there will always be certain individuals who will not grasp all the details. As the teacher moves among his students, he will discover where difficulties are being encountered. Individual help with a repetition of certain parts of a previous demonstration is the only solution to this problem.

A word of caution may be in order. Individual help should rarely take the form of teacher work on the student's project, unless sufficient opportunity remains for the student to master the problem or technique which is giving trouble. The instructor should keep constantly in mind that student growth and behavior changes—not the projects—are the ends in view. *Only in a case where the project is likely to be spoiled, or where the student's self-confidence and sense of achievement are endangered, is the instructor justified in taking the work temporarily into his own hands.*

Fig. 10–2. Individual demonstrations are follow-ups to group demonstrations. (Courtesy, South Bend Lathe, South Bend, Indiana)

Except for the preliminary preparation involved, the several types of demonstrations are practically identical; thus the class demonstration will be used as a basis for discussing more fully the techniques used in the demonstration method of teaching.

ADVANTAGES OF THE DEMONSTRATION

For the industrial arts teacher the class or group demonstration *conserves time and effort.* By demonstrating to several students or to the entire class, considerable saving in teacher time and repetitive effort can be achieved. These savings can better be spent in individual instruction, i.e., in following up on the class demonstration, in checking student planning sheets, and the like.

The class demonstration provides *uniformity of instruction,* that is, all students receive the same basic exposure. With minor exceptions all students see the same operation being demonstrated and hear the same accompanying word description by the teacher.

In the demonstration *verbalism is minimized;* the emphasis is not on words, but on objects and things. Since most people are visually oriented, they can readily "see" and understand what is being demonstrated. The "seeing" language of the demonstration is universal and one that all can understand. It is a special boon to those slow in comprehending verbal instruction alone.

Another unique advantage of the demonstration involves the old adage, "seeing is believing." When a pupil sees the teacher perform with the same tools and materials that he will use, he is confident that *since it can be done, he can do it too.* His doubts and uncertainties vanish. He knows that what man has done once, man can do again. By removing this psychological block to learning, the student is assured of success in performing the skill, operation, or experiment just demonstrated by the teacher. And once he makes a serious effort, progress will be noted which will tend to strengthen his chances of further success.

A demonstration usually has *high student appeal.* Most students like demonstrations because interesting action takes place with actual objects and things. Demonstrations involve tangibles with a minimum of verbalization or abstractness. Students know that soon they may be manipulating the action themselves and for this reason they derive satisfaction from the teacher's demonstration.

MAKING THE DEMONSTRATION EFFECTIVE

To be educationally effective, a demonstration must not only be well planned, but also skillfully performed. There is no better way in which a teacher can build prestige with his students than through his ability to prepare and present skillful demonstrations. Conversely, there is nothing that will lower a teacher in the esteem of the class more quickly than lack of finesse in demonstrating or even an occasional failure. It is of utmost importance, therefore, that every teacher master thoroughly the art of giving a demonstration and then practice until near perfection is achieved.

It must be remembered that mastery of subject matter alone is no guarantee for giving a successful demonstration. Both mastery of subject matter and mastery of the art of teaching are prime requisites which go hand in hand. While merely knowing *what to do* can never ensure a good demonstration, it is equally true that without such knowledge it is improbable that an effective lesson will result.

The principles underlying a good demonstration are the same everywhere; only the area of application is different. The art of demonstrating divides itself naturally into three major parts: (1) preparing for the demonstration, (2) presenting the demonstration, and (3) terminating the demonstration. In addition, while strictly not a part of the demonstration proper, the follow-up to the demonstration is of great importance. Each of these parts is vital to effective learning and each will be considered in turn.

PREPARING FOR THE DEMONSTRATION

A good demonstration does not begin when the class has assembled and the teacher starts the lesson. For the demonstration to be effective, much teacher preparation and planning must have taken place previously. Some of the more important preliminary steps are as follows:

1. Keep Objectives in Mind

The mere teaching of a process or an operation is not sufficient justification for giving a demonstration in the laboratory. A demonstration is given to help fulfill the objectives of the course, which, if realized, will help bring about certain desired student behavior changes. The teacher should have these objectives and the expected student outcomes clearly in mind while planning the demonstration.

2. Limit the Demonstration

It is important to limit both the content and the time devoted to presenting a demonstration.

Limit the Content. In scope the demonstration should be limited to a single concept, topic, process, operation, or experiment. This can be accomplished by carefully analyzing the material to be covered and dividing it into small segments each of which may become a demonstration lesson. It is better to plan several short demonstrations than a single, long drawn-out one.

Thoughtful planning at this point will reduce the temptation for the teacher to include superfluous information in the demonstration. No matter how interesting it may be, related information should be excluded from the demonstration. Likewise, the teacher should resist any impulse to include lengthy references to his personal experiences in industry. If such information is so important, it may well become the subject of a separate informative lesson. A demonstration should include only that information necessary for understanding the demonstration.

Limit the Time. In length a demonstration should be long enough to attain the objective of the lesson. If the presentation is too long or

The Demonstration

"drags," students will become bored, lose interest, and probably forget some of the key operations. Student-interest span is an important factor to be considered in determining the length of a demonstration. While no standard time limits can be prescribed, the practices of successful teachers do provide useful guidelines. With seventh-and eigthth-grade students, the demonstration should be limited to 10 or 15 minutes. Even with older pupils the demonstration probably should not exceed 20 minutes. If these time limits are exceeded, serious consideration should be given to dividing the demonstration into shorter presentations.

3. Make a Demonstration Plan

It is important that every step in the demonstration be planned carefully for its subsequent performance in proper sequence. A written lesson plan is recommended for each class demonstration to be given in the course. Details on lesson planning have been provided in Chapter 7. Also included in that chapter was an illustration of a sample lesson plan for a demonstration (see Fig. 7–7).

As mentioned previously, one successful class demonstration will substantially reduce the number of group and individual demonstrations required and will effect great savings in teacher time and effort. *A carefully-made plan is the key to a successful demonstration.* Like a road map which helps the traveler reach a given destination, a demonstration plan helps the teacher achieve desired outcomes. It is indispensable in providing uniform instruction to all classes and is valuable in evaluating student behavior changes.

Many teachers like to file their demonstration plans in some convenient place. This makes it unnecessary to prepare a new plan each time the demonstration is to be given. If the teacher writes suggestions on the sheet after the demonstration has been given, they will be useful in improving future performances. By reviewing the lesson plan in advance of the demonstration, the teacher can quickly refresh his memory of the lesson content, note the time required, and make certain that all necessary tools, machines, supplies, equipment, instruction sheets, and audio-visual media will be ready.

Considerable effort is often required on the part of the teacher to assemble everything that will be needed for a particular demonstration. Some teachers like to keep their demonstration plan filed with selected materials, visual aids, and other items used in the demonstration. In this way much teacher effort can be spared in preparing for a demonstration.

4. Plan for Student Application

Psychologists state that learning occurs best when the learner responds immediately to the stimulus. For this reason it is important to provide for student implementation and application as soon as possible after the demonstration is completed. If the students are expected to be able to perform the operation or process, then the demonstration should be timed so that an early opportunity will be provided for student practice.

It will be recalled from the previous chapter that learning is thought to be even more effective if the student response receives immediate reinforcement. Thus, as a student begins to practice the manipulative aspects, the teacher should reinforce student behavior accordingly. In some cases reteaching and redemonstrating will be required. However, this is part of the guided-performance role of the teacher.

5. Obtain the Materials Needed

Unless the demonstration is one that includes the getting out of stock, the teacher should see that all materials are in readiness.

6. Check the Machines

If machines are to be used, the teacher should make certain that they are in safe working order, properly adjusted, and ready for use. All safety guards which students will be expected to use should be in place on the machine. Eye protection devices should be on hand.

7. Ready All Hand Tools

Nothing is more likely to spoil a demonstration than for the instructor to get part way through and find that some essential tool has been forgotten. *This is the cardinal sin of any demonstration.* While the teacher is away from the class hunting for the needed tool, much of the effectiveness of the demonstration is lost; class control may suffer through excessive talking, jostling, and even horseplay. There is really little excuse for forgetting a tool during a demonstration since required tools should be listed on the demonstration plan. Besides, if the instructor makes a "dry run" of the demonstration prior to class presentation, all of the needed tools will be on hand. If, however, a tool is inadvertently forgotten, it is better for the instructor to stay with the group and send a student for the tool needed. In this connection an officer of the

The Demonstration

student personnel organization, perhaps the Tool Clerk, may be pressed into service.

The instructor should use the same hand tools which the students are expected to use. A class will be mightily discouraged if the instructor uses tools which are reserved chiefly for his demonstrations. They will suspect that the difference between their performance and that of the teacher lies in the tools used. All edge tools in the shop should be sharp at all times and ready either for regular student use or for the instructor's demonstration.

8. Obtain or Prepare Instructional Aids

These are natural supplements to the demonstration and should be used wherever necessary to promote clear understanding by the students. Some demonstrations lose their impact because components are so tiny that students cannot see them or because certain actions are so rapid that students are unable to comprehend exactly what happened. In such cases, "stopped-motion" still pictures, enlarged models of small parts, or mockups can play a vital role in aiding student comprehension.

Illustrated handout material is often useful and sketches may be made on the chalkboard. Drawings prepared beforehand are likely to be better than a sketch hurriedly made during the progress of the demonstration. Better yet, the use of transparencies for overhead projection is recommended (see Chapter 15). Once prepared, these visuals will save the teacher countless hours of repetitive drawing effort.

9. Have Partially Completed Work at Hand

Some demonstrations cannot be carried to completion because a given period of time must elapse for conditioning the materials. For example, leather must be moistened and allowed to stand prior to tooling or carving, glue must dry before parts can be worked on, or wood finishes must dry thoroughly before further work can proceed. In these cases the best solution is to have several pieces of the work in various states of completion. The demonstration may then move swiftly from major point to major point within the allotted time.

10. Practice the Demonstration

The instructor should be absolutely sure that the demonstration will "click." The only way to make certain of this is to run through the procedure before the class arrives. Without this preparation, even the most experienced teacher occasionally finds himself in difficulty. It is much better to make sure than to find oneself in that most embarrassing of situations where things fail to go as anticipated and the demonstration becomes a failure. At such a time excuses never cover the situation. A few minutes spent in previous practice will pay large dividends in results. Besides, going through the demonstration before presenting it to the class will help ensure that no tool, audio-visual, or other item has been forgotten or overlooked.

PRESENTING THE DEMONSTRATION

With every possible preparation made, the instructor is ready to meet his class. To make the lesson most effective, however, certain techniques will need to be followed. Some of these are:

1. Give the Demonstration When It Best Fits the Needs of Most Pupils

Some instructors seem to think that demonstrations must *always* be given at the beginning of the class period. It is true that the beginning of a period is often convenient because the students are assembled as a group and therefore individuals have not yet begun separate activities from which they will dislike to be called later on. However, a demonstration may be better timed to student needs if given at some time other than the beginning of the period. It must be remembered that immediate student application is an important aspect of the demonstration method of teaching. Therefore, sufficient time must be allowed following the demonstration for the students to apply what they have learned. This precludes the giving of demonstrations during the closing moments of the class before cleanup.

2. Let the Students Know What It Is All About

It is good practice to tell the class before the demonstration actually begins what it is all about, what is going to happen, and *why* the demonstration is important. Greater learning will occur when the students know the purpose of the lesson. The teacher should make a point of telling the class about the points of particular importance. He should make it abundantly clear that all students are expected to learn from the demonstration. Research indicates that this procedure will result in increased learning.

The Demonstration

3. Arouse the Desire To Learn

It is important for the teacher to try to stimulate pupil interest in the demonstration. One incentive to accomplish this is to relate the demonstration to the student's own work, if possible. If he is going to have to perform a manipulative operation himself later, a student may be more interested and inclined to pay closer attention to the demonstration. In short, the teacher should try to motivate the pupil by linking the demonstration with previous learning and to relate it to his future activities.

4. Arrange the Students So That All May See and Hear

There is always a tendency for students to crowd close around the instructor during the demonstration. This frequently makes it impossible for those in the rear to see what is taking place. It is well for the teacher to arrange the class in a large arc (not a circle) with the taller students at the back. The teacher should check occasionally as the lesson progresses to see that a gradual crowding-in does not occur. Under no circumstances should a teacher ever permit the class to encircle him. If this does happen, the instructor's back will be toward some students thus blocking their view of the demonstration. This situation also invites student inattention and even horseplay.

A large overhead mirror, mounted permanently over the demonstration bench, is often useful in helping students see during a demonstration. This device also has the unique advantage in that students will observe the action head on just as does the instructor. Closed-circuit television and videotape recording also help students see and hear during a demonstration. Applications of these devices to the class demonstration will be discussed later in this chapter.

5. Perform the Demonstration So That Students Can See Exactly What Is Taking Place

A demonstration must be skillfully performed, but the instructor should never utilize the demonstration purposely to impress the class with his skill. He must remember, too, that because of his skill and experience what is easy for him to do may appear to be not so simple from the pupil's point of view. Because he already possesses considerable skill and manual dexterity, the experienced teacher may have a tendency to demonstrate too rapidly for the student to follow. He must remember that often the hand is quicker than the eye. This is especially true when working with younger pupils. Pacing the work too slowly will make the lesson drag. Yet, if the work is done too rapidly, many students will fail to grasp all the essential points. Weaver and Cenci suggest that

> Sometimes it is a good idea to go through the demonstration once at normal speed in order to give the learners an overview of the operation, then repeat it more slowly for the learner's benefit.[2]

When giving a demonstration in front of a class the instructor must realize that the work and his movements will appear in reverse action to the class. For example, when he moves his right hand, this appendage and its movements are on the left side of the students. It may well be that certain students will have difficulty in reversing this order in their minds and in practicising "conscious imitation" as they emulate the teacher's actions.

Furthermore, those students who are off to one side will see the demonstration from a different angle from those on the other side or in front of the teacher. Later, when the student tries to perform the operation himself, he will see the action from a still different point of view. Shemick puts it this way:

> It appears that when a student changes his position from viewer to operator, the change of orientation is sufficiently varied to cause confusion and lack of confidence on the part of the student.[3]

Whenever possible the instructor should try to present the demonstration so that the student will see it exactly as if he were doing it himself. In some cases it may be possible to turn the work around or invite the class to step closer to view the action head on.

6. Maintain High Standards of Craftsmanship

In the demonstration the instructor should maintain the same level of craftsmanship that he expects of his students. One can expect from students no greater degree of skill or accuracy than the teacher achieves. A demonstration is one of the best opportunities for stressing the importance of fine workmanship. Under no circumstances should the instructor excuse poor workmanship in a demonstration, whether due to lack of time, dull tools, or any other reason.

---

[2]Gilbert C. Weaver and Louis Cenci, *Applied Teaching Techniques* (New York, N.Y.: Pitman Publishing Corporation, 1960), p. 77.

[3]John M. Shemick, *A Study of the Relative Effectiveness in Teaching a Manipulative Skill—a Multi-Media Teaching Program versus Classroom Demonstration with Printed Instruction Sheets.* Title VII, Project no. 1597, U.S. Department of Health, Education and Welfare, Washington, D.C. (University Park, The Pennsylvania State University, 1964), p. 1.

7. Talk to the Class

The instructor should talk to the students while he is performing the demonstration for the express purpose of directing their attention to important phases of the action taking place. Thus the students will not only see the work being done, but will hear how it is being accomplished. The teacher need not keep up an incessant flow of words; short periods of silence are forceful in that they help maintain student attention.

The instructor should look at and talk directly to the students, not to the work or the equipment. He must remember that he is teaching individuals and not things. He needs to observe his listeners as he speaks to get some idea if his words and actions are being understood. Talking directly to the students will often prevent their interest and attention from wandering. If it is necessary for the instructor to concentrate on the demonstration itself, he should pause occasionally to explain just what is taking place.

8. Ask Questions during the Demonstration

A lesson in which the class does not participate, at least to the extent of answering some questions, is of questionable value. An instructor can never be sure that his lesson is "getting over" unless he checks continually as it progresses. Answers to questions provide the teacher with clues to student understanding as the demonstration proceeds. Questions requiring a simple "yes" or "no" response should be avoided. Instead, the teacher should ask questions designed to test understanding or requiring the student to think. For example, the teacher may ask: "Why do I set this lathe tool below center?" or "Why do I plane in both directions on the end of this board?" Such questions tend to keep the attention of the class focused on what the teacher is doing and lessen the risk of losing one or more students during the course of the demonstration.

It is obvious, of course, that the teacher should not resort to questions thought up on the spur-of-the-moment during the demonstration. Rather, he should plan good questions well in advance of the demonstration, that is, during the preparation of the demonstration plan. These questions should be designed to measure student understanding at various points during the demonstration and should be recorded on the demonstration plan.

As in any good class discussion, order should prevail when questions are posed to the class. Unison answers should be avoided. It is good practice to ask the question, pause, look around the class, and then call on a particular student.

9. Involve the Students Manipulatively

In some demonstrations at certain grade levels it may be possible for the teacher to plan for manipulative participation in the demonstration by one or more students. Wherever it is necessary to repeat the same series of operations more than once, a student may be asked to do them after the first time. Besides contributing to class interest, this procedure gives the instructor an opportunity to judge the effectiveness of his instruction. The teacher should be cognizant of time, however, and not permit the demonstration to drag out unnecessarily.

10. Show Interest in the Demonstration

Students are affected not only by what they see and hear but also by the manner in which it is presented. Sincerity, interest, and enthusiasm on the part of the instructor will tend to induce similar qualities in students.

11. Stress Safety Practices

The most effective way to teach safety is through a positive approach by demonstrating the safe way to work. Take every precaution that you expect the student to observe. Use safety guards and eye protective devices wherever needed. Never miss an opportunity to stress the *safe* way to perform a given operation. Emphasize the safety responsibility each individual has both to himself and to others in his work area.

12. Repeat Parts of the Demonstration if Necessary

If questioning shows that certain details of the lesson have not been grasped by a majority of the class, go over that part of the demonstration again.

13. Where Alternatives Exist Demonstrate One Way

Where there are several ways, techniques, or methods to perform a given operation, the teacher should demonstrate only one to beginning classes lest the pupils become confused. For advanced classes the matter becomes one for teacher judgment.

TERMINATING THE DEMONSTRATION

The teacher's responsibility does not end with the completion of the manipulative aspects of the demonstration. Certain terminating steps must be taken to make sure that the students have grasped all the details and that they will remember the salient points. Therefore the instructor should be sure to clinch the demonstration by taking the following steps:

1. Summarize Important Points

Briefly review the high points of the demonstration, placing particular emphasis on the more important details. It is very important to leave the class with a clear-cut impression. In this connection an operation sheet or other written handout material is sometimes useful in providing the class with a clear summary. Such material may become an important part of a student's notebook and may be useful in preparing for tests.

2. Ask Summary Questions

A weak way to conclude a demonstration is for the teacher to ask: "Are there any questions?" Usually there are none. Asking questions is a good way for the teacher to terminate a demonstration, but of course, these should be prepared in advance. Questions of the recall type, or those testing student understanding, or those involving student application are preferred. The teacher should make sure that the students understand the demonstration by asking questions which will indicate such knowledge.

3. Provide an Early Opportunity for Students to Apply Their New Knowledge

This is not always possible, especially in a general shop, but every effort should be made to shorten the time as much as possible between demonstration and student application, that is, between stimulus and response. The desirability of immediate response to a stimulus backed by teacher reinforcement has been mentioned previously.

4. Put Away Tools and Materials

Putting things away is an opportunity for the instructor to set an example for the students. If he expects students to take care of their tools and equipment, the teacher should be willing to do the same.

FOLLOW-UP TO THE DEMONSTRATION

The demonstration cannot really be considered successful or finished until the process or skill demonstrated has been performed by the students and their work has been checked by the instructor. As the teacher moves around the laboratory he should compare the observable results of his demonstration in terms of expected student outcomes with those outlined on his demonstration plan. He should record the number of group and invidiual demonstrations needed following a class demonstration and seek to improve his plan and subsequent presentations accordingly.

Before presenting another demonstration the teacher should allow sufficient time for students to practice the skill just demonstrated.

THE DEMONSTRATION AREA

It is necessary, of course, for the demonstration to be given at the most appropriate location. Some demonstrations will be given at various machines and work centers throughout the shop. In these instances students will be standing and grouped in an arc at the demonstration site. Other demonstrations, however, can be performed better at a demonstration bench in the area designated for assembly and dismissal of the class as well as for presenting informative lessons. Here the class may be seated comfortably in front of a demonstration bench or table. There is the added advantage that at the beginning of the period students will already be seated in this area (for assembly purposes), so that some demonstrations may begin without delay. An area such as this is convenient for both students and the teacher because it is especially laid out for demonstrations. It will usually contain movable chairs, demonstration bench, chalkboard, screen, several types of projectors such as slide, movie, opaque, and overhead, as well as storage facilities. Where closed-circuit television is being used to facilitate the demonstration, the camera, viewing screen, and monitor will also be present.

NEW DEVELOPMENTS

Enterprising industrial arts teachers everywhere have always been willing to try the new and to modify the old. Their teaching is characterized by a spirit of inquiry as they seek to improve the process of learning. Whenever a new idea or teaching innovation comes along, these instructors have sought to adapt it to industrial arts. It is not surprising

The Demonstration

therefore to find outstanding teachers of industrial arts actively engaged in classroom applications of the newer educational developments, such as closed-circuit television, team teaching, programed instruction, and the like. To them, this is all part of the great adventure of teaching.

Still facing today's teachers is the age-old problem: How best can we teach a motor skill? The traditional method, and the one described in this chapter, has been based on conscious imitation. The teacher demonstrates the skill while the student watches; then the student tries to perform the skill by imitating what he has just seen and heard. The situation is complicated still further by large classes in the laboratory.

It is not possible in the space available to describe the numerous experimental studies and innovative practices now going on in industrial arts which are designed to improve the teaching of skills. However, brief consideration will be given to selected advances along three fronts: (1) using television in the demonstration, (2) teaching skills with programed instruction, and (3) related research findings.

USING TELEVISION IN THE DEMONSTRATION

1. Single-Room Closed-Circuit Television

The use of closed-circuit television is not new in education. Its principal use has been to serve multisection classes where growth in student enrollment has outstripped that of the teaching staff. A more limited application has been to supplement teacher demonstrations in a single room or laboratory. More will be said of closed-circuit television in Chapter 15; at this point, reference is confined to its use in connection with the demonstration.

There are two difficulties inherent in the class demonstration, both of which have been referred to previously. One of these is that both the work and the instructor's movements appear in reversed action to the class. The other difficulty is that frequently certain parts, readings, or procedural details are so small that students cannot see them even if only a few feet away. Closed-circuit television appears to offer at least a partial solution to each of these problems.

Essentially a system of single-room closed-circuit television consists of a television camera connected by coaxial cable to one or more large-screen television sets. The camera may be fixed in one position and controlled by the instructor (see Fig. 10–3), or as is shown in Fig. 10–4, the camera may be mounted on a dolly and controlled by a cameraman. Figure 10–3 shows a typical single-room closed-circuit television system in operation. In more sophisticated set-ups, two television cameras may be employed.

Fig. 10–3. A typical example of single-room closed-circuit television. (Courtesy, D. P. Barnard and H. A. Herbert, University of Wisconsin—Stout)

In using television in the classroom, the students watch and listen while the teacher demonstrates and talks in the usual manner. At the appropriate moment he directs the class to watch the television screen. As shown in Fig. 10–3, the teacher continues to talk in his regular manner because audio amplification is not ordinarily employed. In this instance television is used only when the teacher wishes to show how the action will look to the operator or when he wishes to magnify some small part or detail. Note the degree of magnification shown in Fig. 10–3. At all other times the center of interest remains on the actual demonstration or demonstrator. Other demonstrations have been televised in which the class watches the entire demonstration on television. Such televised demonstrations might be especially useful where danger from spray, sparks, or flying chips may exist. Sometimes shop machinery is located in cramped quarters thus adding to the difficulty of a teacher's presentation. Figure 10–4 shows how closed-circuit television can be used advantageously under such conditions.

The Demonstration

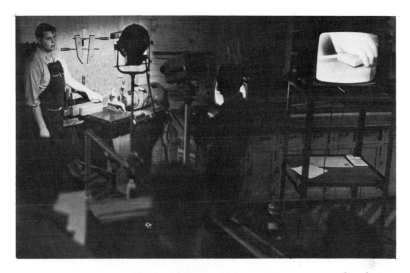

Fig. 10-4. Television facilitates a group demonstration on woodworking machinery in cramped quarters. (Courtesy, D. P. Barnard and H. A. Herbert, University of Wisconsin—Stout)

Figure 10-5 shows how a zero camera angle permits the television learner to see the demonstration exactly as the teacher sees it; the figure also illustrates the degree of magnification possible.

2. Videotape Recording and the Demonstration

Videotape recording is similar to single-room closed-circuit television with the exception of one additional component—a recorder which records both sight and sound on a magnetic tape. Instant playback is possible. Both video and sound tracks may be erased at will or preserved permanently on the tape. Equipment is available to record the video in either black and white or color.

Industrial arts teachers are beginning to experiment with videotaping especially since the cost of the equipment has now become reasonable and since knowledge of its use is more widespread. Teacher demonstrations, experiments, and informative lessons can be "canned" for later use. There is probably some merit in presenting exactly the same demonstration techniques and information to each section of pupils which is easy to do with videotape equipment. Figure 10-6 shows a teacher setting up the camera, recorder, and monitor in preparation for videotaping a demonstration in the industrial arts laboratory.

244   The Demonstration

Fig. 10–5. With a zero camera angle, television learners see the demonstration as the teacher sees it. (Courtesy, D. P. Barnard and H. A. Herbert, University of Wisconsin—Stout)

Fig. 10–6. This teacher is setting up the videotape recorder in the construction laboratory preparatory to taping a demonstration. (Courtesy, Frank Wilson, East Carolina University)

The Demonstration 245

Videotapes are valuable in teacher education programs for taping student-teacher presentations in the field as well as in providing opportunities for students to practice demonstrations and informative lessons in methods classes. Figure 10–7 shows a teacher and a teacher education student critiquing a demonstration which the latter has just given in a teaching methods class.

Fig. 10–7. A videotape recording enables this teacher education student to observe exactly how he looked and sounded as he gave his demonstration in a methods class. He and the teacher can note weaknesses and strengths in the presentation in this follow-up conference. (Courtesy, Frank Wilson, East Carolina University)

TEACHING SKILLS WITH PROGRAMED INSTRUCTION

1. The Hofer Study[4]

In 1963 Hofer sought to teach a skill by means of programed instructional material. His study of 57 seventh-grade students compared the conventional teacher demonstration with programed instruction consisting of written instructions illustrated with a set of photos. Figure 10–8 is illustrative of the programed instructional material used in the study. The experiment involved four operations new to students including: (1) making a sand foundry mold, (2) copper enameling, (3) drilling and counterboring a hole, and (4) cutting internal threads with a tap. Those students who observed the conventional demonstration performed the operation immediately afterward; those who followed the written program performed the operation as they read.

[4]Armand G. Hofer, "An Experimental Comparison of Self-Instructional Materials and Demonstrations in the Teaching of Manipulative Operations in Industrial Arts" (unpublished doctoral dissertation, The University of Missouri, 1963).

Based on his study, Hofer concluded that:

1. There was no difference in the quality of student performance in favor of either method.

2. Students using programed material learned and retained slightly more information on terminology and procedure as measured by paper-pencil tests given immediately after performance and one week later.

3. Students using programed instruction required 7 percent more time to receive instruction and to perform the operation.

4. Students in the conventional demonstration group required 69 percent more individual assistance than those using programed material.

5. As far as intelligence is concerned, students in the upper half of the group required about the same amount of time for instruction and performance with both methods of instruction.

Although students in the lower half of the group took 12 percent more time to receive programed instruction, they learned more than through teacher demonstration.

Students in the upper half required 74 percent more individual assistance during performance when the instruction was presented by the demonstration; students in the lower half required 64 percent more individual assistance.

6. Reading ability was not a serious handicap in the use of programed materials.

Although this investigation indicated that programed materials appear to be effective in teaching manipulative operations where the criterion is one successful performance, Hofer speculates that better results might be achieved if programed materials were combined with teacher demonstrations. He feels that there is still a place for the inspiration of a skilled demonstration. While recognizing the obvious advantages of written programed materials, Hofer also points out several limitations which may not be apparent to the casual observer. These are (1) lack of sound and motion, and (2) students following programed material would need teacher supervision at critical points where student safety is involved.

2. The Shemick Study[5]

In the past several years teaching machines have been used successfully to teach simple motor skills such as assembly and disassembly. In a 1964 study which was supported by the Office of Education, Shemick

---

[5]Shemick, *op. cit.*, pp. 1–20.

The Demonstration

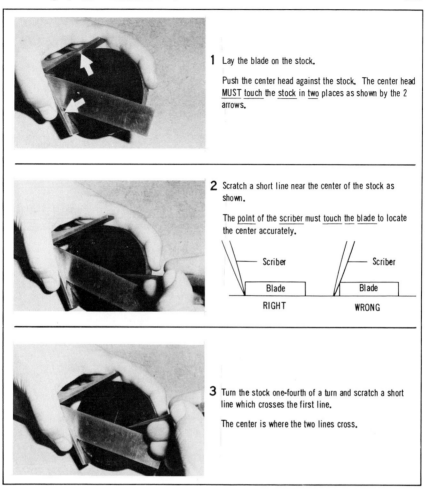

Fig. 10–8. An example of the programed instructional method used in the Hofer study. (Courtesy, Armand G. Hofer, University of Wisconsin—Stout)

sought to teach by machine a skill involving a high level of coordination.

The teaching machine selected was the Audio-Graphic (see Fig. 10–9) which provides the learner with both visual and sound stimuli. The visuals are 2 × 2-inch slides projected onto a frosted screen. Magnetic tape is used to provide sound which may be heard either through a speaker or earphones. The machine may be controlled by either push buttons or foot pedals. In this teaching machine, however, the learner is not required to make a response in order to advance from frame to frame in the program.

Metal spinning was the skill selected to be taught by machine. This

Fig. 10–9. The Audio-Graphic teaching machine. (Courtesy, The Pennsylvania State University)

skill was first analyzed into ten operational units or instructional sequences as follows: (1) personal preparation and safety rules, (2) machine nomenclature and orientation, (3) centering the disk, (4) initial forming, (5) tool action and remedial procedures, (6) finish spinning, (7) planishing, (8) trimming edge to size, (9) rolling the bead, and (10) polishing. The programed instruction consisted of 34 slides with 15 minutes of taped instruction.

Although the slides in each instructional unit were changed automatically, the learner could stop the machine at any time for a longer examination of the slide. The machine was programed to stop at the end of each operational unit, whereupon the learner was then to put into practice what he had just seen and heard. Before attempting to do the work himself, he could also replay the operational unit. When the learner had completed that portion of the spinning, he activated the machine to begin the next instructional unit.

Ten college-level students were taught metal spinning by teacher

The Demonstration

demonstration supplemented with instruction sheets while an equal number learned spinning by means of the Audio-Graphic. Figure 10 shows a student in the experimental group using the Audio-Graphic. The task to be performed was to spin a 6-in. bowl with rolled edge from 18-gauge aluminum.

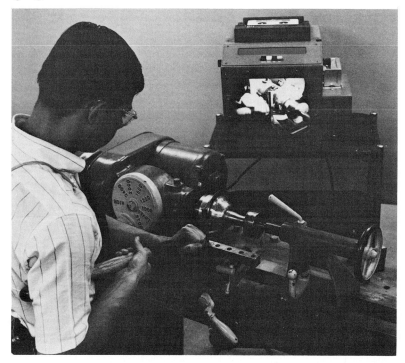

Fig. 10-10. The Audio-Graphic in use. (Courtesy, The Pennsylvania State University)

Based on the data gathered, Shemick concluded that:

1. In terms of the amount of material consumed in spinning and the quality of the finished product, there was no difference in student performance in favor of either method.

2. Students using the teaching machine required significantly more time to receive the instruction and to perform the operation. This group also requires significantly less teacher assistance.

3. In the teacher-demonstration group, there was the usual delay between instruction and performance ranging, in this study, from five to fourteen days. Analysis of the data, however, revealed no pattern in favor of either extreme in delay.

4. The noise level in a typical metal shop did not seem to affect adversely either method of teaching.

Although his study indicates that the Audio-Graphic appears to be effective in teaching metal spinning insofar as the amount of materials consumed and the quality of the finished product are concerned, Shemick makes the following observations:

1. The Audio-Graphic lacks the action necessary for quickly learning a motor skill.

2. As prepared, the program did not provide an overall orientation and the learners seemed to move blindly from one operational sequence to another.

3. If used just before student performance, machine instruction probably would make an excellent review of a teacher demonstration.

RELATED RESEARCH FINDINGS

Efforts to improve the industrial arts demonstration have received the attention of numerous investigators in recent years. Space permits reference to only the following two selections.

1. The LaRue Study[6]

In 1968 LaRue compared the effects on initial learning and retention of both positive and negative instruction in industrial arts demonstrations. He had noted that the professional literature of the field for the past 38 years had consistently advocated that demonstrations be presented in a straightforward, positive manner with a flawlessly correct procedure. At the same time he had observed that some experienced teachers purposely carried out certain operations incorrectly for the dramatic effect that resulted. These teachers believed that students subjected to such negative instruction would be impressed and so would learn more and remember longer.

The study tested the main hypothesis that the use of dramatically negative teaching technique in presenting demonstrations in industrial arts yields greater initial learning and retention than the traditional (positive) method.

The design of the experiment included five equivalent groups of sixth-grade pupils. Group A received dramatically negative instruction

---

[6]James P. LaRue, "Dramatically Negative Teaching Technique as Supplemental to Positive Instruction Compared with Traditional Teaching Technique in Industrial Arts Demonstrations" (unpublished doctoral dissertation, The Pennsylvania State University, 1968).

consisting of an educational film and a colored slide/tape presentation; Group B received a positive version of the film and the slide/tape presentation; Group C received a live demonstration which was dramatically negative; Group D received a live demonstration which was traditionally positive in character; and Group E took the criterion tests (immediately after treatment and four weeks later) to establish how an uninstructed, but similar, group would perform. It is to be noted that the groups receiving dramatically negative instruction also received instruction showing correct use or procedure. In other words, the dramatically negative technique necessitates that incorrect or negative usage be shown either before or after the correct procedure.

Although the analysis of data revealed insufficient evidence to support the main hypothesis of the study, the findings did suggest that dramatically negative techniques may be as effective in promoting initial learning and retention as the traditional positive manner of presenting industrial arts demonstrations. Based on his findings, LaRue concluded that (1) it is better to use actual demonstrations of either a dramatically negative or traditionally positive character than films and slides in presenting basic instruction in common hand tools in industrial arts, and (2) the dramatically negative technique, though not quite so effective as traditionally positive instruction, was nevertheless a very effective instructional technique as far as initial learning and retention were concerned.

2. The Kruppa Study[7]

In 1968 Kruppa sought to test the effectivity of three teaching techniques related to the industrial arts demonstration. His study was in the area of introductory woodworking and involved 128 non-major industrial arts college students who were subjected to the following treatments: (1) class demonstrations only, (2) overview-film demonstrations followed by identical class demonstrations, and (3) overview-film demonstrations followed by optional student use of concept films with supplementary study guides. The relative effectiveness of these three methods was measured in terms of (1) the amount of technical knowledge gained, (2) student performance of constructed products, and (3) teacher time required for presenting individual demonstrations.

Based on an analysis of the data, Kruppa concluded that the three different teaching techniques were equally effective in providing tech-

[7]J. Russell Kruppa, "A Comparison of Three Teaching Techniques Using Instructional Films in Selected Units in Industrial Arts" (unpublished doctoral dissertation, The Pennsylvania State University, 1968.).

nical knowledge and in student performance of constructed products. Although no statistically significant difference in teacher time was noted for giving individual demonstrations when the three teaching techniques were compared, a substantial savings of teacher time was attributed to the use of the overview-film demonstrations plus concept films technique. This time was a resultant of the fact that two of the experimental groups (class demonstrations only and overview-film demonstrations followed by class demonstrations) required teacher time in preparation, presentation, and administration, while the group using the overview-film plus concept films required little or no teacher time since the latter equipment was operated by the students. As noted previously, this saving in teacher time appeared to be made with loss neither in student acquisition of technical knowledge nor in student performance in constructing products. Another advantage of this technique is that the filmed demonstrations are not subject to variation as may be the case with teacher demonstrations.

In concluding his study, Kruppa states:

> The implications of this study are not that teachers should be replaced with overview-film demonstrations and concept films, but that some of the time which teachers are now devoting to class demonstrations might be utilized by providing additional individual help, and better classroom supervision. This valuable saving in time might also be wisely used for course enrichment through additional experiences and information.[8]

SUMMARY

This chapter pointed out the importance, advantages, and uses of the demonstration as a method of teaching industrial arts classes. Three types of demonstrations (class, group, and individual) were discussed. The three natural divisions of the class demonstration with the cardinal features of each division are as follows:

1. *Preparing for the demonstration*
    a. Keep objectives in mind.
    b. Limit the demonstration in terms of content and time.
    c. Prepare a demonstration plan.
    d. Plan for student application.
    e. Obtain the materials needed.
    f. Check the machines.
    g. Ready all hand tools.
    h. Obtain or prepare necessary visual aids.

[8] *Ibid.*, p. 110.

The Demonstration

        i. Have partially completed work at hand.
        j. Practice the demonstration.
2. *Presenting the demonstration*
        a. Give demonstration when it best fits needs of most pupils.
        b. Let the students know what it is all about.
        c. Arouse the desire to learn.
        d. Arrange students so that all may see and hear.
        e. Perform demonstration so students can see exactly what is taking place.
        f. Maintain high standards of craftsmanship.
        g. Talk directly to the class.
        h. Ask questions during the demonstration.
        i. Involve the students.
        j. Show interest in the demonstration.
        k. Stress safety practices.
        l. Repeat parts of demonstration, if necessary.
        m. Where alternatives exist demonstrate one way.
3. *Terminating the demonstration*
        a. Summarize important points.
        b. Ask summary questions.
        c. Provide early student application.
        d. Put away tools and materials.
        e. Follow up on the demonstration.

Demonstrations are given at various machines and work stations throughout the laboratory as well as in special demonstration areas where seating and audio-visual media are available.

Single-room closed-circuit television and videotape recording appear to be useful in the industrial arts demonstration. Large magnification of small parts is possible and the action within the demonstration appears to the viewer exactly as it does to the demonstrator.

Two experimental studies show that, with certain limitations, skill can be taught by means of programed instruction and teaching machines. Both researchers seem to feel that these new media may produce optimum results if used supplementary to, rather than as a replacement for, the teacher demonstration.

Another research investigation showed that (1) it is better to use actual demonstrations of either a dramatically negative or traditionally positive character than films and slides in presenting basic instruction in common hand tools and (2) the dramatically negative teaching approach used to supplement positive instruction is as effective in promoting learning and retention as the traditional manner of presenting an industrial arts demonstration.

Another experimental study found that demonstrations presented by means of overview-films and concept films resulted in a saving of teacher time which might be used to provide additional individual help and better classroom supervision.

DISCUSSION TOPICS AND ASSIGNMENTS

1. Describe the three types of demonstrations.
2. Make a list of ten demonstrations which you would expect to give to a beginning class. Justify these in terms of objectives.
3. Make a lesson plan covering a selected demonstration (see Chapter 7).
4. Prepare and present in class a demonstration of a simple process.
5. Make a sketch of the ideal demonstration area for a junior high school industrial arts laboratory. Prepare a list of equipment which you recommend.
6. What are the pros and cons of using programed instruction and-teaching machines in teaching motor skills?
7. Examine the literature in the field for investigations and experimental studies on the demonstration and report your findings.

SELECTED REFERENCES

Hofer, Armand G. "'Teaching Manipulative Operations With Programed Materials,"
    Industrial Arts and Vocational Education, vol. 53, no. 8 (October 1964). pp. 49–51.
———. "An Experimental Comparison of Self-Instructional Materials and Demonstrations in the Teaching of Manipulative Operations In Industrial Arts" (unpublished doctoral dissertation, The University of Missouri, Columbia, 1963).
Kruppa, J. Russell. "A Comparison of Three Teaching Techniques Using Instructional Films in Selected Units in Industrial Arts" (unpublished doctoral dissertation, The Pennsylvania State University, University Park, 1968).
LaRue, James P. "Dramatically Negative Teaching Technique as Supplemental to Positive Instruction Compared with Traditional Teaching Technique in Industrial Arts Demonstrations" (unpublished doctoral dissertation, The Pennsylvania State University, University Park, 1968).
Leighbody, Gerald B. and Donald M. Kidd. *Methods of Teaching Shop and Technical Subjects.* Albany, N. Y.: Delmar Publishers, Inc. 1966.
Miller, Rex and Fred W. Culpepper, Jr. "How to Give Effective Demonstrations," *Industrial Arts and Vocational Education*, vol. 60, no. 6 (September 1971), pp. 24–25.

Miller, W. R. "Reconciling Basic Teaching Methods and Problem Solving," *School Shop*, vol. XXIV, no. 5 (January 1965), pp. 13–14, 31.
Pankowski, Dallas. "A Guide for the Development of Motor Skills," *Industrial Arts and Vocational Education*, vol. 54, no. 2 (February 1965), pp. 24–26, 62–64.
Pendered, Norman C. "What's Your KGD?" *School Shop*, vol. XVII, no. 4 (December 1957), p. 34
_____. "Shop Demonstration Check List," *School Shop*, vol. XVII, no. 7 (March 1958), p. 32.
Shemick, John M. "Teaching a Skill by Machine," *Industrial Arts and Vocational Education*, vol. 54, no. 8 (October 1965), pp. 30–31.
_____. *A Study of the Relative Effectiveness in Teaching a Manipulative Skill —A Multi-Media Teaching Program versus Classroom Demonstration with Printed Instruction Sheets.* Title VII, Project No. 1597, U. S. Department of Health, Education and Welfare, Washington, D. C. University Park: The Pennsylvania State University, 1964.
Silvius, G. Harold and Estell M. Curry. *Teaching Successfully in Industrial Education.* Bloomington, Ill.: McKnight and McKnight Publishing Company, 1967.
Spence, William P. "The Demonstration," *Industrial Arts and Vocational Education*, vol. 52, no. 8 (October 1963), pp. 46–52.
Weaver, Gilbert G. and Louis Cenci. *Applied Teaching Techniques.* New York, N. Y.: Pitman Publishing Company, 1960.

Chapter 11

Laboratory Management — Starting the Class

Managing the laboratory is an important responsibility of the industrial arts teacher. It is one to which he must devote full effort, for his success as a teacher may well hinge on his management abilities. The importance of management as a factor in the quality of an industrial arts program was reported by Pendered some years ago.[1] His study yielded a valid, reliable rating scale to evaluate the quality of an industrial arts program. The instrument considered the following four major factors: (1) *Instructional Program* (what is taught), (2) *Physical Conditions* (where it is taught), (3) *Methods and Shop Management* (how it is taught), and (4) *The Teacher* (who does the teaching). The quality of a program was defined as the summation of scores on these four factors.

The findings of this research revealed that the highest correlation existing between any pair of variables was between *Quality of Program* and *Methods and Management* ($r = .924$). The factor (Methods and Management) was also found to be a highly reliable predictor of the quality of an industrial arts program. In fact, alone it predicted quality as well as the combined scores of all other factors. For this reason a final recommendation in the study urged teachers:

> Place greater emphasis on teaching methods and techniques of shop management since these so greatly influence the quality of the industrial arts program.[2]

It must be pointed out, however, that the factor, *Methods and Shop Management*, consisted of items relating both to methods of teaching

---

[1] Norman C. Pendered, "An Evaluative and Comparative Study of Industrial Arts Programs in Selected Junior High Schools of Pennsylvania at Various Levels of Financial Expenditure" (unpublished doctoral dissertation, The Pennsylvania State University, 1951).

[2] *Ibid.*, p. 129.

Laboratory Management—Starting the Class

and to techniques of shop management. Therefore, it would be incorrect to attribute the full influence of this factor on the quality of a program to shop management alone. However, management items doubtless did contribute significantly to the overall effect of the factor.

This chapter and the two following are devoted to a consideration of selected aspects of laboratory management including (1) starting the class, (2) student rotation, (3) pupil personnel organization, and (4) records and record keeping. The keys to successful shop management are careful planning and sound organization which are emphasized throughout these chapters.

STARTING THE CLASS

Starting an industrial arts class is an activity in which the teacher engages at least once, perhaps oftener, during the school year depending on the organizational pattern of the school. If the duration of the industrial arts course is 36 weeks (an entire school year), then the teacher will need start the class but one time. On the other hand, if a given section of pupils is rotated on a nine-weeks basis through art, music, home economics, and industrial arts, then the teacher will be faced with starting four different classes in one school year.

To the student the beginning of a course in industrial arts should be a challenging as well as an interesting experience. It is an opportunity to "size up" the teacher, to gain a preview of future activities, and to become part of the class organization.

To the teacher the activities connected with starting an industrial arts class are of vital importance. It is essential that the teacher make a good beginning by "getting off on the right foot" and by starting the class in the direction in which it is to go. In short, it is imperative for the teacher to gain and hold control of the learning situation. This is much easier to accomplish on the first day rather than try to do it later in the year. Much of the teacher's future success in the course depends upon a good beginning. It is important, therefore, that special attention be given to preparation for the first class meeting.

It is doubtful if two teachers will use identical techniques for starting an industrial arts class. This is as it should be for each is a unique individual and will necessarily put something of himself into the teaching situation. However, there are certain guidelines related to starting a class which have proven their worth and which may be used to advantage by all teachers.

Since the method of starting a class will vary with the type of laboratory organization as well as with the grade level of the pupils, it seems pertinent to recall the several ways in which industrial arts activities are commonly organized. As noted in Chapter 3, the three general bases for organizing industrial arts activities are (1) the unit shop, (2) the comprehensive general shop, and (3) the general unit shop.

1. Starting the Unit Shop

It is generally conceded to be somewhat easier for a teacher to start a class in a unit shop than one in a general shop. Two reasons for this are (1) it is simpler to start a class in only one area (e.g., wood, metals, or electricity) than in several of these areas simultaneously, (2) ordinarily all students in a unit shop are started on the same or similar projects which greatly facilitates teacher demonstrations.

Before the arrival of the class in the laboratory, the well-prepared teacher will make careful plans for the first period. Special attention should be given to a number of details including the following:

1. Prepare a lesson plan (see Chapter 7).

2. Check tools and equipment for safe operating condition.

3. Plan a shop tour (if one is to be used) and set up selected machines or equipment for brief operation.

4. Prepare instructional materials to teach planning, including handouts, overhead transparencies, and the like.

5. Prepare the demonstration, if one is to be used. Ready all materials, tools, equipment, instruction sheets, and audio-visual aids (see Chapters 14 and 15).

With the actual arrival of the students in the laboratory, the teacher who wishes to gain early control of the situation will take the initiative in starting some activity immediately. This may take the form of directing students to seats in the instructional or demonstration area and distributing information cards to be filled out. These cards may be used for checking the roll and will serve as a basis for guidance interviews later in the course. After these formalities have been taken care of, the process of getting acquainted may begin. *The importance of learning to know each student individually cannot be overestimated.* Some instructors have successfully used the practice of having each student state his name and tell something about himself including, perhaps, his interests or hobbies. This serves to introduce him to the teacher and to his classmates as well. If the teacher so desires, students may be assigned at this time to regular seats in the demonstration area, to work areas, or to work stations; an alphabetic arrangement is often made merely for the convenience of the teacher.

When these preliminaries have been taken care of, the next step usually is to give students a preview of the course. Some teachers have found that a brief outline of the course content, a display of typical projects, and an opportunity for questions from the class constitute good overview material. The importance of shop safety should be stressed, especially the 100 percent eye protection program. If each class uses the same set of safety glasses, the teacher should explain the procedure to be followed in using the lens cleaning tissue and the sanitizing equipment. One type of safety glass sanitizer used in industrial arts laboratories is illustrated in Fig. 11-1.

Fig. 11-1. Safety glass monitor. Courtesy, Sellstrom Manufacturing Company)

If each student is issued his own safety glasses (this is preferable), the rule on forgetting to bring them to class should be pointed out: *No safety glasses—no work in the laboratory.* Although extra safety glasses should be available in the laboratory, especially for visitors, it is not good practice to condone student frailities in forgetting their safety glasses by permitting them to use the shop glasses, unless some good reason exists. It is desirable, too, in this orientation phase to lay the groundwork for a democratic class organization or student personnel system. Additional details on this topic will be presented in Chapter 12.

Following such a presentation, several procedures are open to the teacher. He may (1) take the class on a shop tour, (2) introduce the manipulative activities by planning the first project with the class, (3) present a demonstration, or (4) present an informative lesson. The latter probably should be avoided on initial contact with the class. Most students enjoy activity and "doing things;" they are anxious to get to work. A long talk, especially during the first class period, will contribute little to class interest and expectancy.

The Shop Tour. There is much to recommend the use of a well-planned shop tour as a means for initiating students to a new industrial arts laboratory. The major purposes of such a tour are to acquaint the students with the organization and possibilities of the shop and to arouse their interest and enthusiasm for the new experience. To accomplish these ends fully, careful planning and preparation are necessary.

A good shop tour will include the following features:

1. Provide for an overview—The teacher should provide an overview of the whole laboratory by pointing out the major items of interest that are to be included on the tour. For example, in a unit woodworking shop, he may wish to call attention to the specialized tool panels, to the stock room and finishing room as well as to the power equipment available, such as the lathes, surfacer, circular saw, and so on. Figure 11-2 shows a teacher whetting the interest of students on a shop tour by briefly comparing an auger bit with an oversized model.

Fig. 11-2. This enlarged model of an auger bit captures the interest of these junior high school students who are on a shop tour. (Courtesy, Julian Cleveland, E. B. Aycock Junior High School, Greenville, North Carolina. Photo by Walter F. Deal, III)

Laboratory Management—Starting the Class

2. Briefly operate machines or equipment—Various machines and work stations may be set up to provide the class with a quick glimpse of the possibilities in the shop. For example, a piece of stock may be between centers in the wood lathe ready for turning; the Uniplane or circular saw may be set for taking a cut; or self-instructional devices, such as film loop projectors or sound-on-slide presentations may be ready in the carrels. Students will be more impressed if some of the equipment is used and certain machines operated by the teacher, even if only for a moment or two, than if they merely look at them. The sight and sound of equipment and machinery in operation tends to arouse student interest.

3. Keep the tour moving—The tour should be planned so that the group moves quickly from one point of interest to the next. The trip is to provide a general picture of the laboratory and should not consume too much time.

4. Encourage student questions—During the tour frequent opportunity should be allowed for questions from the class.

5. Stress safety—Safe working practices which are of *immediate* importance, such as eye safety may be stressed.

6. Point out good housekeeping features—General organization and housekeeping practices within the laboratory may be pointed out on the shop tour.

Under ordinary circumstances a shop tour should not require more than one-half hour for its completion. If the trip is drawn out much longer than this, the enthusiasm which has been aroused may tend to lag. The teacher should remember to keep the trip *brief, informative,* and *interesting* and not try to make it a full course.

Planning the First Project. Students in industrial arts classes are interested primarily in getting to work quickly. There is much to be said for capitalizing on the zeal or eagerness of a new class by getting the tools into their hands at the earliest possible moment. Before this can be done, the students must know what they are to do and how they are to do it. It becomes necessary, therefore, for the teacher to outline the first project and to demonstrate at least the first operations involved.

Much stress has been laid on the necessity for pupil planning. If this technique is to be used effectively, it should be started with the first project. Delay makes the initiation of planning that much more difficult. A detailed discussion of planning has already been made in Chapter 8 where it was emphasized that *planning must be taught.* The instructor will consider it time well spent to plan the first project with the class. Two purposes are served by this early introduction to planning: (1) the class will understand what is involved in making the project; and more important, (2) the fact will be made clear right from the start that planning is an indispensable prelude to construction. It is at this stage when students are first learning to plan that a project chart may be useful. Two sample project charts are illustrated in Fig. 11-3.

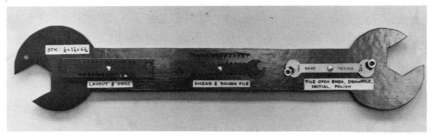

Fig. 11-3. Project charts help students visualize steps in constructing a project. (Courtesy, George Lartz, Sheboygan Falls, Wisconsin)

It must be assumed that students will have but little knowledge of how to plan when they are beginning their industrial arts work. The teacher should make every effort to draw from the class the procedure to be followed in making the project. Such cooperative effort in planning will lay the groundwork for individual planning which will follow later.

With beginning students it is not necessary to plan the entire project completely before starting actual construction. In fact, many instructors have found that considerable development must take place before students can be expected to plan an entire project by themselves prior to construction. If the first few steps can be outlined, this part of the procedure probably should be considered as acceptable at this time.

The Demonstration. After the first few steps of the project have been planned, both the class and the instructor should be ready for a demonstration of how the work is to be accomplished. As stated previously, this first demonstration is of the greatest importance to the instructor. It is his opportunity to make a favorable impression on the class. If his work is well planned and skillfully executed, he will gain the confidence of the class and he is well on the road to success. However, if he runs into difficulty or if his performance is awkward and shows lack of planning or skill, his stock immediately goes down in the estimation of the class. It behooves the teacher, therefore, to take every possible precaution to ensure that this first demonstration is successful in all respects. Chapter 10 provided detailed treatment of the steps to be followed in preparing for and in giving a class demonstration.

Following the demonstration, the class should be allowed to begin work immediately. This is the moment for which the students have been waiting! It is likely that the teacher may wish to have rough stock prepared for the students' use in making this first project. If so, then the material should be distributed at once and the students should be allowed to begin construction. The cleanup for this first work session may be handled simply by asking each student at the end of the period to clean around the place where he has been working and to put away the tools he has been using.

Other Activities. On subsequent days the teacher will want to introduce the student personnel plan and to demonstrate additional operations and processes. The sooner these activities are introduced, especially the personnel plan, the smoother the shop organization will function. It is important that an informative lesson be presented before the class has progressed too far. The first of such lessons probably should be given during the early days of the course. From that day forward, the use of informative material should become a regular part of the class routine.

2. Starting the General Shop Class

There is one significant difference between starting a class in a comprehensive general shop and starting one under a unit type organization. In the ideal general shop, students are at work in several different material areas (e.g. wood, metals, graphic arts, et al.) at the same time. The problem is, therefore, how can one teacher start a class in several areas simultaneously? Numerous techniques have been suggested, but in general, these can be resolved into two basic approaches: (1) the unit method approach, and (2) the materials-centered approach with either a multiple demonstration or a single demonstration.

The Unit Method Approach. The unit method of teaching was described in detail in Chapter 7, *Organizing Learning Activities*. The point was emphasized that there is a need in industrial arts for a central core or theme which will integrate the informative aspects and the manipulative experiences. The unit method of teaching provides this integration and is especially applicable to general shop activities at the middle school and the junior high school levels. As applied to industrial arts activities, the unit method does not ensure similar experiences for each student in each of several different materials areas. Projectwork may involve several areas, but the emphasis is on the theme of the unit rather than on selected or predetermined experiences in the materials areas themselves.

The enterprising and resourceful teacher of industrial arts may well give serious consideration to starting a general shop class with the unit method approach. To do this some of the steps outlined previously for starting a unit shop should be employed. This would include the following:

1. Prepare a detailed lesson plan for the first meeting of the class.
2. Check tools and equipment for safe operating condition.
3. Plan a shop tour.
4. Prepare instructional materials to teach planning, including handouts, overhead projectuals, and the like.
5. Begin the first step in the unit method, that is, the introduction and orientation phase (see Chapter 7 for details).

The Materials-Centered Approach. Some general shop teachers prefer to have several students at work at all times in each of the different materials areas offered in the laboratory. For example, in a class of 15 students at any given time during the course, one-third of the group may be in wood technology, another third in metals technology, and the last third in visual communications. The students then rotate through the areas so that by the end of the course, each student will have covered all three areas.

At once the teacher is faced with this question: How shall students be placed *initially* in a given area? Three techniques for solving this problem are (1) calling for volunteers to work in a given area (and hoping the student choices come out evenly!), (2) using some chance means, such as, drawing student names or numbers out of a hat, or (3) using an alphabetic assignment with the simple explanation that each student will eventually get to work in each area.

The next question facing the teacher is: How can several groups of students be started in as many different materials areas simultaneously? Two solutions to this problem are (1) the multiple demonstration and (2) the single demonstration.

The Multiple Demonstration. This technique involves several teacher demonstrations each of which covers a beginning experience in each of the materials areas. This approach has the advantage that the class is immediately started in several areas and the spirit of the general shop organization is functioning from the start. Several disadvantages to this approach are evident. Among the more serious of these are the following:

1. Since two, three, or more demonstrations must be given before

any student begins work, actual construction activities may be delayed for several class periods. In some cases considerable time may elapse before the students actually use the shop tools and equipment. This may adversely affect the initial enthusiasm of the class.

2. Since all students will be expected to watch all demonstrations, there is a good chance that they will become confused. It is also likely that they will remember vividly the details of the last demonstration only.

3. Since the students realize that they will be making only one of the projects demonstrated, they may not give full attention to all demonstrations.

In spite of these disadvantages, however, many instructors feel that the great advantage gained by having the entire class started in various areas is sufficient to warrant the use of the multiple demonstration. This approach has been used successfully by many teachers who are persuaded to its effectiveness.

Several variations of the multiple demonstration approach have been developed which tend to minimize the disadvantages cited. In one variation the teacher plans a short demonstration on an introductory project for each area. Obviously all demonstrations cannot be given at the same time, so the teacher plans some worthwhile activity for each group waiting its turn at the demonstration bench. These activities may involve self-instructional materials, such as assignment sheets, programed instruction sheets, or the use of single-concept film loops. As soon as the teacher completes one demonstration to the first group, he moves quickly to give the second demonstration to the second group of students in another area, and so on, until the entire class has been covered. In this manner the whole class is started in a multiarea shop within the first class period. This particular technique works well in both the comprehensive general shop and in the general unit shop.

Another variation of the multiple-demonstration approach is to use a project which involves work in two or more materials areas. Under some circumstances it is possible to give more than one demonstration and still maintain interest of the group, since each student knows that he will need to perform the operations in order to complete the project. The difficulty in this approach lies in finding suitable projects involving activities in the desired areas.

The Single Demonstration. In attempting to avoid some of the defects of the multiple-demonstration approach, some teachers prefer to start a general shop as if it were a unit shop. It is evident that this method can be used only if there is sufficient equipment in some one area of the general shop to permit an entire class to work in that area without excessive waiting for tools or equipment. To be specific, the average general shop has only a limited number of work stations available, say, in the metals casting area, and it would be impractical to try to start an entire class in this one area. In some general shop laboratories, however, sufficient tools may be available, for example, in the woodworking area, to permit the entire class to make a successful start here.

This method has the obvious disadvantage in that a long period is required prior to the separation of the class into groups working in the various areas. In other words, considerable time may elapse before the shop is operating on a general shop basis. The plan has several advantages, however, including the following:

1. Since only one demonstration is required, the class as a whole may be started sooner than with the multiple-demonstration approach.

2. The attention of the class will be centered on the single demonstration because each student knows that he will be expected to perform the operations which are being presented.

3. Introduction to student planning is facilitated because all students will be planning the same or a similar project.

4. The teacher needs to prepare for one demonstration only, rather than several demonstrations.

If it is decided to use the single-demonstration approach, the initial steps are similar to those outlined for the unit shop as follows:

1. Get acquainted.
2. Arrange a tour of the shop.
3. Give an introduction to planning (plan the first project together).
4. Have a demonstration.
5. Start class to work.

In the days which follow these first activities, however, the program will differ greatly from that outlined for the unit shop. Although the problem of dividing the class into the various other areas has been deferred, it still must be solved. To reach the desired goal of having several materials areas operating concurrently, the teacher should begin at once to give lessons and demonstrations in areas other than the one in which the class was started. For example, if the first demonstration were given in woodworking, the second might be in metalwork, the third in crafts, and so on, until all the areas have been covered. By the time the first student has completed the introductory project, demonstrations and lessons in several areas will have been presented. Student interest will have been aroused in one or more of these new areas. By skillful guidance the teacher can have his students working in several areas as they begin their second project.

ROTATING PUPILS THROUGH AREAS

Regardless of the type of organization utilized (unit, comprehensive general, or general unit), the industrial arts teacher is faced with the problem of moving students through the various areas or activities to provide broad exploratory experiences. In the case of the unit shop, each student should gain some experience with all of the machines and equipment; in the general shop he should gain experience in each materials area (e.g. woods, metals, et al.). An exception to this occurs where the unit method is employed. The reason for this is that emphasis is not on providing like experiences for each student in a variety of materials areas.

The two ways of student rotation through the offerings of a given industrial arts program are self-evident—by groups or by individuals.

1. Group Rotation

For many years teachers have been dividing a class into small groups and placing each in one of the machine areas or materials areas. At the end of a certain period of time, each group is moved to a new machine or area. The working time in a given area is a factor of total time available for the course and the number of areas or activities to which students are to be exposed. For example, in a 36-weeks course for a six-area general shop, each student may be permitted to work six weeks in each area. This example assumes that each area is equally important. Some teachers, however, weight the working time of an area in relation to its relative importance or to the amount of tools and equipment available.

Obviously a weakness of this plan is that not every student completes his work at the exact time to change to a new area with his group. Many unfinished projects are a common result. To offset this weakness, some teachers rotate students through all desired areas, but allow a common time at the end of the course (several periods or weeks) to complete any unfinished work. This variation provides a workable solution to the problem.

2. Individual Rotation

At first, the idea of rotating students through various activities on an individual basis appears to be the ideal solution. While this plan is pedagogically defensible, it does pose certain problems for the teacher. For example, it is unlikely that a group of students will be transferring to a new area at the same time. For this reason the teacher will be called upon to present fewer group demonstrations, but will give more individual demonstrations and instruction. Then too, there is the problem of the laggard student who may never get exposed to all the areas within the total allotted time, if the plan of individual rotation is followed. Despite these shortcomings, however, individual rotation is a plan that has worked well over the years for many industrial arts teachers.

SUMMARY

Managing the industrial arts laboratory is an important responsibility that contributes greatly to the success or failure of the teacher. The keys to successful shop management are planning and organization. Much of the success of a course depends upon a good beginning. It is imperative for the teacher to gain and hold control of the learning from the first day forward. A good beginning requires careful planning.

Starting an industrial arts class is related to the type of shop organization present. Three general bases for organizing industrial arts activities are: (1) the unit shop, (2) the comprehensive general shop, and (3) the general unit shop.

The steps usually followed in starting the unit shop include (1) planning the first lesson, (2) getting acquainted, (3) taking a shop tour, (4) planning the first project, (5) giving a demonstration, (6) starting students to work, (7) establishing the personnel organization, and (8) giving an informative lesson.

A shop tour is designed to acquaint students with the organization and possibilities of the laboratory and to arouse their interest and enthusiasm. The tour should be brief, informative, and interesting.

A summary of the steps to be followed in starting a general shop class includes: (1) planning the first lesson, (2) getting acquainted, (3) taking a shop tour, (4a) if the unit method approach is used, begin the orientation phase; or (4b) if a materials-centered approach is used, perform a multiple demonstration, divide the class into groups corresponding to the materials areas and set them to work. An alternative procedure within the materials-centered approach is to perform a single demonstration, set the class to work in one area and later demonstrate in other areas into which individuals are directed after finishing the introductory project.

If the unit method is employed, rotation of individuals through the several materials areas of the laboratory is unnecessary. If the materials-centered approach is utilized, rotation may be accomplished either by groups or by individuals. Each plan has advantages and disadvantages and both have been used effectively.

DISCUSSION TOPICS AND ASSIGNMENTS

1. Review the unit shop, the general shop, and the general unit shop with regard to (a) physical facilities, (b) teacher preparation, (c) types of projects, and (d) methods of teaching.

2. Can a good teacher be a poor manager? Explain your answer.

3. Why is it easier for a teacher to start a unit shop than a comprehensive general shop?

4. What features characterize an effective shop tour?

5. How can the unit method be justified as a means of starting a general shop class?

6. Make lesson plans covering the first three class meetings for (1) a unit shop and (b) a general shop.

7. Do you think experienced teachers make lesson plans? Why?

8. Compare and contrast the recommended procedures for starting a general shop class.

9. Discuss the factors which limit the effectiveness of (a) the multiple-demonstration approach for starting a general shop class and (b) the single-demonstration approach.

10. What variations can you suggest for using the multiple-demonstration approach for starting a class in a general unit shop?

11. Why is rotation of students not a part of the unit method?

12. What are the advantages and disadvantages of group rotation and individual rotation?

SELECTED REFERENCES

Adams, John H. "How to Start the New Year," *Industrial Arts and Vocational Education*, vol. 50, no. 7 (September 1961), pp. 35–37.

Anderson, W. Carlisle. "Into the Third or Fourth Week," *Industrial Arts and Vocational Education*, vol. 52, no. 1 (January 1963), p. 25.

Bernhardy, George W. "Organizing Your Shop for the School Year," *Industrial Arts and Vocational Education*, vol. 56, no. 7 (September 1967), pp. 26–29, 68.

Dunn, Adam E. "Make a Safety Inspection of Your Shop," *School Shop*, vol. XXVII, no. 5 (January 1968), pp. 48–52.

Miller, Wilbur R. "Starting the School Year with a Beginning General Shop Class," *Industrial Arts and Vocational Education*, vol. 51, no. 7 (September 1962), pp. 30–32.

Pendered, Norman C. "An Evaluative and Comparative Study of Industrial Arts Programs in Selected Junior High Schools of Pennsylvania at Various Levels of Financial Expenditure" (unpublished doctoral dissertation, The Pennsylvania State University, University Park, 1951).

Price, John S. "Multiple Areas Project," *School Shop,* vol. XXVI, no. 5 (January 1967), pp. 32–35.

Rathbun, Jess. "Another Look at Eye Protection," *Industrial Arts and Vocational Education,* vol. 56, no. 3 (March 1967), pp. 78–79.

Ruley, M. J. "How to Begin the School Year," *Industrial Arts and Vocational Education,* vol. 58, no. 7 (September 1969), pp. 22–23.

Silvius, G. Harold and Estell H. Curry. *Teaching Successfully in Industrial Education.* Bloomington, Ill.: McKnight and McKnight Publishing Company, 1967.

Swan, John T. "How to Organize the Shop for the First Day," *Industrial Arts and Vocational Education,* vol. 51, no. 7 (September 1962), pp. 34–35.

Chapter 12

Laboratory Management — Personnel Organization

The pupil personnel organization is recognized as a valuable means for helping students achieve some of the objectives of industrial arts. Certain of the behavior changes related to the organization of industry and the development of desirable social relationships can best be attained through cooperative planning and participation in the personnel organization.

The term *personnel organization* apparently has been borrowed from industry where it is an important feature of the managerial system. As applied to the industrial arts laboratory, the personnel plan may be defined as *the organization of students for the purpose of achieving certain behavior changes and other desirable outcomes.*

Most students enjoy role-playing activities. The personnel organization provides opportunities for students to role-play important administrative and supervisory posts in their world of industry. For this reason students usually enjoy participation in the personnel plan and approach their assignments with interest and enthusiasm.

PURPOSES OF THE PERSONNEL ORGANIZATION

The purposes of the personnel plan, which may not be the same for each school, may range all the way from a mere clean-up schedule (which is in reality no personnel system at all) to an important facet of the instructional program. A true personnel plan will have at least three major purposes: (1) to train for leadership and followership, (2) to explore industry, and (3) to relieve the instructor of routine duties.

1. To Train for Leadership and Followership

The rather informal atmosphere of the school laboratory makes possible an organization which can contribute effectively to training in leadership and its concomitant followership. Only through opportunities to assume and discharge responsibility can leadership be developed. In like manner, it is only through identification with a group undertaking and an understanding of group purposes that effective followership can be secured. It has frequently been noted that the troublesome boy who is always in difficulty with both the teacher and his fellow students becomes an effective and accepted leader when given specific and well-defined responsibilities. On the other hand, those who seem to have little ability for self-direction sometimes become efficient workers when their activities are directed by others. If the personnel organization had no other purpose than to give opportunities for training in leadership, its place as part of the industrial arts· program would be fully justified.

2. To Explore Industry

An important aspect of industry relates to its organization. If one is to understand fully the means by which an industry operates, he must know how its organization functions. An effective method for providing these understandings is through the development and operation of a personnel organization in the industrial arts laboratory.

In order that the personnel organization may be most effective, however, students must realize the likenesses and the differences between their organization and the organization within an industry. The duties and responsibilities of comparable positions should be pointed out and discussed. Only in cases where such understandings exist can the full value of the personnel organization be realized. Chart VII depicts a typical industrial arts personnel system.

To lend an air of realism to student personnel systems, Greer[1] proposes that they be organized like industry on a labor-management basis of collective bargaining. He describes a pilot study in which through letters of application students were selected by the instructor for management posts such as shop superintendent, safety engineer, and publicity manager. All other students were considered "labor" and were assigned to various jobs in the personnel organization. The labor force then organized into a local union with elected officers while management formed a manufacturer's association. The instructor met with

---

[1] Arthur G. Greer, "Let's Organize the School Shop," *School Shop*, vol. XXI, no. 2 (October 1961), pp. 16–17.

CHART VII
Flow Chart for a Typical Industrial Arts Personnel System

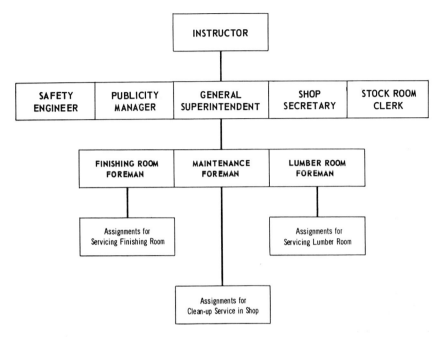

each faction at the bargaining table to discuss duties and responsibilities of personnel jobs as well as minimum pay standards (grades), absences and tardiness (payroll deductions), shop discipline, and overtime benefits. Every student in the class participated in some manner. In evaluating this personnel organization Greer notes:

> It was very enlightening and gratifying to observe the intense interest, cooperation, and managerial abilities that resulted from this project. Keen competition developed among class members, not only scholastically, but also in ingenuity, responsibility, and cooperation with each other. The drastic change in attitude from complete indifference for education to warm interest and respect for ability and knowledge was an unexpected bonus, which alone was worth more than all of the effort expended to start the program.[2]

3. To Relieve the Instructor of Routine Duties

Many routine duties in the school laboratory demand from the teacher an amount of time which might be better spent in more important aspects of teaching. Most of these activities have some educational

[2] *Ibid.*, p. 17.

value when performed by students. Typical of such duties are the dispensing of small supplies, keeping routine records, taking class attendance, checking on ventilation and lighting, inspecting the condition of tools (see Fig. 12–1), the charging out of library books and other items. Through a successful personnel system a large proportion of such duties can be turned over to responsible members of the class.

Fig. 12–1. This maintenance foreman is performing a valuable service to the program by checking on the condition of tools and equipment in the laboratory. (Courtesy, Donald Prescott, Snow Hill Junior High School, Snow Hill, North Carolina. Photo by Robert McElroy)

The teacher can add considerably to overall efficiency in the laboratory by coordinating the efforts of the pupil personnel organization with his records-keeping system (see Chapter 13).

ORGANIZING A PERSONNEL PLAN

It was noted in the previous chapter that clean-up activities for the first work session of a class could be handled simply by requesting each student to return his tools to the panel and to clean up around his work site. This suggestion was made in the spirit of capitalizing on student interest and enthusiasm for a new activity. It was pointed out, however, that the groundwork for a democratic personnel organization should be laid during the beginning days of the course.

The teacher should give prompt attention to this matter and get the personnel organization functioning at the earliest possible moment. It is imperative to do this because a class of students can leave a laboratory in quite a mess after only one period of work. When the next class enters they are not highly motivated either to work in a dirty shop or to clean up a laboratory left untidy by others. By the end of a single day the physical conditions in the laboratory could approach sheer chaos. Worse than this, bad impressions and poor work habits may already have been formed which will at best be difficult to overcome. Unless the teacher takes the initiative immediately to get a personnel organization underway, he has only himself to blame for the chaos, confusion, disorganization, and extra work he may be called upon to do to restore the laboratory to normalcy.

Personnel systems are basically of two types when considered from the standpoint of the method of organization. They may be either teacher-developed or class-developed.

1. Teacher-Developed Plans

The easiest way to establish a personnel system is for the teacher to develop the plan in all its details and then present it to the class. This method has the advantages of being quick, direct, and authoritative. It has the further advantage of being applicable to all classes and of being usable year after year.

On the debit side, however, a teacher-imposed plan is quickly recognized as such by the students. It can never become the students' own system. Participation under such circumstances becomes a matter of doing what the teacher suggests because it is expected. A more democratic method of class organization would seem more desirable.

2. Class-Developed Plans

It is quite possible and feasible to have a personnel plan developed by a class or by several classes working together. The process of establishing such a plan provides excellent motivation for the study of various types of organizations within an industry. The attention of the students may first be directed to the function of the personnel plans for various types of local or nearby plants. From such studies the need for a similar organization in the school shop can be developed. An industrial visit (see Chapter 14) may be planned not to view the manufacturing process, but to study the organizational plan of the company. Resource people, such as superintendents or foremen, may be invited to the class to present their views. If scheduling presents a problem, the use of telelecture or videotape recording may be considered.

After researching the problem, a committee may suggest a student personnel plan to the class or the teacher may work with the entire class in developing a suitable organization. Proceeding along these lines will require considerably more time than would be required for the teacher to develop the plan himself. When the plan is completed, however, the students will know that it is *their* plan and they will be much more interested in seeing it work. The extra time spent in the democratic development of a student personnel organization is usually more than compensated for by the high degree of cooperation that results.

FEATURES OF A PERSONNEL ORGANIZATION

A personnel organization should be designed especially to fit the conditions of a particular situation. Standard plans will not work in all circumstances. Inasmuch as an important objective of any personnel plan is to give students some idea of the manner in which industry is organized, it might be well if the type selected bore some relationship to the industry represented by the school organization. For example, the personnel plan for the general metals laboratory should be to some degree typical of plans commonly found in the metals industries.

Salient features of a pupil personnel plan should include the following:

1. Develop a Plan to Reach the Objectives of Industrial Arts

Certain behavioral changes outlined for Objective 1 and Objective 7 (see Chapters 5 and 6) can best be achieved by student planning and participation in a pupil personnel organization. To develop leadership and followership qualities requires that each student be given an opportunity to exert leadership roles and to act as a cooperative follower. The only way to teach honesty is to give students a chance to be honest. Likewise, the only way to develop leadership abilities is to give students opportunities to lead. Some students who may be reluctant to assume leadership roles will need to be "pushed forward" by the instructor who should insist that they assume the responsibilities assigned to them.

2. Use a Class-Developed Plan

The plan of organization should be developed from within the group (with the help, advice, and approval of the instructor), rather than be imposed upon the class by the teacher. This need not mean that a totally new organization must be developed each term. It does mean, however, that each class should be given the opportunity to review and, if desirable, revise the personnel system used by previous classes.

3. Avoid Making the Organization a Clean-up Schedule

In most cases where a personnel plan has failed to function, the teacher has attempted to use it merely as a clean-up technique. Any such organization worthy of the name should subscribe to the three major objectives previously cited. To accomplish these objectives, students must sense that the plan really has a definite purpose and must feel that they are important to its proper functioning. They must also be convinced that it has a more worthwhile purpose than merely keeping the laboratory clean. To provide an exploratory situation paralleling to some extent the personnel system of industry cannot be accomplished if the prime purpose of the organization is to keep the laboratory in order.

If only clean-up is emphasized, the instructor will be called upon to perform many routine tasks, such as roll call, dispensing of supplies, controlling library loans, and the like. He may better use this time to the advantage of the instructional program.

Familiarity with the organization of industry will become more meaningful after students have had first-hand experiences with the personnel organization in the shop.

4. Fill Offices Democratically

Offices may be filled in a democratic manner by election or by a system of promotion once elected. In some situations appointments may be made by the instructor. In either case, the instructor should make the decision in this matter or should abide by the class's decision as to which method is to be employed.

5. Define Duties and Responsibilities

A definite statement should be made of the duties and responsibilities of each member of the personnel organization. These duties should be described fully and should then be posted where they may be referred to at any time. A chart which simply lists the names or titles of the offices is not sufficient. A complete description of each job should be developed, preferably by the class or by a committee appointed by it.

6. Give Officers Selected Responsibilities with Commensurate Authority

There should be a definite assumption of responsibility by certain members in the organization. It is wasted effort to designate a certain student as superintendent or foreman and then give him no opportunity to assume responsibility. Furthermore, the authority exercised by each officer in keeping with his responsibility should be backed up by the instructor.

7. Change Officers

Officers should be changed when they have exhausted the educational possibilities of a given position. Keeping a student in an office after he has mastered all the details connected with its operation, simply because he is efficient, tends to discourage initiative. Besides, it reduces the opportunities for providing leadership training to others in the class.

8. Rotate the Clean-up Jobs

To keep things moving along lively in the organization plan, clean-up and maintenance jobs should be rotated at regular intervals, perhaps weekly. Frequent rotation of assignments prevents students from becoming discouraged or dismayed with a given task, particularly if it is a menial or unpleasant one. On the other hand, if the rotation occurs too frequently some students will be confused and will be unable to master their assignment.

9. Provide Class Time for Officers to Meet

Major officers should be given an opportunity to meet and discuss problems of the personnel organization during class hours. By this means weaknesses in the organization may be caught early and remedial measures taken. Frequent meetings of this type are good morale builders. They give the officers a feeling that their work is essential and important. This lays the foundation for a strong, efficient organization with high morale. Figure 12-2 shows the officers of a student personnel organization at the junior high school level conferring on policy matters.

Fig. 12-2. These officers of the student personnel organization are meeting on class time to discuss a problem which has just materialized in the laboratory. (Courtesy, Donald Prescott, Snow Hill Junior High School, Snow Hill, North Carolina. Photo by Robert McElroy)

10. Provide a Suggestion Box

Not all the good ideas for improving the laboratory or the personnel system will come from the instructor or the officers. A way should be provided for receiving ideas and suggestions from any member of the class. Grievances can also be aired anonymously. A suggestion box will provide this avenue of communication.

TYPICAL EXAMPLES OF PERSONNEL ORGANIZATIONS

It has been indicated that the exact form which a personnel plan will take depends on local conditions and needs. No fixed pattern can be prescribed which will be equally applicable to all situations. Examples of types that have been used successfully in various schools may serve as a starting point for the development of a plan to meet the requirements of a specific situation. The following types have been selected for inclusion here on the basis of having met the needs of the particular schools in which they were used.

CHART VIII
A Personnel Organization Plan – Sample No. 1

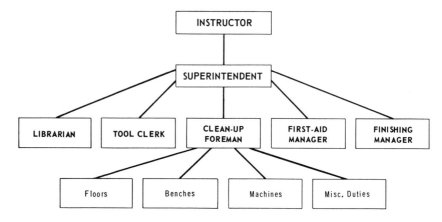

The duties of the personnel shown on Chart VIII are as follows:

Superintendent's Duties

1. Supervise entire personnel system. Take charge if instructor has to leave laboratory temporarily.
2. Call roll and keep roll book.
3. Assist teacher in preparing for class demonstrations (e.g., get tools out).
4. Keep a list of needed materials and supplies so that the instructor will be able to maintain sufficient quantities on hand at all times.
5. Receive ideas and suggestions from the class. Check the Suggestion Box daily.
6. Call clean up. The signal may be several notes on a marimba or a quick blinking of the overhead lights.
7. Dismiss class.

Librarian's Duties

1. Keep the planning center and library in good order.
2. Check project files.
3. Charge out and receive returned library books and periodicals.
4. Maintain a supply of student planning sheets.
5. Maintain the library of 8-mm single-concept film loops and other audio-visual materials.

Tool Clerk's Duties

1. Check all tool panels at the end of the period and report any missing tools to the superintendent.

Laboratory Management—Personnel Organization

2. Maintain *Tools Borrowed* sheet to check out tools on loan to students, faculty, or other staff members.
3. Inspect all hand tools and report dull, broken, or unsafe ones to the superintendent.

Clean-Up Foreman's Duties

1. Oversee all clean-up personnel.
2. Make substitutions for absentees.
3. Check with other clean-up officers to see if their portion of the laboratory is taken care of at clean-up time.
4. Report to the superintendent when the clean-up is satisfactorily completed so that the class may be dismissed.
5. Receive ideas from the class for improving the clean-up.

First-Aid Manager's Duties

1. Check on first aid supplies and keep sufficient quantities on hand.
2. Report all injuries to the instructor immediately.
3. Report safety suggestions or safety hazards to the superintendent and the instructor.

Finishing Manager's Duties

1. Take charge of all finishing supplies.
2. Check on all paint, varnish, and shellac brushes to make sure they are in good order.
3. Keep the finishing cabinet and area in neat order.
4. Report needed finishing supplies to the superintendent.

The following data supplement Chart IX.

Personnel Duties

Superintendent:
1. Supervises the general cleaning of the laboratory.
2. Reports to the instructor anything out of the ordinary, such as lost or broken tools and inoperative machines.
3. Suggests possible shop improvements gained through experiences as superintendent; also acts as class representative.
4. Works with the maintenance man in supervision of maintenance work in shop.

Office Manager:
1. Assists instructor in keeping shop library and office records.
2. Cleans up shop planning center, office, and library.
3. Checks shop library books and periodicals daily.

CHART IX
A Personnel Organization Plan – Sample No. 2

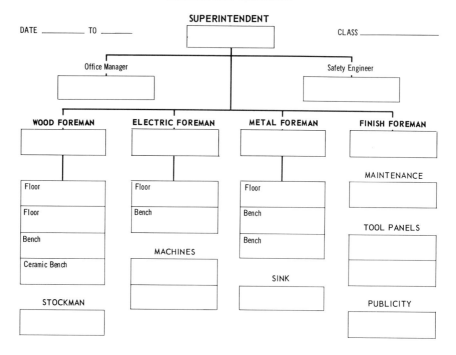

Safety Engineer:
1. Checks shop at beginning and end of each period for proper ventilation and lighting; turns off all gas outlets.
2. Makes safety check of machines and equipment.
3. Is responsible for general shop safety program and its improvement.
4. Posts safety posters and information; replenishes first-aid cabinet when necessary.

Foremen:

Wood and Ceramics:
1. Is responsible for all tools, benches, and machines of that department.
2. Supervises and assists in clean-up of wood and ceramics area with students assigned.
3. Closes all wood vises and checks with tool men at end of class period.

Laboratory Management—Personnel Organization

Electric:
1. Cleans benches, machines, and window sills of department.
2. Turns off all electrical connections and stores test instruments at close of period.
3. Collaborates with safety engineer to see that all electrical connections are in proper condition.

Metals:
1. Appoints assistant in charge of metals casting area with whose assistance he is responsible for all benches, tools, and equipment of the department.
2. Closes all vises and cabinets and sees that tools and supplies are properly stored at end of period.

Finishing:
1. Is responsible for proper storage and condition of all supplies and equipment in finishing room.
2. Cleans up and closes all finishing cabinets and containers; checks supplies and replenishes when necessary.
3. Cleans room and drying racks with vacuum cleaner.
4. Places oily rags and papers in safety-type waste containers.

Stockman:
1. Is responsible for arrangement and storage of supplies in bins and racks.
2. Reports shortages to representative of Supply Committee.

Machines:
1. Checks on condition and safe operation of all machines and tools of shop.
2. Cleans and dusts machines at end of period.

Sink:
1. Cleans sink at close of period. Uses powdered soap and, if necessary, solvent.
2. Checks on supply of paper towels and soap; replenishes as needed.

Maintenance:
1. Lubricates machines according to lubrication schedule.
2. Reports all necessary maintenance and supervises it during tour of duty.

Tool Panels:
1. Checks all tools at beginning of period; reports broken or missing tools.

2. Collects and replaces tools on panels at end of period.

Publicity:
1. Is responsible for displays both within shop and in hall cases.
2. Writes for school paper articles on pupil activities in shop.

DEVICES FOR DEPICTING PERSONNEL ORGANIZATIONS

Over the years industrial arts teachers have produced ingenious charts and contrivances to depict the personnel system operating in the laboratory. These devices have the important function of quickly showing the students exactly what they are to do. Each of these aids usually provides a place for each student's name or number as well as the title of his assignment in the organization. Detailed duties to be performed or responsibilities to be assumed are also outlined. More often than not these devices provide a means of rotating the assignments, especially among the labor or clean-up force.

Two common means for depicting pupil personnel organizations are the clock type and the vertical-columns type.

1. The Clock Type

This is a large circle or disc which is divided into as many segments as the number of assignments to be rotated in the organization. These are called job segments and are labeled alphabetically to correspond with an accompanying description of the jobs which is posted nearby. Just off the perimeter of the disc (on the backing panel), but facing the segments are numbers arranged like the numerals on a clock face. Each student is assigned a number and performs the task or duty as indicated by the alphabetized segment of the disc opposite his number. Rotation of assignments is accomplished simply by rotating the disc, say, clockwise so that a given segment will face the next higher number. In Fig. 12-3 several students are shown discussing their personnel assignments.

2. The Vertical-Columns Type

Another popular way to show the personnel organization and to rotate duties is the vertical-columns type (see Fig. 12-4). In its simplest form this device consists of two vertical columns each of which is divided into matching panels by a series of horizontal lines. The number of panels corresponds to the maximum number of students, jobs, or responsibilities to be rotated. Each job title or duty is labeled on one of

Laboratory Management—Personnel Organization

Fig. 12-3. These students are checking their clean-up assignment on this clock-type chart. (Courtesy, William J. Wilkinson, Nether Providence High School, Wallingford, Pennsylvania)

the panels of the first column. In the second column each panel is overlaid with separate, movable cardboard strips. Students' names or numbers are placed on these strips. This column must be constructed with tracks which not only hold the individual strips in place, but also allow them to be movable. Rotation of assignments is accomplished by removing the cardboard strip from the bottom of column two. All other strips will then slide down leaving room at the top of the column for the displaced card. It is obvious, of course, that additional columns with movable strips can be added to the chart to accommodate other classes. Attractive labels for this type of chart can be made by the use of (1) a primary or large letter typewriter and gummed labels, (2) instant or rub-on letters, or (3) a tapewriter producing self-adhesive embossed letters on plastic tape.

SUMMARY

The personnel organization is an important means for helping students reach some of the objectives of industrial arts. Its three basic functions are: (1) to train for leadership, followership, and cooperation; (2) to explore the organization of industry; and (3) to relieve the instructor of certain routine duties. The plan may be organized by the instructor or developed from within the group, but it should be started as soon as possible after the course begins. Success of the personnel system

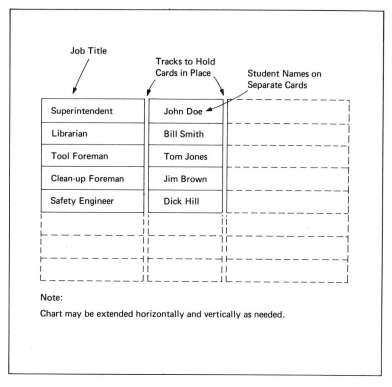

Fig. 12-4. The vertical-columns type of personnel chart.

depends on its being accepted by students as "their plan" and on complete cooperation among officers and the instructor.

It is important that the personnel organization not become merely a device to clean up the shop. Rather, the officers should be given selected responsibilities with commensurate authority. Offices may be filled by election from the class, appointed by the instructor, or by a system of promotion. Positions in the personnel organization should be rotated when the educational benefits have been exhausted. The clock-type and the vertical-columns type are two common devices to depict the personnel organization and to provide for rotation of duties and responsibilities.

DISCUSSION TOPICS AND ASSIGNMENTS

1. Make a complete bibliography of all articles dealing with the personnel organization which have appeared in professional journals during the past five years.

Laboratory Management—Personnel Organization

2. Make a detailed lesson plan in which you present and develop the need for a personnel plan to a seventh-grade group.

3. Compare the advantages and disadvantages of the teacher-made personnel plan with the student-developed plan.

4. Make a list of routine tasks which are sometimes performed by the teacher, but which could be effectively done by students under a personnel system.

5. Obtain from some industry an outline of its personnel system. Compare it with one of those presented in this chapter.

6. Plan a contrivance to show the personnel organization for a junior high school laboratory section of 18 students. Do not use the devices mentioned in this chapter.

SELECTED REFERENCES

Boone, James L. and Gus Baker. "Push Cleanup Panic Button," *Industrial Arts and Vocational Education*, vol. 55, no. 2 (February 1966), pp. 40–41.

Friedlander, David. "Put a Pupil Personnel Plan Into Action," *School Shop*, vol. XXVIII, no. 6 (September 1968), pp. 54, 56.

Greer, Arthur G. "Let's Organize the School Shop," *School Shop*, vol. XXI, no. 2 (October 1961), pp. 16–17.

Gumtow, E. M. "Quick Clean-Up Chart," *Industrial Arts and Vocational Education*, vol. 56, no. 3 (March 1967), p. 75.

Hoyt, Carlton W. "Organization in the High School Shop," *Industrial Arts and Vocational Education*, vol. 57, no. 3 (March 1968), pp. 69–72.

Husak, John. "Everybody Sooner or Later Is in Charge of Cleanup," *School Shop*, vol. XXIII, no. 2 (October 1963), p. 31.

Huss, William E. "Is Your Personnel Organization Democratic," *School Shop*, vol. XI, no. 9 (May 1952), pp. 7–8.

Marpet, Lee. "Rotating Shop Assignments the Easy Way," *School Shop*, vol. XXVII, no. 2 (October 1967), p. 71.

McKnight, Donald L. "The Value of Foreman Assignments," *School Shop*, vol. XXV, no. 3 (November 1965), p. 45.

Pelikan, Robert A. and Rolland C. Wolfe. "This Clean-up System Works," *Industrial Arts and Vocational Education*, vol. 54, no. 3 (March 1965), p. 81.

Pompeo, Anthony J. "Student Personnel Organization," *Industrial Arts and Vocational Education*, vol. 57, no. 3 (March 1968), pp. 73–74.

Rinehouse, James. "A Disc for Rotating Student Personnel in the Shop," *School Shop*, vol. XXVIII, no. 10 (June 1969), pp. 33–34.

Visca, Norman. "Motivating Students and Organizing Your Shop," *Industrial Arts and Vocational Education*, vol. 60, no. 6 (September 1971), pp. 28–29.

Chapter 13

Laboratory Management—Records and Record Keeping

The professional role of the industrial arts teacher has a number of facets. The teacher is responsible primarily for guiding the educational growth of his pupils, but he is responsible also for the proper management of a laboratory facility and for many thousands of dollars worth of machines, tools, equipment, and supplies. This stewardship carries with it the additional responsibility of accounting for that which is entrusted to his care. Since it is virtually impossible for the teacher to remember each and every one of the details connected with his daily work, it becomes necessary for him to resort to some form of written records.

A good system of shop records is an indispensable asset to the teacher. Educational records are valuable in the instructional process while management records facilitate the routine operation of the laboratory. This chapter will seek to do the following: (1) stress the importance of keeping records, (2) describe the general types of records useful in industrial arts, (3) discuss methods of record keeping, and (4) provide illustrations of typical record forms. In addition, the salient features to be considered in designing a records system for the industrial arts laboratory will be emphasized.

SCOPE OF THE PROBLEM

The scope of record keeping in industrial arts may be realized if one considers the great number of things which the teacher must keep track of or about which he may be called upon to furnish information. The industrial arts teacher may need to keep records of the following: lesson plans, unit plans, and courses of study; class attendance; personal data on pupils; quiz scores, examinations, and final grades; demonstrations and informative lessons given; homework assignments; student planning sheets and project grades; anecdotal notes of student behavior and work habits ratings; records of student and/or parent conferences; pupil progress charts; accident reports; pupil locker numbers; permission to leave the laboratory; requisitions for the purchase of tools, machines, equipment, and supplies; running and/or yearly inventories of tools, machines, equipment, and supplies; preventive maintenance check lists; parental permission slips to operate machines; student safety cards; requests for repairs or construction; personnel organization plans and student assignments; public relations records; loan records for library books, tools, and equipment borrowed; community resources survey forms; financial records of student payment for supplies; general laboratory fees; and safety inspection records. While the above listing of records may appear somewhat frightening, especially to the beginning teacher, it must be remembered that not all industrial arts teachers keep all of these records. The list simply represents the various kinds of records that have been kept by some teachers.

It is readily apparent that the problem of records and record keeping is compounded where the teacher has multi-sections of students at several different grade levels. This is true also in the comprehensive general shop where a group of students are busily engaged in several different materials areas simultaneously.

THE IMPORTANCE OF RECORDS

Laboratory records are important for the teacher for several reasons including: (1) inventorying and budget making, (2) assaying instructional progress, (3) making student evaluations, (4) managing the laboratory, and (5) providing evidence to protect himself from charges or criticism.

1. Inventorying and Budget Making

One of the most important reasons for maintaining shop records involves inventorying and budget making. Regardless of educational objectives, no industrial arts program can be any stronger than the financial support it is accorded. An efficient records system can provide the school administrator with the input necessary for making the industrial arts budget. A well-organized accounting system is often indicative of the kind of teacher and, perhaps, the quality of teaching going on in the laboratory. Records provide tangible evidence to help the administrator muster strong financial support for the industrial arts program.

2. Instructional Progress

Probably the most important function of records, when considered from all points of view, is the influence they have on the improvement of instruction. Records should provide a clear index of just what lessons have been taught, what demonstrations have been given, and what each student has achieved. Without such records no teacher carrying a full teaching load could be expected to recall exactly what he had taught to each section, to say nothing of what each student had accomplished or had failed to accomplish.

Continual study of the instructional records will assure the teacher that each student has received the essential instruction. These records will reveal student strengths as well as be helpful in diagnosing learning difficulties.

3. Student Evaluation

Records are important in student evaluation. Attendance records, behavorial analyses, test scores, grades for projects, extra work, and participation in the personnel organization should be recorded together with the final course grade. In a sense, records serve to protect the student. For example, if a student has done good work, the records should show it. There should be no question of one poor project or a poor quiz grade immediately before the end of the marking period having undue influence on the composite grade.

4. Laboratory Management

There are numerous forms and records which simplify the routine management of the industrial arts laboratory. For example, accurate inventory records of supplies will tend to minimize awkward situations of running out of stock and also will save much teacher time in requisitioning replenishments. Records are also useful in keeping track of library books and magazines as well as tools and other equipment which may be borrowed by students, custodians, and faculty members. Safety records and permission slips form an essential part of the safety program. Check lists facilitate regular safety inspections of machine tools and other equipment and are important in the preventative maintenance program. In short, *the key to successful laboratory management is organization* and records play a vital role in that organization.

5. Teacher Protection

Finally, it is important for the industrial arts teacher to keep records for personal reasons, that is, he keeps records for his own protection. First of all, the teacher needs to be protected against any question or inference in regard to financial matters of the laboratory. Records should show all money received and exactly what became of it. Secondly, the teacher needs to be protected against any charge of favoritism or poor judgment in assigning grades or marks. His records should be explicit enough to permit him to show any student or parent exactly why a certain grade was earned. Finally, the industrial arts teacher needs to be protected from any charge that certain subject matter may not have been taught adequately. This may become a significant factor in matters of safety instruction. Charges of negligence have been determined on the basis of whether or not adequate safety instruction had been given by the teacher.

GENERAL TYPES OF RECORDS

Probably each industrial arts teacher will develop a set of records to meet local conditions, but each record can be classified under one of the following three general categories: (1) administrative, (2) instructional or (3) financial.

ADMINISTRATIVE RECORDS

Under this heading are grouped those records dealing with the administration or management of the laboratory. Certain records of this type are often kept at the request of the school administration. Typical administrative records include the following: (1) attendance records, (2) inventory records, (3) requisitions, (4) safety records, and (5) loan records.

1. Attendance Records

While the attendance of students in school is usually noted in the homeroom, it is desirable and often required by the principal's office to check and record the attendance in each class. This is especially true if the industrial arts laboratory is located in a separate building or in a wing at some distance from other classrooms. More importantly, however, the industrial arts teacher will want to record absences and tardiness so that he will know who missed a class demonstration, informative lesson, or other learning activity.

As a manager of shop activities, the teacher will also want to control student departures from the laboratory. Students who have permission to leave the laboratory should sign out on a slip or chart near the entrance and sign in on their return. Such a form should require the student's name, destination, time of leaving, and time of re-entering the laboratory. The teacher should know at all times exactly what students are in his class and where are those who are not in the room at the moment. It is embarrassing for the teacher to be called upon by the office to find a particular student assigned to an industrial arts class and not know where he is or what he is doing.

2. Inventory Records

The teacher is responsible for the machines, tools, equipment, and supplies assigned to his laboratory. It is essential, therefore, that an accurate record be kept of what is currently on hand. It is impossible to anticipate when the need for an exact accounting will arise, but the prudent teacher will keep his inventory accurate and up-to-date at all times. Common practice in schools seems to be to require an annual inventory.

Records of Machines, Tools, and Equipment. Figure 13-1 shows one method for keeping a card inventory for machines. Similar inventory forms can be prepared for small hand tools, equipment (furniture, benches, cabinets, etc.), and portable machines.

Records of Supplies. In a doctoral study concerned with an analysis of inventory systems for expendable supplies in industrial arts, Etsweiler noted:

> Capital in the form of money and securities is usually very carefully guarded, while capital in the form of materials, equipment, tools, and supplies is rarely so carefully protected. Accounting of expendable supplies is often inefficient. Frequently there is little or no record as to when supplies were used, by whom they were used, and for what purpose they were used.[1]

In this investigation selected features of a good inventory system were validated by a nationwide jury of leaders in industrial arts and by another jury of school administrators. These criteria were then used to

---

[1] William H. Etsweiler, Jr., "An Analysis of Inventory Systems for Expendable Supplies in Industrial Arts, With Recommendations for a Simplified Perpetual Inventory System" (unpublished doctoral dissertation, The Pennsylvania State University, 1956), p. 1.

MACHINES

Name _____ Serial No. _____

Size or capacity _____ Mfg. _____

Cost _____ New or used _____ Date _____

If power, fill in following:

Make of motor _____ Type _____ Phase _____ Volts _____

Condition: (Code: N-new; G-good; U-usable)

1965	1966	1967	1968	1969	1970	1971
N	N	N	N	N	N	N
G	G	G	G	G	G	G
U	U	U	U	U	U	U

Side 1

Accessories (guards, attachments, tools, etc.)

Name	Make	Size	Inventory Remarks

Side 2

Fig. 13–1. An inventory card for shop machines.

analyze the inventory systems of 207 industrial arts teachers in Pennsylvania. Upon statistical analysis the data revealed the following twelve items to be most important to a simplified perpetual inventory system for expendable industrial arts supplies:

1. Provide for an annual inventory of supplies.
2. Classify all expendable supplies, i.e., hardwoods, ferrous metals, abrasives, welding supplies.

3. Furnish an index of all classified supplies.
4. Indicate quantity of each item to be ordered.
5. Include brief, but adequate, specifications with each item to be ordered.
6. Provide a current cost record per ordering unit.
7. Maintain a record of back-ordered items.
8. Provide a record of receipts of supplies.
9. Arrange in a manner so as to provide for inclusion of new classes of supplies.
10. Provide for subdivisions of all classified index material as necessary for individual items.

EXPENDABLE SUPPLY INVENTORY RECORD

Name of Item: Flint paper, extra fine   Code: A-5-a

Complete Specifications

Flint paper, extra fine
Ordering unit: 100 sheet package
Sheet size: 9" x 10"
Grit size: 4/0 - 220, 3/0 - 180

Cost Record

Date	Unit Cost
9/1/52	$1.30 / pkg.
9/2/54	$1.50 / pkg.

Annual Inventory	Ordered			Received		Back-Order Record	Annual Consumption
	Date	Quant.	Vend.	Date	Quant.		
6/1/52 10 pkg.	7/1/52	5	3	9/2/52	5	—	8 pkgs.
6/3/53 7 pkgs.	7/2/53	10	1	9/1/53	10	—	11 pkgs.
6/3/54	7/5/54	5	2	9/2/54	5	—	9 pkgs.
6/2/55	7/5/55	15	2	9/3/55	15	—	

VENDORS

No. 1 Brodhead Garrett Co.   Address: Cleveland, Ohio
No. 2 William Dixon   Address: Newark, N. J.
No. 3 Patterson Brothers   Address: 15 Park Row, N.Y.C.
No. 4 _____   Address: _____

Fig. 13–2. Expendable supply inventory record form after a 4-year period of service. (Courtesy, W. H. Etsweiler, Jr.)

Laboratory Management—Records and Record Keeping

11. Provide space for including names of reliable vendors of each class of items.

12. Include a record of dates when orders are placed.[2]

On the basis of these twelve criteria, Etsweiler developed an inventory form. Figure 13-2 illustrates this form after four years of service.[3] The illustration contains data relevant to one item of stock, namely, Flint Paper, Extra Fine, A-5-a. To satisfy criteria numbered 2, 3, 9, and 10 (see above), a classification plan was developed which is shown in Fig. 13-3. This code and identification scheme are needed for use with the expendable supply inventory record form (Fig. 13-2).

Major Group Headings for Expendable Supplies	Group Subheadings	
A - Abrasives	A - 1 Aluminum Oxide Paper	
B - Automotive Supplies	A - 2 Crocus Cloth	
C - Ceramics Supplies	A - 3 Emery Cloth	
D - Electrical Supplies	A - 4 Emery Polishing Paper	
E - Graphic Arts Supplies	A - 5 Flint Paper	
F - Finished Materials and Solvents	A - 6 Garnet Cabinet Paper	
G - Ferrous Metals	A - 7 Garnet Finishing Paper	
H - Nonferrous Metals	A - 8 Speed-Wet Garnet Paper	
I - Hardware	A - 9 Metalite Cloth	
J - Softwoods	A - 10 Speed-Wet Durite Paper	
K - Hardwoods	A - 11 Oscillating Sander Sheets	
L - Plywood and Veneer	A - 12 Cloth Belts for Portable Sanders	
M - Soldering Supplies	A - 13 Sanding Discs	
N - Welding Supplies	A - 14 Sanding Sleeves	
O - Drafting Supplies	A - 15 Pumice and Rotten Stone	
P - Plastics		
Q - Wood Fasteners		
R - Metal Fasteners		
S - Glues		
T - Lubricants	Subsubheading	
U - Art Metal Supplies		
V - Textiles		
W - Textile Supplies	Flint Paper, Extra fine (4/0, 3/0)	A-5-a
X - Leather Supplies	Flint Paper, Fine (2/0, 0)	A-5-b
Y -	Flint Paper, Medium (½, 1)	A-5-c
Z -	Flint Paper, Coarse (1½, 2)	A-5-d
AA -	Flint Paper, Extra Coarse (2½, 3)	A-5-e

Fig. 13-3. Code for use with the expendable inventory record form. (Courtesy, W. H. Etsweiler, Jr.)

3. Requisitions

Written requisitions for the purchase of tools, supplies, and equipment constitute another important category of laboratory records. In addition to quantity, price, and vendor, such forms usually contain

[2] *Ibid.*, pp. 82–83.
[3] *Ibid.*, p. 78.

details or specifications of the desired item. Most teachers like to retain a copy of all requisitions submitted to the department chairman, local administrator, or purchasing agent. To facilitate ordering, it is desirable for the inventory form to match the requisition form in content and sequence.

4. Safety Records

Records of demonstrations, lessons, and tests on the safe operation of tools and machines comprise a vital segment of the shop safety records. Reference has already been made to their possible use in refuting charges of negligence in the event the teacher becomes involved in legal action. In addition, parental permission slips are used in some schools. While these do not absolve the teacher of responsibility, they are useful in helping to control the use of power tools and machinery and in promoting safety consciouness on the part of the student.

If an accident does occur in the school laboratory, a report must be filed. The industrial arts teacher should make certain that the accident report form includes at least the minimum content as recommended by the National Safety Council. In regard to minimal content, the Council states:

> First and foremost, it is recognized that there is no one report form that will satisfy the needs of every school system. There is no one format that can be said to be better than any other. But there is a required body of information which is basic to the analysis and utilization of accident and injury data, if the information is to have any value for accident prevention purposes.[4]

The minimum data required for an accident report form as recommended by the National Safety Council includes the following items:

MINIMUM CONTENT FOR AN ACCIDENT REPORT[5]

1. Name
2. Address
3. School
4. Sex
5. Age
6. Grade/Special Program

---

[4] *Student Accident Reporting Guidebook* (Chicago, Ill.: National Safety Council, 1966), p. 8.
[5] *Ibid.*, p. 29.

Laboratory Management—Records and Record Keeping

7. Date and Time of Accident, Day of Week
8. Nature of Injury
9. Part of Body Injured
10. Degree of Injury
11. Number of Days Lost
12. Cause of Injury
13. Jurisdictional Classification of Accident
14. Location of Accident
15. Activity of Person
16. Status of Activity
17. Supervision
18. Agency Involved (apparatus, equipment, etc.)
19. Unsafe Act
20. Unsafe Mechanical-Physical Condition
21. Unsafe Personal Factor
22. Corrective Action Taken or Recommended
23. Property Damage
24. Description
25. Date of Report
26. Report Prepared by (signature)
27. Principal's Signature

Optional Data

As required by local school system: as applicable, information on first aid, doctor, hospital, notifications, insurance, witnesses, etc.

The National Safety Council has no standard format for the arrangement of these 27 items. However, the form should be simple, easy to read, easy to complete, and with sufficient room for supplying the data required. Figure 13–4 illustrates a sample arrangement of the recommended minimum content.

5. Loan Records

Faculty members, custodians, students and others frequently need to borrow items for which the industrial arts teacher is responsible. This includes hand and power tools, equipment, and library materials. The latter can be easily charged out "library style" if each book or periodical is equipped with a card and pocket. Other items to be used on a loan basis can be charged out on a ruled sheet showing item, date, and borrower's signature. It is prudent practice for the teacher to arrange a tentative date for the return of the item.

RECOMMENDED STANDARD STUDENT ACCIDENT REPORT

(check one)
- ☑ School Jurisdictional
- ☐ Non-School Jurisdictional

(check one)
- ☐ Recordable
- ☑ Reportable Only

School District: Ranier Valley
City, State: Ranier Valley, Pa.

General

1. **Name:** James Brown
2. **Address:** 102 Adams St., Ranier Valley
3. **School:** Reynolds Junior High
4. **Sex:** Male ☑ Female ☐
5. **Age:** 12
6. **Grade/Special Program:** 7th
7. **Time Accident Occurred**
   - Date: 1/14/70
   - Day of Week: Monday
   - Exact Time: 1:06
   - AM ☐ PM ☑

Injury

8. **Nature of Injury:** Burn
9. **Part of Body Injured:** Left hand, index finger/thumb
10. **Degree of Injury** (check one)
    - Death ☐
    - Permanent ☐
    - Temporary (lost time) ☐
    - Non-Disabling (no lost time) ☑
11. **Days Lost**
    - From School: —
    - From Activities Other Than School: —
    - Total: —
12. **Cause of Injury:** Touched hot soldering copper

Accident

13. **Accident Jurisdiction** (check one)
    - School: Grounds ☐ Building ☑
    - Non-School: Home ☐ Other ☐
    - To and From ☐
    - Other Activities Not on School Property ☐
14. **Location of Accident:** Soldering bench, I.A. Laboratory
15. **Activity of Person:** Soldering a tin box
16. **Status of Activity:** Regular lab. period
17. **Supervision** (if yes, give title & name of supervisor): Yes ☑ No ☐ — I.A. Teacher
18. **Agency Involved:** Hot soldering copper
19. **Unsafe Act:** Used equipment unsafely—copper slipped
20. **Unsafe Mechanical/Physical Condition:** None
21. **Unsafe Personal Factor:** None
22. **Corrective Action Taken or Recommended:** Retaught soldering demonstration; observed student at work
23. **Property Damage**
    - School $ —
    - Non-School $ —
    - Total $ —
24. **Description** (Give a word picture of the accident, explaining who, what, when, why and how):

    James Brown was soldering at the bench with two other students. As he moved along a seam with the hot soldering copper, he applied incorrect pressure. The copper slipped and made contact with his index finger and thumb of his right hand.

Signature

25. **Date of Report:** 1/15/70
26. **Report Prepared by** (signature & title): N. C. Pendy
27. **Principal's Signature:** J. B. Smith

This form is recommended for securing data for accident prevention and safety education. School districts may reproduce this form adding space for optional data. Reference: *Student Accident Reporting Guidebook*, National Safety Council, 425 N. Michigan Avenue, Chicago, Illinois 60611. 1966. 34 pages.

Fig. 13–4. Recommended standard student accident report. (Courtesy, National Safety Council)

INSTRUCTIONAL RECORDS

Every teacher recognizes the need for records which will define clearly the status and progress of instruction. Some of the more important types of these records include the following: (1) informative lessons, (2) demonstrations, (3) tests, (4) progress chart, (5) projects, and (6) final grades.

1. Informative Lessons

The teacher should maintain a record of the informative lessons which have been taught. It is particularly important that the record show not only what lessons, but also the sections or classes to whom the lessons were given. The reason for this is apparent. In a busy school schedule it often happens that one or more sections of a given grade do not appear for industrial arts at the assigned period because of some rearrangement in the schedule. This may be due to a special assembly, a class meeting, a field trip for another class, or a hundred and one other reasons. Unless an accurate record of lessons is kept, it is easy for the teacher to forget that a certain lesson was missed by a class. He may then hold the students responsible for instruction which they never received.

2. Demonstrations

All that has been said concerning the necessity for a record of informative lessons holds equally true for class demonstrations. It is especially important where matters of safety instruction are concerned.

3. Test Records

A record of quizzes, tests, and examinations with the grades assigned is requisite to later evaluation of the student's total accomplishment.

4. Progress Chart

A progress chart is a cumulative record which shows the status of an individual as he works toward the completion of the requirements of a course. The traditional progress chart consists of a list of manipulative operations to be performed as well as the names of each student in the class. As a student performs an operation, it is "checked off." By this means both teacher and student can see what has been done and what remains to be completed (see Fig. 13–5). The progress chart appears to have been inherited from vocational education along with the trade analysis technique for selecting subject matter.

Sometimes progress charts of large size are posted on the bulletin board. Instead of wall charts, some teachers prefer to keep the same information on notebook-size paper. The record of an entire class can be kept on a single sheet or a separate page for each student may be maintained. In the latter case an individual progress record of accomplishment can perform a useful function if forwarded to the industrial arts teacher in another school if the student should transfer. Or, such a record is useful if the student continues for a second or third year in the same unit-type shop.

There is considerable difference of opinion on the effectiveness of the progress chart for industrial arts classes. Some teachers hold that such visible evidence of accomplishment tends to stimulate the slower students to greater effort in order to keep up with the group. Others contend that students with less ability become discouraged and give up trying. There appears to be no objective evidence to support either of these views. The use of progress charts thus becomes a matter of teacher judgment.

Another criticism of the conventional progress chart is that it places too much emphasis on the mere performance of manipulative processes. This is a logical objection to the type commonly used today. It is possible, however, to construct a progress chart on the basis of desirable behavior changes that were outlined in Chapter 6. An example of how this might be done is shown in Fig. 13–6. Such a chart would tend to place the emphasis where it logically belongs and, at the same time, keep before the class the objectives of the course. It is readily conceded, however, that it is much more difficult to obtain objective evidence on the attainment of behavior changes than evidence on the performance of a manipulative process. Nevertheless, if the industrial arts teacher is sincere in his desire to measure growth toward his objectives, he must be willing to prepare and use some such record. Whether or not it will be in the form of a progress chart is a matter that each teacher must decide for himself.

5. Projects

While projects are not considered as ends in themselves, but rather as a means toward an end, nevertheless, the character and number of projects completed are probably some indication of the extent to which some of the objectives of the course are being attained. After all, unless

ACTIVITIES	WOODWORKING																					
DIVISIONS	LAYOUT									SAWING							PLANING				
PROCESSES AND OPERATIONS	Layout from Drawing	Layout Pattern on Stock	Read a Rule	Test with a Try-Square	Pencil Gauge a Line	Layout with Marking Gauge	Layout with Dividers	Layout with T-Bevel	Layout with Knife	Crosscut to a Line	Rip to a Line	Fine Cuts with Back Saw	Saw Curves with Coping Saw	Cut Angles in Miter Box	Cut Irregular Shapes on Jig Saw	Smooth Cuts with Cab File	Plane a Surface Square	Plane an Edge	Plane an End Grain	Shape with Spokeshave	Plane with Router Plane
1 Anthony, Charles		✓	✓	✓	✓					✓					✓	✓	✓	✓	✓		
2 Bareham, Donald			✓	✓						✓					✓	✓	✓	✓	✓		
3 Carter, David						✓					✓										
4 Hellard, Paul																					
5 Johns, Robert		✓	✓	✓	✓	✓		✓	✓	✓	✓	✓	✓					✓	✓		
6 Osmond, Robert		✓	✓	✓	✓			✓		✓	✓		✓				✓	✓	✓		
7 Moss, Donald		✓	✓	✓	✓					✓				✓				✓	✓		
8 Scott, James	✓	✓	✓	✓	✓					✓								✓	✓		

Fig. 13-5. A common type of progress chart.

COMPREHENSIVE GENERAL SHOP GRADE 7B	RECREATIONAL AND AVOCATIONAL											APPRECIATION OF CRAFTSMANSHIP AND DESIGN							
	Ask Advice on Constructive Activities	Interested and Engages in Constructional Hobbies	Spends Spare Time in Shop at School or at Home	Reads Magazines, such as Popular Mechanics, Popular Science	Asks Questions and Talks about Hobbies	Consults Catalogs for Information about Hobbies	Contributes to Class Discussions from Reading along Lines of Interests	Takes Initiative in Visiting Industries	Develops Home Workshop			Recognizes Good Design and Applies it in Construction of Projects	Shows Appreciation of Design in Artifacts by Speech and Actions	Recognizes and Appreciates Period Pieces	Uses "Streamlining" Correctly in Development of Projects	Redesigns Projects to Improve Appearance and Utility	Selects and Develops Projects Suitable to Material Used	Recognizes and Avoids Poor Design and "Over Decoration"	
Albert, Howard																			
Aford, Ben																			
Barnes, William																			
Buntz, Herman																			
Carl, Robert																			
Conway, Arthur																			
Crossman, Daniel																			
Culver, Neal																			
Dale, Arthur																			
Dare, John																			
Doxie, Harvey																			
Dunn, Arthur																			
Eaton, Harold																			
French, John																			
Fuller, Kenneth																			
Hill, James																			

Fig. 13–6. A suggested type of progress chart based on behavior changes.

projects contribute to the objectives of the course, they should not be constructed by the pupils. It is expected, therefore, that some record of projects will be kept. Such a record probably should indicate the name of the project, the degree of skill exhibited, and the amount of time required for completion. Also, it may be desirable to record the materials cost and whether the student has paid (if a charge is to be made).

Another phase of record keeping closely related to the project itself involves the instructor's checking of student planning sheets. A separate record or chart may not necessarily be required; the instructor's initials and date at the check point may suffice. Or, this information can be recorded with the project information.

6. Final Grades

While final grades are usually recorded on report cards or individual grade slips, it is desirable for the teacher to keep his own record for later reference. Experience has demonstrated that one can never tell when or for what purpose these grades will be required.

FINANCIAL RECORDS

Teachers should be particularly careful in keeping accurate records of all financial matters related to the industrial arts laboratory. Frequently, a teacher's reputation and standing in the school and the community may depend on his ability to demonstrate clearly the disposition of funds which have been collected for projects or materials. Whatever else a teacher may do or leave undone, his record of financial dealings should be clear and accurate. All money received should be immediately recorded and, if possible, the entry should be made in the presence of the student from whom the money is collected. When any money goes into the teacher's pocket or cash box, there may be doubt in the minds of some students as to whether it actually is turned over to the proper fund.

It is better for the teacher not to handle any student money whatever. Arrangements can usually be made for the office to accept incoming monies to the credit of the industrial arts account. There are several ways to collect money from students for general laboratory fees, projectwork, or extra materials used. One popular method is for the student to purchase from the office a "Materials Ticket" for several dollars (see Fig. 13-7). Denomination marks from one cent to twenty-five cents or a dollar are printed along the edges of the card. When the student "pays" for materials used, the teacher punches out the proper denomination marks with a hand punch. The card may be kept on file in the laboratory or retained by the student. To encourage the use of "Materials Cards" rather than cash for each transaction, some schools give the students a bonus by selling a card worth, say $3.50 for $3.00

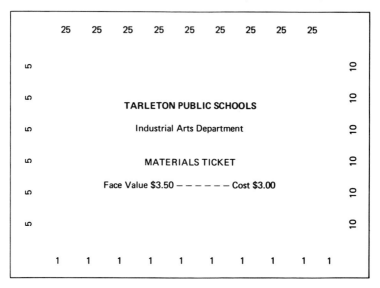

Fig. 13-7. Sample Materials Ticket.

Another method involves triplicate copies of a material's bill. The student fills out the bill, gains the instructor's approval, and then takes the bill to the office with his payment. The office receipts the bill and retains one copy. The student keeps his copy and gives the instructor a copy for his records. Pressure-sensitive pads with interleaves of three different colors are useful for this purpose. Or a printed card of appropriate size can be perforated into three sections.

DESIGNING A RECORDS SYSTEM

1. Kinds of Records Needed

Before preparing any forms or charts for a records system, the teacher will do well to analyze his needs carefully. He should list in this order: (1) information required by the school, (2) essential information, that is, minimal data necessary to carry on the program, and (3) desirable information or nice-to-have data. Records are likely to be effective and accurate to the extent to which they can be kept without excessive expenditure of time and effort. The number and character of the records becomes, therefore, a compromise between the ultimate that would be desirable and the number that can be kept accurately and up-to-date. The teacher should bear in mind it is better to maintain a minimum number of well-kept records than to plan such an ambitious system that it eventually falls of its own weight. As a general rule it may be stated that any record system is inefficient if it requires more than one half-hour of the teacher's time each day. Such a system should be examined critically and either refined or discarded for a more efficient one. A records system should be the servant of the teacher and not his master.

2. Longevity of Records

After selecting the kinds of information to be recorded for the program, the teacher should consider the longevity of each record needed. Some data, such as inventory records on tools, machines, and equipment will need to be both cumulative and permanent. On the other hand instructional records may have only semi-permanent or transient value. Regardless of the nature of the permanency of its records, however, any system should be characterized by *flexibility,* that is, provision should be made for the addition of new items and entries and the deletion of obsolete ones without disruption to the system.

3. Closed versus Open Records

Another factor to consider in designing a records system is for the teacher to decide which information will be closed or private and which may be open for pupil inspection. Relative to this decision is one concerning the personnel who will maintain a given record. The pupil personnel system offers fine opportunities to help students develop in responsibility and in leadership qualities. So long as educational benefits accrue to students, the teacher is justified in utilizing their services in record keeping. Specific suggestions for utilizing the student personnel organization will be offered later in this chapter.

4. Mode of Response

The mode of recording pertinent data is an important factor in designing a records system and merits the teacher's full consideration. Written responses should be brief and concise. Objective measures may be recorded by a check mark or a punched hole. The latter is useful in that it provides a measure of control in an open-records system by the one responsible for keeping the record.

5. Record Format

The format for each record form is quite important. The teacher must decide whether to use a large wall or desk chart or whether single notebook pages or cards will suffice. He will have to decide, too, which records are to be individualized and which can be kept for a class or group of objects. In any event the teacher should give thought to combining several records on one form. Whenever this can be accomplished, time and effort in handling data will be saved. Figure 13–8 shows how the records for one student may be kept on a 5 × 8-inch card. Instead of cards some teachers may prefer to keep similar information for the entire class on a sheet of notebook paper.

BLANKSVILLE HIGH SCHOOL

Industrial Arts Department
Individual Record

Name ______________ Grade ________ Section ______ No. ______

Projects:

	Grade	Begun	Finished	Cost	Pd.
1					
2					
3					
4					
5					
6					
7					
				Total	

Note Book
1st quarter
2nd "
3rd "
4th "

Test Marks

Date		Date	
1		4	
2		5	
3		6	

Outside Preparation

	Job	Time						
1			5			9		
2			6			10		
3			7			11		
4			8			12		

Fig. 13–8. A comprehensive record blank.

6. Record Storage

Another consideration in planning a records system is the matter of storing and housing the records. A limiting factor of immediate concern is the size of the card or sheet containing the data. More important, the teacher should give top priority to his own preferences. He may well consider the possibilities of housing records by asking himself these and other pertinent questions: (1) Should I keep individualized records in manila folders in a file cabinet? (2) Or is the same information better recorded on small cards and stored in a card file box? (3) Should I adapt my class or roll book or should I prepare duplicate forms for each section or class? Any solution to the housing of records should, of course, always be in terms of easy accessibility.

7. Record Preparation

After all factors have been weighed in terms of the needs of the teacher, thought should be given to the preparation of the forms themselves. Hand-lettered charts can be attractive, but unless well done, some other means of presentation should be considered. Besides, such charts usually represent a time-consuming effort for the teacher. The virtues of the common printing and duplicating processes should be tempered by consideration of cost and the equipment which may be readily available to the teacher in the industrial arts department.

KEEPING RECORDS

Record keeping can be either a pleasant task or a disagreeable burden depending upon the nature of the teacher himself and how well he has organized this particular responsibility. Some teachers regard record keeping as a necessary evil. They feel that at best it is drudgery which does not merit serious attention. But, to the conscientious teacher, some form of record keeping is regarded not only as desirable, but as absolutely essential to the operation of a successful program of industrial arts.

The pupil personnel organization can be utilized to great advantage in record keeping. For example, the superintendent can be given the authority in certain financial matters, such as punching out the costs of a project on a student's "Materials Ticket." A records clerk may be made responsible for such matters as class attendance records and the recording of related lessons and demonstrations given to his section. In some cases an officer of the personnel organization or even the student concerned can be charged with the responsibility of keeping a progress chart up-to-date (if one is used). If a progress chart based on behavior changes were prepared (see Fig. 13-6), this would certainly give students an opportunity for critical self-evaluation which would, perhaps, be as valuable as the record itself. The librarian and his assistant can contribute significantly to the record-keeping system by checking shelves and by charging out and receiving returned library books and periodicals, audio tapes, film loops, and the like. The safety engineer can complete routine safety inspections of tools and machines (see Fig. 13-9). Of course, the teacher will need to supervise these activities, but the burden can be placed on the students themselves. The maintenance of laboratory records can provide excellent opportunities for members of the student personnel system to assume responsibility and to develop leadership qualities as well as other desirable personal traits. It goes without saying, of course, that the teacher should insist all forms have neat entries and be kept up-to-date lest carelessness breed the impression of unimportance.

Fig. 13-9. This teacher is showing the Safety Engineer of the student personnel organization how to record his routine safety inspections of power tools. (Courtesy, California State Department of Education)

SUMMARY

The industrial arts teacher plays a dual professional role by serving both as teacher and manager of the laboratory. Records are necessary in the operation of a successful program. They are valuable in (1) budget making, (2) noting the progress of instruction and in student evaluation, (3) facilitating the daily routine management of the laboratory, and (4) protecting the teacher against charges or implications.

Laboratory records are many and varied, but each can be classified under one of the following general categories: (1) administrative, (2) instructional, or (3) financial. *Administrative records* include (1) attendance, (2) inventory records for machines, tools, equipment, and supplies, (3) requisitions, (4) safety, and (5) loan records. For *instructional records* the teacher needs to record data on (1) informative lessons, (2) demonstrations, (3) tests, (4) student progress (manipulative or behavior change), (5) projects, and (6) final grades. *Financial records* include general laboratory fees and charges for supplies and materials.

Methods for keeping laboratory records efficiently were discussed with emphasis on the pupil personnel organization. Sample forms were illustrated. Attention was given to factors to be considered in designing a records system including the following: (1)kinds of records needed, (2) record longevity, (3) closed versus open records, (4) mode of response, (5) record format, (6) record storage, and (7) record preparation.

DISCUSSION TOPICS AND ASSIGNMENTS

1. Why should accurate and up-to-date records be kept by the industrial arts teacher?

2. Make as complete a list as possible of all records which an industrial arts teacher might keep. Underline those you feel are essential.

3. For the list (above) indicate those records which could be kept by members of the student personnel organization.

4. Design an instructional record card for a pupil to include as much essential information as desirable.

5. In parallel columns compare the advantages and disadvantages of progress charts.

6. Prepare a progress chart based on a listing of behavior changes for any selected objective of industrial arts.

7. What factors should be considered in designing a records system for an industrial arts laboratory?

SELECTED REFERENCES

Blegen, August H. *Records Management—Step-by-Step.* Stamford, Conn.: Office Publications, Inc., 1966.

Broadbent, V.E. "Improved Check-Off Record," *Industrial Arts and Vocational Education*, vol. 54, no. 2 (February 1965), pp. 40–41.

Dean, C. Thomas. "Records and Management," *Industrial Arts and Vocational Education,* vol. 58, no. 3 (March 1969), pp. 76–81.
Dean, C. Thomas and Irvin T. Lathrop. "Specifications and Requisitions," *Industrial Arts and Vocational Education,* vol. 53, no. 3 (March 1964), pp. 66–67.
Dudley, Arthur J. "On Keeping Management and Educational Records," *Industrial Arts and Vocational Education,* vol. 52, no. 3 (March 1963), pp. 66–70.
Etsweiler, Jr., William H., "An Analysis of Inventory Systems for Expendable Supplies in Industrial Arts, with Recommendations for a Simplified Perpetual Inventory System" (unpublished doctoral dissertation, The Pennsylvania State University, University Park, 1956).
Hoyt, Carlton W. "Organization in the High School Shop," *Industrial Arts and Vocational Education,* vol. 57, no. 3 (March 1968), pp. 69–72.
Jones, Donald R. "The Industrial Arts Bank," *Industrial Arts and Vocational Education,* vol. 55, no. 3 (March 1966), p. 88.
Jones, James H. "Student Safety Test Card," *Industrial Arts and Vocational Education,* vol. 54, no. 5 (May 1965), p. 54.
Kassay, John A. "Maintenance Checklist for the Wood Laboratory," *Industrial Arts and Vocational Education,* vol. 54, no. 3 (March 1965), pp. 73–74.
Krueger, Frederic. "A Cash Receipt Designed with the School Shop in Mind," *School Shop,* vol. XXVIII, no. 5 (January 1969), p. 52.
McLoney, L. Jason. "A New Student Progress Sheet," *Industrial Arts and Vocational Education,* vol. 54, no. 3 (March 1965), pp. 82–83.
Murphy, Jr., John O. "Grade Card System," *Industrial Arts and Vocational Education,* vol. 57, no. 3 (March 1968), pp. 74, 76, 78, 80.
O'Brian, John L. "A System for Inventorying and Budget-Making," *School Shop,* vol. XXVI, no. 6 (February 1967), pp. 48–49, 71.
Organization and Administration Forms for Industrial Arts, Series 1000, Instructional Materials Service, P.O. Box 244, Kennedy, Minn. 56783.
Silvius, G. Harold and Estell M. Curry. *Teaching Successfully in Industrial Education.* Bloomington, Ill.: McKnight and McKnight Publishing Company, 1967.
_____. *Managing Multiple Activities in Industrial Education.* Bloomington, Ill.: McKnight and McKnight Publishing Company, 1971.

Chapter 14

Instructional Media—Part I

In times past the typical teacher was content to use a single instructional aid, such as the chalkboard or perhaps an object or a specimen to illustrate his presentations. Sometimes an occasional motion-picture film or filmstrip was employed. Later other instructional aids were fitted into the lessons, but generally they were used to supplement or enrich the teacher's presentation to the class. Today as a result of better understanding of the learning process, our thinking concerning the role of sensory aids to learning has changed radically. The terminology is changing. New terms are appearing in the literature, such as, instructional or educational media, polysensory learning, cross-media kits, multimedia approach, total self-instructional system, and a host of others. Underlying this changing terminology is a more basic change involving our concept of audio-visual instruction and its relationship to the learning process. It seems that three main thrusts appear to be slowly emerging. These are (1) the concept of polysensory learning through an integrated multimedia approach, (2) total reliance on portions of the "instructional system" for the achievement of clearly-defined educational objects, and (3) growing emphasis on student responsibility for his own learning through self-paced and even self-instruction.

1. Polysensory Learning through a Multimedia Approach

As the term suggests, polysensory learning involves the use of a variety of human senses in the learning process. The sensory organs are the portals of human learning. It is only through these senses that learning stimuli can be received. Of the five senses, most learning (approximately 88 percent) comes through the combined senses of sight and hearing. It is said that 75 percent of what we learn comes through the sense of sight; about 13 percent comes through hearing; about 6 percent through the senses of smelling and tasting; and about 6 percent through the sense of touch. Since one sense tends to complement the others, the idea in polysensory learning is to utilize as many as possible of these avenues of learning to gain the skills, understandings, attitudes, and appreciations deemed desirable. To attain this end the teacher utilizes a multimedia approach, that is, a variety of appropriately-selected media is employed, such as single-concept film loops, overhead projections, aural recordings, closed-circuit television, and video tape recordings, to mention only a few. Since most humans are visually oriented, it is not surprising to discover that a multimedia approach is also a highly visual approach.

The multimedia approach does not mean necessarily that selected sensory aids are used one at a time. It may well be that in special instances the teacher may utilize twin slide projectors projecting dual images on one large screen or on separate screens with an aurally-recorded description. Or, it may be that a wide-screen presentation is needed. This may involve three individual images projected with three slide projectors on a wide (9×21-ft) screen. The resulting multiple images blend together on the screen to form a continuous panorama. Or, to achieve the desired end, the teacher may make simultaneous use of a slide projector and an overhead projector accompanied by an audio narration via tape recorder or audio cassette. The point is that the multimedia approach does not limit the teacher to sequential use of educational hardware. The possibilities of effective combinations of audio-visual devices seems to be limited only by the teaching situation and the imagination of the teacher.

It is important to recognize that the multimedia approach does not use sensory aids randomly, haphazardly, nor simply because they are available. The mere manipulation of educational hardware does not constitute a true multimedia approach to education. Rather, each media is selected because of its distinct contribution. When used in combination, selected media tend to reinforce each other and promote greater effectivity. Thus, the teacher plans the instructional unit carefully and fuses together the vehicles of instruction into an integrated whole—the purpose of which is to effect desirable behavior changes and thus reach educational objectives previously established.

2. Reliance on the Instructional System

An instructional system usually incorporates self-paced or self-instruction where an individual moves through planned learning activities at his own speed. These activities may involve exposure to several educational media designed to elicit student response as a measure of achievement or behavioral change. An instructional system is pupil-centered—that is, attention is focused not on teacher activities, but on the responses of the student and their immediate reinforcement.

Under such an instructional strategy the role of the teacher changes. No longer is he on center stage; instead, he stands ready to motivate learners, diagnose learning difficulties, evaluate behavorial changes, and plan new learning activities. It is important for the teacher to have confidence in the instructional system and refrain from "interfering" during the learning period. It has been shown on numerous occasions that certain media are effective and reliable in helping students attain selected desirable behavorial patterns. One research finding (see Chapter 15) recommends that the teacher let the film do the teaching. In addition to motion-picture films, programed instruction and slide-tapes are among those instructional media which have been used effectively to teach certain skills, knowledges, attitudes, and appreciations.

3. Student Responsibility for Learning

This new approach to the learning process via self-paced instruction, independent study, and self-instruction results in faster, greater learning which is more meaningful and more interesting to the student. It implies, too, that students should be encouraged to learn *where* and *when* they can best do so. In other words, in the future more learning probably will take place outside the school. But even today many media, such as programed texts, audio cassettes, and even lightweight, portable, battery-powered videotape recorders may be carried home for perusal by the student. While this new approach to learning does utilize hardware, it is not intended that its use will replace the teacher nor lessen his responsibility. It will, however, provide some additional time for planning instructional strategies, for diagnosing individual learning difficulties, for working both with individuals and small groups, and for evaluating behavior changes.

It must be admitted that the foregoing discussion reflects current thinking as supported by innovative practices on a limited scale. While this may well be indicative of things to come, only the future will reveal whether these ideas will ever become a part of the daily routine of the average teacher. It is certain, however, that as educational technology advances, teachers will endeavor to keep up-to-date and to implement new ideas as their situation and community support permit.

Audio-visual sensory aids are the building blocks which are needed for polysensory instruction through a multi-media approach. They are needed for instructional systems or strategies wherein the student becomes more responsible for his own learning. This chapter and the one to follow seek to describe these building blocks or audio-visual sensory aids, leaving to the teacher the decision on how to use them in the learning activities he plans for his students. This chapter will consider in turn the following instructional media: (1) industrial visits; (2) demonstrations; (3) tape recorders and audio cassettes; (4) objects and specimens; (5) models and mock-ups; (6) project charts and process charts; (7) bulletin boards, chalkboards, and flannelboards; (8) free and inexpensive materials. Other major types of educational media dealing mostly with projected materials or based on photographic processes will be discussed in Chapter 15.

INSTRUCTIONAL MEDIA IN INDUSTRIAL ARTS

Industrial arts teachers are in an enviable position in being able to utilize in the learning process the senses of sight, touch, hearing, smelling, and on rare occasions, the sense of taste. It is not surprising, therefore, to discover that these instructors always have sought to capitalize upon this inherent advantage of industrial arts education. It is common to find school laboratories rich with audio-visual materials, but more important, to note that they are being used in everyday instruction. The literature of the field is replete with descriptions, plans, and photographs of demonstration equipment and other audio-visual aids especially adapted for improving industrial arts instruction.

The industrial arts teacher has found many ways to improve his teaching through the use of audio-visual media. Among the more important contributions which such educational media can make to industrial arts are the following:

1. Aid to Industrial Exploration and Orientation

Industry and technology must be made real and vital to youth. Words alone cannot accomplish this. Every audio-visual device at the teacher's command must be pressed into service to make clear just what is meant by industry—its raw materials, workers, processes, organization, and finished products. Verbalism and abstractions must be eliminated or reduced to a minimum. The student also needs help in bridging the gap between the processes as they are performed in the school laboratory and methods used in industry. To develop industrial concepts and understandings, a variety of sensory aids must be employed in the learning process.

2. Aid in Providing Information

If a program of industrial arts is to achieve its objectives, it must be much more than mere manipulative work alone. An extensive body of information must always accompany the manipulative activities because both comprise the learning activities. The informative aspects should cover not only the exploratory phases of industry (as indicated in the preceding heading), but should also include (1) important concepts concerning the evolution and development of industry; (2) applications of science and mathematics to industrial processes; (3) appreciation of design in manufactured products; (4) consumer information as applied to the purchase, use, and maintenance of products; and (5) recreational aspects of craft and industrial activities. Much of this content will need to be illustrated by means of audio-visual aids. The use of a wide and careful selection of instructional media, for the purpose of vitalizing and motivating the informative learning activities, is one of the most effective methods for maintaining student interest in this important phase of the industrial arts program.

3. Aid in the Teaching of Skills

Visual aids of some type are almost indispensable in the teaching of skills and manipulative activities. It is practically impossible to learn how to perform a series of operations without reference to audio-visual aids, such as pictures, diagrams, demonstrations, and films. The extent to which the teacher relies on such devices may be realized if one attempts to describe verbally how to perform even a simple manipulative act. Research findings over the years coupled with the widespread adoption of audio-visual materials in the training programs of the military service indicate their effectiveness in the teaching of skills. Since audio-visual aids are of such importance in the teaching of skill, they should be used even more than they now are in industrial arts. The use of newer educational media, such as programed instruction sheets, teaching machines, and 8-mm single-concept films point the way to improved audio-visual methods for teaching skills.

4. Aid in Creating Atmosphere

Another function of instructional media which may not directly affect instruction, but which certainly has an important indirect effect, is their use to create a favorable learning atmosphere in the laboratory. In particular, the development of the ability to think which is the central purpose of industrial arts education requires a setting that is conducive to thinking. It is possible to transform a bare and unattractive room into a stimulating environment by the use of color and attractive sensory aids, such as pictures, charts, models, and mock-ups. Materials of this type tend to induce an atmosphere for work, study, and experimentation. There is some evidence which seems to suggest a relationship between the atmosphere of the school laboratory and the quality of learning which takes place there.

TYPES OF INSTRUCTIONAL MEDIA

Today more than ever before, the industrial arts teacher has available a wide array of instructional media. Research and development are continuously developing new media as well as improving the old. Several types of instructional media that can be used by the industrial arts teacher have already been mentioned. These and other aids to instruction will be considered in detail under the following headings: (1) industrial visits; (2) demonstrations; (3) tape recorders and audio casettes; (4) objects and specimens; (5) models and mock-ups; (6) project charts and process charts; (7) bulletin boards, chalkboards, and flannelboards; and (8) free and inexpensive materials.

1. Industrial Visits

The industrial trip is one of the most valuable forms of instructional media utilized in industrial arts education. It is especially effective because of its realism. The student is not confronted with a picture or a verbal representation of industry, but observes the real thing. His impressions are first-hand; he hears, smells, and sees industry and technology at work. Depending on the nature of the trip, his senses of touch and taste may be involved as well.

It has been stated previously that casual observation of industrial processes by individuals (especially youth) is generally discouraged by industry today. However, many industrial plants are receptive to teacher requests for visits by their shop classes. Some companies regard industrial visits as an important facet of their public relations programs and go to great lengths to make their tours interesting and educationally beneficial. Such firms often provide special tours with trained guides who distribute attractive, worthwhile literature; sometimes they offer specimen sets or samples of their product and occasionally provide lunches in the plant cafeteria. Figure 14–1 shows a group of school children on an industrial visit to an automobile plant.

Research findings either supporting or refuting the educational val-

Instructional Media—Part I

Fig. 14-1. School pupils riding a tour train. (Courtesy, Pontiac Motor Division, General Motors Corporation)

ues of the field trip seem conspicuous by their absence from the professional literature. However, one research study by Hillson, Wylie, and Wolfensberger[1] found that a field trip experience did effect a small, but highly significant, change in the direction of a more favorable attitude of a group of college students as measured by a standardized attitude scale.

Then in 1969 Goldsbury made a study of field trips taken vicariously through slide-tapes.[2] The investigation involved 251 third-grade students who were divided into three groups. One group experienced a field trip directly; the second group experienced the same field trip vicariously through a slide-tape presentation; the third group saw the slide-tape presentation and then experienced the field trip. Parallel forms of a test covering facts, concepts, and attitudes were administered in pre- and posttesting.

The findings revealed that the combination of slide-tape and direct experience was the most effective of the three approaches used. Vicarious field trips (slide-tape presentations) proved more effective than the

[1] Joseph S. Hillson, Alexander A. Wylie, and Wolf P. Wolfensberger. "The Field Trip as a Supplement to Teaching: An Experimental Study," *Journal of Educational Research*, vol. 53, no. 1 (September 1959), pp. 19-22.
[2] Joseph W. Goldsbury, "A Feasibility Study of Local Field Trips Taken Vicariously Through Slide-Tapes" (unpublished doctoral dissertation, The Ohio State University, 1969).

direct experience of the field trip itself. These results suggest that teachers may improve the educational effectiveness of a field trip through the use of a slide-tape presentation as a supplement to the actual visit. The implication is, too, that vicarious field trips via slide-tapes may be a good substitute, at least in some cases, for real field trips.

Before considering any industrial visit and especially before arousing the interest of the class, the teacher should be sure the trip has a definite purpose and is an important learning activity in the unit being studied. He should compare the educational benefits with the costs of the visit. The latter includes both monetary costs as well as time losses by pupils if other classes are missed. The teacher should be convinced the proposed visit is the best way to achieve the desired results, that is, no other educational media will do as well or better.

When properly planned and carried through, the industrial visit (or plant visitation) represents the ultimate in effectiveness for industrial exploration and orientation. Its value will depend chiefly on how well the class has been prepared for the trip and the extent to which experiences and impressions are discussed after the excursion is completed.

Trips to industries should never be taken on the "spur of the moment." Careful thought should be given to the following details: (1) timeliness of the visit in terms of the unit of instruction, (2) approval of the school administration, (3) consent of the parents of the pupils involved, (4) provision for those pupils in the class who cannot make the trip for one reason or another, (5) plans for adequate adult supervision from start to finish, (6) detailed plans for reaching the plant, especially if bus transportation is involved, (7) pre and postvisit assembly (if the class is divided into small groups), (8) advance preparation of the students for what they will see and hear during the visit, and (9) plans for a class lesson and discussion upon returning to school.

If possible, the instructor should visit the industrial plant beforehand and determine what the students are likely to see. He should then discuss with them the points to be observed. At this time he should acquaint the class with the details of transportation, assembly instructions, safety rules to be observed, and expected pupil conduct during the tour. The need for good conduct should be stressed because the class, the teacher, and the school itself will be judged by the actions of individual pupils while on the tour. During this preparatory session the students should be urged to ask questions while on the tour about any phase of the work which is not fully understood.

If the industry to be visited is large, it may be advisable to divide the class into sections and to make each group responsible for observing some particular phase or process. In this case an adult or a student leader should be appointed for each section or group. Instruction should be given for groups to stay together and the necessary safety rules should be stressed.

From an educational point of view, the most important part of an industrial trip is the class discussion which follows. Such topics as the organization of the industry, what is manufactured, what occupations are involved, what raw materials are used, what finished products are produced, and the social and economic contributions of the industry to the community should be considered. If the class has been divided into groups with special assignments, each group should make its report to the class.

Although the term *industrial visits* has been used in this section, it is to be understood that other field trips may be equally appropriate. Such visits could include museums, construction sites, garages, photographic labs, and other places where skills may be observed and information gained.

It is recommended that the industrial arts teacher survey the locality to ascertain what industries, firms, museums, etc., are available. A suggested form for this purpose is shown in Fig. 16–4 (see Chapter 16). Such a survey may well reveal that a number of worthwhile sites lie within walking distance of the school and may be visited within the class period.

2. Demonstrations

The demonstration as an industrial arts teaching method has been discussed fully in an earlier chapter (see Chapter 10). Included also was a discussion of certain of the newer instructional media, such as, closed-circuit television, programed instruction, and the Audio-Graphic teaching machine, as these have been used experimentally in pilot programs to teach manipulative skills in industrial arts. The demonstration is mentioned here simply to emphasize the fact that it is an instructional medium of first importance. It ranks with the industrial visit in the matter of realism, since real materials, real tools, and actual operations are presented for student instruction. Figure 14–2 depicts a demonstration of an important industrial process in the metals laboratory.

Since the major value of the demonstration depends on its presentation through the audio-visual senses, every possible effort should be made to ensure that all students can see and hear exactly what is being demonstrated. This requires careful placing of students and also such skillful demonstrating that the important movements and operations are clearly evident to each student in the group.

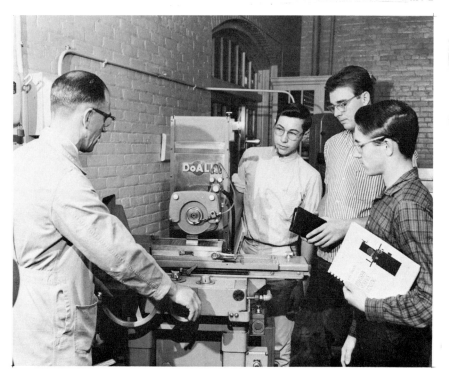

Fig. 14–2. The industrial arts demonstration is highly effective because of its realism. (Courtesy, Do All Company)

3. Tape Recorders and Audio Casettes

This equipment seems to have limited usefulness in industrial arts programs—which may explain, at least in part, the dearth of material in the literature of the field. The principal uses for these aural aids appear to be confined to (1) the addition of sound to sets of slides or to filmstrips, (2) use with the telelecture, (3) presenting prerecorded lectures or instructions, and (4) providing background music in, say, an electronics laboratory or a drafting room. Other instances are noted where tape recorders have been used in presenting the steps in project construction, in reviewing the material covered in a given period of time, and in student seminars where the research and experimentation approach to industrial arts is being employed (see Fig. 14–3).

Figure 14–4 illustrates a novel use of the tape recorder in which taped instructions are being sent to two students at work on metal lathes. Students do not operate the lathe while receiving the audio signals, rather the instruction is restricted to nomenclature and to mak-

Instructional Media—Part I 321

Fig. 14-3. In this laboratory a tape recorder is being used during the student seminar which is an important phase of the research and experimentation approach to industrial arts. (Courtesy, Alan P. Keeny, Montgomery Hills Junior High School, Silver Spring, Maryland)

ing setups for the instructor's approval. Figure 14-5 shows a close-up view of the master console by which the instructor controls the tapes and the microphone which may be cut in at any time for giving special instructions. In this laboratory a similar audio system, but employing speakers instead of earphones, is used in the graphic arts area.

Fig. 14-4. A tape recorder is being used to send set-up instructions to students at the metal lathes. (Courtesy, Robert Jackson, Tamanend Junior High School, Warrington, Pennsylvania)

Fig. 14–5. Close-up view of the console showing tape recorders, microphone, and control switches. (Courtesy, Robert Jackson, Tamanend Junior High School, Warrington, Pennsylvania)

The Educators Guide to Free Tapes, Scripts, Transcriptions[3] is a major source of taped material. Although there is no subject heading called *industrial arts*, certain materials under the following headings may be useful in the industrial arts program: home safety, forestry, labor conditions, electrical appliance safety. Another prime source of taped material is *Audio Tape Recordings*.[4] Listed are many tape reels and tape cassettes on industry, inventions, communications, natural resources, science, television, transportation, and wood technology which are of great interest to industrial arts teachers. The use of taped material of this nature can add a new dimension to teaching and learning in industrial arts.

4. Objects and Specimens

Although they are really quite similar, the fine distinction between objects and specimens is largely a matter of use. Both are "real things," but if the item is used to represent the class or group of objects to which it belongs, it is more precisely called a *specimen*. Thus, specimens are examples or samples representing a class of objects. For example, a piece of white pine is an object, but it is more exactly termed a specimen if it is exhibited as a sample of, say, softwoods.

Johann Pestalozzi, the "father of manual training," has been called the "great reformer in the development of all manual and industrial

---

[3] *Educators Guide to Free Tapes, Scripts, Transcriptions* (Randolph, Wis.: Educators Progress Service, Inc.), published annually.
[4] *Audio Tape Recordings* (Ann Arbor, Mich.: University of Michigan), 1971.

education."[5] One of his greatest contributions to education was object teaching, that is, the use of real objects as a means of teaching subject matter. Where real objects were not available, he then used well-made models. Little need be said here about the extensive use of object teaching practiced by the industrial arts teacher over the years. The world of industry is a world of objects; likewise, in the laboratory the industrial arts teacher daily leads the student into meaningful contact with palpable, tangible things. As they work with the tools, materials, processes, and products of industry, students gain valuable insights and understandings through first-hand contacts and close-range observations of "real things."

Students should be encouraged to bring objects and specimens to class not only to enrich the curriculum and to improve the shop atmosphere, but also to illustrate their reports and class assignments as well as their research and development topics.

When objects and specimens are used to illustrate a learning activity, it is not advisable for the teacher to pass them around the class. Such practice invariably causes distraction and even confusion especially if a number of objects are circulating around the class at one time. Instead, the objects may be displayed in front of the class and then made available for closer examination or self-study after the lesson. Sometimes the opaque projector can be used with small objects and specimens during the presentation.

Like object teaching, objects are also the basis for a useful form of testing in industrial arts. Object tests have been widely used by industrial arts teachers for many years.

5. Models and Mock-ups

The "real things"—that is, objects and specimens make the best teaching aids. When these are not available the teacher can often use to great advantage models which are scaled reproductions or miniature samples of the "real thing."

Model making in America has become a big business. Two of the largest users of models in this country are the military services and the aerospace industries. In addition, models are used extensively in inventions and patent applications, in home building and city planning, in machine and product design, in plant layout, in the automobile, boat, and airplane industries, and in movie and television productions.

Over the years industrial arts teachers have produced countless

---

[5]Charles A. Bennett, *History of Manual and Industrial Education up to 1870* (Peoria, Ill.: The Manual Arts Press, 1926), p. 107.

models for teaching purposes. Generally they are constructed by the teacher, but valuable educational benefits do accrue for students who help produce models, cut-aways, and other instructional devices for use in the laboratory. This is particularly true in the drafting laboratory where student-constructed 3-D models may be an integral part of the course. By constructing models, students gain insights and understandings of the principles of drafting.

Common practice usually has been to make enlargements of small objects in the laboratory. Some examples are giant-size models of planes (see Fig. 14-8), chisels, marking gauges, auger bits (see Fig. 11-2), lathe tool bits, tool holders and gauges, composing sticks, gauge pins, foundry type, halftone screens, meter faces, electrical resistors, electron tubes, rules and architect's scales, micrometer and vernier scales, and so on—the list is nearly endless. In other instances teachers have made scaled working models of a variety of machines including several types of early printing presses, veneer cutting lathes, steam and internal combustion engines, cut-away models, and so on—again a complete listing would be lengthy.

Some of the unique features of the use of models in the learning process include the following:

1. Models are three dimensional, which means that the pupil's tactual sense as well as his visual sense may be brought into play in the learning process. The student can touch, handle, and sometimes operate the model—thus firmly instilling concepts, understandings, and appreciations. Models needed to be studied in detail, not just looked at casually by students. A study assignment sheet (see Chapter 9) based on self-study of a model can be a valuable educational experience. If the model can be taken apart and reassembled, the student can examine each part individually and study its relationship to other parts and to the whole. In this respect, models can make a unique contribution to the learning process.

2. A model can be made to any scale and so provide a three-dimensional version of a size convenient for laboratory observation and study. For example, a scaled model can depict in miniature a huge object, such as a 100 ft. high blast furnace, or it may be an enlargement of a tiny object like a piece of foundry type. The student shown in Fig. 14-6 will have a better understanding of the component parts of the hand saw, particularly the angle and set of the teeth and the resulting kerf because of this enlarged model. Teachers of technical drawing have long used oversize models in the teaching of orthographic projection so that all in the class may see. Paper, cardboard, wood, and sheet plastic materials have been used. Many teachers are now using models cut from styrofoam by means of a shop-made cutter using a heated nichrome wire.

Fig. 14-6. Enlarged models help build sound concepts and clear understandings of the functions of small, but highly important, components. (Courtesy, Julian Cleveland, C. B. Aycock Junior High School, Greenville, North Carolina. Photo by Walter F. Deal, III)

3. Working models have the additional advantage in that they can depict in simplified fashion the principles of complex operations or processes. Usually the nonessential details are left out so that attention is focused on the fundamentals. Notice in Fig. 14-7 the use of a giant-size model of the electrical meter being demonstrated by the instructor. "Work-it-yourself" models are often featured in museums because of their high interest and appeal as well as their educational benefit.

4. Models can be made to be taken apart or in cutaway fashion to show cross sections and to reveal interior details which normally cannot be seen. Such models help the learner to visualize and to understand better just what goes on inside. If the model also has moving parts, the educational value is increased. Take-apart and cutaway models have

Fig. 14–7. Teaching students to read an electrical meter is easy with this oversize model. (Courtesy, William J. Wilkinson, Nether Providence High School, Wallingford, Pennsylvania)

been instructional bulwarks for many years in industrial arts. The efforts of teachers have ranged from models showing the interior of a simple dry cell to those showing a complete automobile engine in action. Figure 14–8 shows a class learning about the jackplane through a giant-size take-apart model. Note the teacher is using a regular-size jackplane with the model to avoid teaching any misconceptions as to size. This is an important point to remember in using large-size models as teaching aids. Some models are made of clear plastic which facilitates observation of the movements and relationships of interior parts. Of special interest to industrial arts teachers are models of an automobile engine and chassis currently on the market.

5. The use of working models to teach industrial processes and to develop insights and understandings of American industry is an exciting prospect and some progress has already been made in this direction. Working models of printing presses using rubber type which print a variety of forms are available. Here the principle of letterpress printing is clearly evident. Industrial processes in the field of plastics can be imitated by means of working models and kits in the school shop. Certain of these actually produce small plastic objects involving the principles of injection molding, extrusion, rotational molding, and other processes. Figure 14–9 illustrates plastic blow-molding in the laboratory. The field of communications is represented with working models of

Fig. 14-8. Note the interest shown by these junior high school boys as the teacher compares a regular-size jackplane with a giant-size take-apart model. (Courtesy, Julian Cleveland, C. B. Aycock Junior High School, Greenville, North Carolina. Photo by Walter F. Deal, III)

Bell's first telephone, Morse's telegraph, and other communication kits.

Instead of using commercially available models, some teachers prefer to construct their own. Figure 14-10 is an example of a ⅜ths-scale working model of an old-time Isaiah Thomas printing press constructed by a graphic arts class. The tremendous satisfaction of making handmade paper which is then printed on this press is an exciting educational adventure for pupils. Other teachers have constructed models which illustrate the principles of offset printing and textile weaving.

6. Pupil-made models can provide educational experiences not possible with any other medium. Models constructed by pupils in one class are highly effective motivating devices for other classes. They also make fine materials for public relations displays and create favorable impressions with both lay and professional groups.

In elementary school industrial arts and in the unit method of teaching industrial arts at the junior high school level, pupil-made models are utilized as individuals and classes study transportation and the evolution of tools and machines. Figure 14-11 shows several ship models constructed by junior high school pupils as part of a unit on the history of transportation. Some teachers favor pupil-drawn plans of models

Fig. 14-9. With this working model the teacher can develop insights and understandings of an important industrial process (plastic blow forming) right in the school laboratory.

Fig. 14-10. A scale model of an Isaiah Thomas printing press. (Courtesy, Ward S. Yorks, Red Lion High School, Red Lion, Pennsylvania)

which are then constructed from materials available in the laboratory while other teachers utilize model kits which are commercially available. There is a wide variety of models especially adapted to the teaching of transportation presently on the market. Some are completed

Instructional Media—Part I 329

models while others are in kit form. Included are replicas of old-time and present automobiles, airplanes and ships, railroad units, submarine and aircraft carriers, helicopters, rockets, space capsules, and moon cars.

Student-built models of homes on landscaped lots have long been

Fig. 14-11. The construction of models of old-time sailing vessels is popular with junior high school students studying a transportation unit.

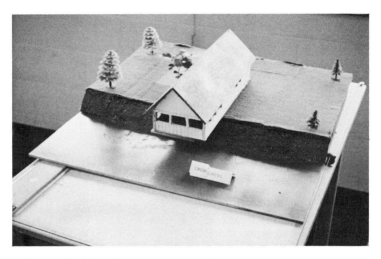

Fig. 14–12. Model home construction has long been a favorite term assignment in senior high school architectural drawing classes. (Courtesy, Robert MacMillan, Bald Eagle High School, Wingate, Pennsylvania)

used in architectural drawing classes at both high school and college levels (see Fig. 14-12).

Mock-ups. A mock-up is similar to a working model in that it simulates the real thing and has moving parts. It differs, however, in that the mock-up is usually made full size and is used to provide practice or training where it is either too costly or is educationally unfeasible to use the real thing. The Link trainer for training airplane pilots is probably one of the best known training mock-ups. Through the use of mock-ups, millions of people have seen the interior workings of space capsules as featured in the nationwide television coverage of America's space flights. In England newly-developed mock-ups are used to teach the art of sailing a boat. In some mock-ups details are purposely left out to focus attention of the learner on fundamentals, while others, such as the Link trainer and the space capsule, are exact replicas of the real thing.

Another kind of mock-up consists of full-size components mounted on a panel. Here student understandings rather than skills are developed. An example of this type of mock-up is a transistor radio circuit. Real electronic components are placed on a panel in proper relationship to each other, connected electrically, and properly identified with labels. Through a well-planned mock-up the functions of various parts and the relationships of one part to another are made easily understandable.

The use of mock-ups as an aid to instruction has been greatly accelerated by their extensive use and development in industrial training programs and in the military services. Significant gains in time required for understanding such complex units as airplane carburetors, fuel systems, and radar hook-ups have been made by displaying these in simplified form on a suitable mounting.

The industrial arts teacher has found many applications for mock-ups in teaching the skills and information of his subject. Typical examples of mock-ups used in the industrial arts laboratory include the wind anchorage system used in house construction; an electrical entrance with meter, distribution panel, and typical circuits (see Fig. 14-13); the ignition system and the braking system of an automobile.

Fig. 14-13. This teacher is using a mock-up to depict common conduit circuits.

The preparation of a mock-up is not a project that can be accomplished in a moment. First it must be carefully planned, then skillfully constructed. Once completed, however, it can be used for many years. Frequently it is both possible and desirable from an educational point of view to use student help in building mock-ups. Certainly the students who work on these projects will develop a depth of understanding of the principles and problems involved which it would be difficult to obtain by any other method.

6. Project Charts and Process Charts

The project chart is a type of visual aid which is particularly well adapted to industrial arts. Usually it consists of a board on which is mounted a project in various stages of completion. For example, a project chart covering the making of a paper knife might include (1) the metal blank, (2) the blank with a design marked upon it, (3) the project rough-sawed to shape, and (4) the finished knife. Project charts may be made for a variety of projects ranging from simple to complex. Their major values lie in helping beginning students to learn project planning and to visualize the steps in project construction. In Chapter 11 two pictures of process charts were presented to illustrate their use in starting an industrial arts class (see Fig. 11-3).

A similar type of chart is a process chart which depicts the steps or processes in the manufacturing of a given object, such as, a hammer, a saw, or an auger bit. These charts are often available from industrial firms and show the major manufacturing steps from raw material to finished product. Student-made process charts are sometimes useful as manufacturing flow charts in planning mass-production projects in the school laboratory.

7. Bulletin Boards, Chalkboards, and Flannelboards

These instructional media are both unique and educationally effective. Since they are used for different purposes, each aid will be considered briefly in the following paragraphs.

Bulletin boards are one of the least expensive instructional aids used in the laboratory. Conventionally, they are rigidly mounted on the wall and usually located near an exit, washing facility, or other area where students congregate. Newer practices suggest the use of mobile bulletin boards supported on folding easels. These are used for special displays at temporary locations throughout the laboratory or school; since these appear "new" or different, they invariably attract immediate attention. Pegboard makes a useful supplementary bulletin board because industrial samples, models, or tools can be easily mounted with hooks, wires, or pegs.

The student personnel organization can be helpful in preparing attractive and colorful bulletin board displays. Art fundamentals of layout, lettering, and color can be taught in this manner. It should be remembered that a good layout should not incorporate too much material, but rather should emphasize a central theme or idea (see Fig. 14-14). Bulletin board displays need to be organized, planned, and kept up-to-date. It is usually desirable to outline a plan for bulletin board displays for a marking period, a semester, or even for the entire year. A well organized filing system such as is suggested in Chapter 15 will greatly facilitate the implementation of such a plan.

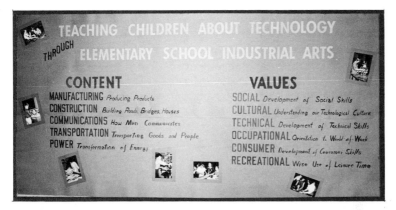

Fig. 14–14. This attractive display illustrates a good rule-of-thumb for bulletin boards: Emphasize a central theme or idea. (Courtesy, William R. Hoots, Jr. Photo by Bureau of Public Relations, East Carolina University)

Chalkboards are one of the most convenient teaching aids readily available. They are especially useful not only for teacher presentations, but for pupil work as well. Some basic rules-of-thumb for the optimum use of the chalkboard include the following:

1. Keep the attention of the class focused on the points being developed by erasing all superfluous or unrelated material from the board.

2. Avoid using the lower portion of the chalkboard because students at the rear of the instructional area may not be able to see the material.

3. Make the lettering and sketches large enough so that they may be seen easily from a distance. Break chalk into short lengths, e.g., ¼-inch or ½-inch and use the sides to produce wide lines. Press firmly on the chalk to produce lines of even width and good contrast.

4. Use colored chalk for headings and divisions as well as to emphasize and clarify portions of sketches or illustrations.

5. Look at and talk to the class—not to the chalkboard—while explaining the material. Remember to stand clear so that students can see the material on the board. Use a chalkboard pointer to focus attention on the point being explained.

Lettering and freehand sketching at the chalkboard are skills which can be acquired only by patient practice. Presentations can be facilitated, however, by the use of hand-held aids, such as ruler, sliding straightedge, protractor, triangle, and compass. While these items are available commercially, they can also be made in the laboratory. Drafting machines are also available for mounting on chalkboards. Shop-made templates of cardboard are sometimes useful.

Some teachers have found it convenient to draw guidelines, isometric axes, and grids, directly on the chalkboard with a felt-tip marking pen or to score lines lightly with a well-sharpened scriber. Although invisible at a distance, such guidelines greatly assist the teacher or student in making a sketch accurately and quickly. Borrowing a technique from the elementary school, some industrial arts teachers transfer drawings onto a chalkboard by means of a perforated pattern made on thin paper. After the drawing is made, it is perforated by a marking wheel or a sewing machine. The transfer is effected by holding the perforated pattern against the chalkboard and patting it with a chalkdust bag or an eraser loaded with chalkdust. The chalkdust passes through the perforations in the pattern and produces a dotted outline on the board which may then be finished as desired. An opaque projector can also be used to advantage in projecting illustrations, charts, and drawings from books onto the chalkboard. Usually the major lines are traced after which the drawing is completed by freehand sketching.

A new kind of chalkboard is the magnetic chalkboard. One type consists of a steel base with a porcelain surface. Chalk, pencil, and crayon take well and can be removed easily with a dry cloth. Magnets will, of course, cling to the metal base so that light objects can be hung. Thin, flat magnets as well as flexible magnetic strips are available commercially; these can be glued to the back of objects to be suspended for display purposes. Or rubber magnetic strips with a pressure-sensitive surface on one side are available for use in mounting felt or cardboard materials. Displays of two- and three-dimensional objects and the rearrangements of parts of mechanisms are but two uses for such chalkboards.

An enterprising industrial arts teacher can easily make his own magnetic chalkboard-bulletin board. A small board can be made of a sheet of tinplate or galvanized iron with a molding around the edges to prevent injuries from possible contact with sharp edges and corners. Or larger boards of plywood can be covered with metal. When given a coat of chalkboard paint, these boards will then serve as both magnetic bulletin boards and chalkboards.

A *flannelboard* is a piece of heavy cardboard, plywood, hardboard, or similar material covered with cotton or felt flannel. A number of materials such as string, yarn, balsa wood, felt, styrofoam, sponge, pipe cleaners, and sandpaper will cling directly to the flannel. If letters,

Instructional Media—Part I

pictures, or other lightweight objects are first backed with strips of these materials, they will become affixed to the flannelboard apparently without any visible means of support. Striking effects are attainable when three-dimensional objects made of balsa wood, cardboard, or styrofoam are used on the flannelboard. These boards are also effective in presenting illustrated lectures and in making titles for movies, colored slides, and closed-circuit television. Colored letters and other cutouts are available commercially which save teacher effort and effect attractive layouts. Signs, notes, or labels for parts can be built up of individual felt or cardboard letters. Or, they may be lettered with a felt-tip marking pen on strips of cardboard backed with sandpaper. Small pieces of hardboard painted with chalkboard paint can be used on the flannelboard as little chalkboards; these, too, make useful signs and labels.

A recent development in flannelboards is provided by a nylon material called "hook and loop" which is used in place of cotton flannel. Hook-and-loop material is also available in strip form with an adhesive backing for easy attachment to the teaching materials to be displayed on the board. When the hook and loop materials of the strip and the board come into contact with each other, they will support objects weighing up to a pound or more in weight.

8. Free and Inexpensive Materials

A wealth of visual aids which are either inexpensive or free for the asking is available to all industrial arts teachers. In form these materials may be booklets, brochures, posters, reports, monographs, samples, charts, demonstration kits, graphs, maps, educational "comic" books, pictures, or exhibit materials. Industrial and commerical firms often vie with each other in trying to produce the most striking and arresting graphic materials.

It must be remembered that commercial organizations are interested in reaching school pupils with their messages. For this reason the teacher should exercise care in selecting and using materials, even if free, to make certain that the information presented is authoritative, up-to-date, un-biased and has a minimum of commercialism. Although these materials are seldom designed especially for laboratory use, many are useful as resource materials for enriching the curriculum and for providing attractive bulletin board displays.

Sources of free or inexpensive materials are to be found literally everywhere, including newspapers and magazines; governmental agencies; state, county, and local organizations; gas, light, railroad, telephone, and water companies; professional associations; and trade associations and industrial concerns. Selected lists of free or low-cost materials are regular reader service features in the industrial arts periodicals. Two major sources of materials are the following: (1) *Elementary Teachers' Guide to Free Curriculum Materials*[6] which lists a number of items under the heading, *industrial arts*. Additional items pertinent to the field may be found under "visual aids," and other headings, such as steel, plastics, or science. (2) *Sources of Free and Inexpensive Educational Materials*[7] which is partially annotated, lists scores of items under the heading, *industrial arts*. Other headings deal with arts and crafts, auto mechanics, welding, etc., as well as materials including paper, plastics, steel, and rubber.

Every industrial arts teacher should have a constantly growing collection of graphic materials that can be used as the occasion demands in connection with specific lessons or to create atmosphere within the laboratory. Careful selection and use of display materials can transform even an ugly room into a cheerful and challenging learning environment. With the vast amounts of such material currently available, there should be no reason for bulletin boards with yellowing pictures, time-worn charts, and tired drawings. The student personnel organization may well be made responsible for assisting in keeping such displays live and up-to-date.

SUMMARY

Due largely to a better understanding of the learning process, the role of audio-visual aids to instruction is changing. New terminology is being used to describe three thrusts which seem to be developing on the educational front: (1) polysensory learning through a multimedia approach, (2) total reliance on portions of the "instructional system" for the attainment of clearly-defined educational objectives, and (3) growing emphasis on student responsibility for his own learning through self-paced instruction.

Instructional media are of special importance to the industrial arts teacher for the following reasons: (1) to aid in exploration and orientation to industry, (2) to teach and illustrate informative aspects, and (3) to assist in teaching skills, and (4) to create an atmosphere in the labora-

---

[6] *Elementary Teachers' Guide to Free Curriculum Materials* (Randolph, Wis.: Educators Progress Service, Inc.), published annually.

[7] *Sources of Free and Inexpensive Educational Materials*. Compiled by Ester Dever (mimeographed). P.O.Box 186, Grafton, West Virginia, 1966.

tory which is conducive to the development of the ability to think in students.

The types of instructional media discussed in this chapter included the following:

1. *Industrial visits* are one of the most valuable forms of instructional media because of their realism. Such trips require careful planning by the teacher and should culminate in a class discussion following the visit.

2. *Demonstrations* rank with industrial visits in realism and constitute an instructional medium of first importance.

3. *Audio recordings* have been used to a limited extent in industrial arts. However, a new dimension to the learning activities can be added via the fine commercial tapes now available in reel and cassette form. These are worthy of investigation by the industrial arts teacher.

4. *Objects and specimens* make important and unique contributions to both teaching and testing in industrial arts.

5. *Models,* especially working and take-apart models, can involve both the tactual and the visual senses and can show in simplified form complex operations as well as interior details not normally seen. Some progress has been made in teaching industrial processes with models.

6. *Mock-ups* are useful devices which simulate the real thing through simplified versions of the relationships of one component to another and to the whole. Mock-ups are used in teaching skills and information.

7. *Project and process charts,* respectively, depict in three-dimensional form the steps in the construction of a project or the steps in manufacturing some industrial product.

8. *Bulletin boards,* especially mobile boards, are useful in displaying two- and three-dimensional objects for student examination. *Chalkboards,* especially of the magnetic type, serve well for illustrating teacher presentations, student sketching, and for displaying objects. *Flannelboards* are effective in illustrating shop talks and in making titles for colored slides.

9. *Free and inexpensive materials* are readily available, but the teacher should exercise discrimination in selecting materials to present information which is authoritative, unbiased, up-to-date, and with a minimum of commercial advertising.

Any industrial arts teacher who fails to involve selected educational media in the learning activities is doing less than his best. Even a good lesson can be greatly improved by the incorporation of the right type of instructional media.

DISCUSSION TOPICS AND ASSIGNMENTS

1. Mention several instances in which the sense of touch and the sense of smell can be utilized in industrial arts teaching.
2. Can you think of a situation in which the sense of taste can be safely used in industrial arts instruction?
3. Do you think that industrial arts teachers utilize instructional media to the maximum in teaching skill? Why?
4. To what uses, other than those mentioned in this chapter, do you think the tape recorder and audio cassette can be adapted in teaching industrial arts?
5. What advantages and disadvantages are there to the use of models in teaching industrial processes?
6. What is the difference between a project chart and a process chart?
7. What are the unique advantages of each of the following: (1) mobile bulletin boards, (2) magnetic chalkboards, and (3) flannelboards.
8. Procure ten names and addresses of sources for free and inexpensive materials in your choice of an industrial arts area.
9. Select an industrial plant with which you are familiar and make a detailed outline for a class visit to that plant.

SELECTED REFERENCES

Allen, David and Bruce J. Hahn, Milo P. Johnson, Richard S. Nelson. *Polysensory Learning Through Multi-Media Instruction in Trade and Technical Education.* Los Angeles, Calif.: Bureau of Industrial Education, California State Department of Education, 1968.

Audio Tape Recordings. Ann Arbor, Mich.: Audio-Visual Education Center, University of Michigan, 1971.

Bakamis, William A. "A Multimedia Model for Total Self-instructional Systems," *School Shop,* vol. XXIX, no. 3 (November 1969), pp. 54–57.

Barnard, David P. "How to Use a Flannelboard," *Industrial Arts and Vocational Education,* vol. 50, no. 9 (November 1961), pp. 26–27.

Brong, Gerald R. "Anatomy of a Multi-Media Accessibility Continuum," *School Shop,* vol. XXVIII, no. 1 (September 1968), pp. 74–76.

Colburn, Dave. "Building a Cassette Tape Bank: Everything You Wanted To Know," *Educational Screen and Audiovisual Guide,* vol. 50, no. 4 (April 1971), pp. 6–9.

Dever, Ester (ed.). *Sources of Free and Inexpensive Educational Materials.* Post Office Box 186, Grafton, West Virginia, 1966 (mimeographed).

Educational Media Index. New York, N. Y.: McGraw-Hill Book Company, Inc., 1964.

Educators Guide to Free Tapes, Scripts, Transcriptions. Randolph, Wis.: Educators Progress Service, Inc.

Elementary Teachers' Guide to Free Curriculum Materials. Randolph, Wis.: Educators Progress Service, Inc.

Fazzini, Phillip A. "How to Get the Most Out of a Field Trip," *School Shop,* vol. XXX, no. 6 (February 1971), p. 66.

Goldsbury, Joseph W. "A Feasibility Study of Local Field Trips Taken Vicariously Through Slide-Tapes" (unpublished doctoral dissertation, The Ohio State University, Columbus, 1969).

Groneman, Chris H. "Supplementary Teaching Materials for Industrial Education," *Industrial Arts and Vocational Education,* vol. 55, no. 6 (June 1966), pp. 58, 60, 62–63.

Guatney, Charles L. and Richard E. Higgins, "Tips on Tape—Its Technology and Sound Use," *School Shop,* vol. XXVIII, no. 6 (February 1969), pp. 54–56, 59.

Hankins, Edward K. "How to Vitalize Instruction With Samples and Specimens," *Industrial Arts and Vocational Education,* vol. 52, no. 2 (February 1963), p. 18.

Hillson, Joseph S. and Alexander A. Wylie, Wolf P. Wolfensberger, "The Field Trip as a Supplement to Teaching: An Experimental Study," *Journal of Educational Research,* vol. 53, no. 1 (September 1959), pp. 19–22.

Hodges, Lewis C. and Gerald A. Silver, "Quickening the Pulse of Industrial Education," *School Shop,* vol. XXVII, no. 8 (April 1968), pp. 68–71, 108.

Lauda, Donald P. "Multimedia Presents," *Industrial Arts and Vocational Education,* vol. 56, no. 10 (December 1967), pp. 36–37.

──────. "Seven Rules for Effective Media Usage," *School Shop,* vol. XXVII, no. 8 (April 1968), pp. 72–73.

Muller, Warden B. "Make and Take, But Don't Forsake, the Field Trip," *Industrial Arts and Vocational Education,* vol. 58, no. 7 (September 1969), pp. 27–29.

Olson, Jerry C. "How to Produce Effective Chalkboard Developments," *Industrial Arts and Vocational Education,* vol. 53, no. 10 (December 1964), pp. 23–24.

Rolbiecki, James J. "Breakthrough With 'See-Through' Models," *School Shop,* vol. XXIX, no. 2 (October 1969), p. 64.

Silver, Gerald A. "Bold Advances in Audiovisual Instruction," *School Shop,* vol. XXVI, no. 8 (April 1967), pp. 71–75, 90.

Wittich, Walter A. and Charles F. Schuller. *Audio-visual Materials: Their Nature and Use.* New York, N. Y.: Harper & Row, Publishers, 1967.

Zanco, M. I. "How to Tape Classroom Lectures," *Industrial Arts and Vocational Education,* vol. 51, no. 7 (September 1962), p. 35.

──────. "How to Individualize Instruction With a Tape Recorder," *Industrial Arts and Vocational Education,* vol. 52, no. 2 (February 1963) p. 19.

Chapter **15**

Instructional Media—Part II

This chapter continues the presentation of instructional media which began in Chapter 14. Discussion here is limited to selected media involving screen projections or based on photographic processes. The specific media to be considered in the following sections include (1) educational television, (2) closed-circuit television, (3) videotape recording, (4) films, including 8-mm single-concept film loops, (5) filmstrips, (6) slides, (7) overhead projections, (8) still pictures, and (9) opaque projections. The chapter concludes with a section entitled "Photography for the Industrial Arts Teacher."

TYPES OF INSTRUCTIONAL MEDIA*

1. Educational Television

The rapid development and acceptance of television has opened to teachers a whole new world of possibilities in educational media. New channels have been designated exclusively for educational purposes and vast transmission facilities, including orbiting satellites, have brought programs which are educationally beneficial and entertaining to hundreds of thousands of school pupils. Some school systems now have their own television studies over which can be broadcast programs specifically designed for particular classes and grade levels. This makes possible the selection of experts in specialized fields to give demonstrations or to teach lessons to large viewing audiences. For example, all of the seventh-grade classes in a given school or city might watch a skilled craftsman, such as a welder, a cabinet maker, or a silversmith at work. By such means every pupil may enjoy the benefit of enriched instruction.

*Continued from Chapter 14.

The Central Virginia Educational Television network has developed a series of 30 telelessons called "Industrial Arts for the 70's" which is intended for students in grades 9 through 12. Mark Delp, an industrial arts instructor, designed the series and serves as studio teacher. The purpose of the programs is to provide enrichment not readily available in the typical industrial arts laboratory. Lesson content is drawn from the cultural, social, guidance, consumer, recreational, and occupational aspects of today's society. Typical lessons are on the history of colonial industry, mass production, and how to get a job; several occupational information lessons on engineering, managerial, and production positions are also presented. The 25-minute lessons are televised weekly in the sequence outlined in the *Teachers' Manuals*. These lessons are not intended to be a course in industrial arts nor are they to be viewed casually by students. The teacher is urged to conduct certain previewing activities. After seeing the televised lesson, post-lesson activities may be carried out as suggested in the manual. Figure 15–1 shows a high school class participating in the previewing session.

Fig. 15–1. This class is about to see a televised lesson from the series, "Industrial Arts in the 70's," sponsored by the Central Virginia Educational Television network. (Courtesy, Mark Delp, Richmond, Virginia)

Industrial arts teachers should be alert to the possibilities of this instructional medium which holds such high appeal for boys and girls.

Selected programs may well be suggested for viewing by the student, much as outside assignments are now made in books. In addition, the teacher should be eager to seek broadcast time for the purpose of furthering a sound program of public relations as well as for providing educational and recreational services to the community drawn from the world of industrial arts.

2. Closed-Circuit Television

This media is a limited form of educational television in which the aural and visual signals as picked up by the camera are transmitted through coaxial cable directly to the television receiver. Where the camera and receiver are in the same room, the system is called single-room closed-circuit television or single-room direct camera-to-receiver television.

A study by Herbert[1] disclosed four principal reasons given by selected industrial arts faculty at Stout State University for using television to supplement their laboratory demonstrations. In order of importance these reasons were (1) to magnify small objects, (2) to present material otherwise inaccessible to the group, (3) to show the action from the demonstrator's viewpoint, and (4) to permit close-up viewings of dangerous operations.

The equipment used in closed-circuit television in industrial arts laboratories ranges from simple to sophisticated. Sometimes a teacher is successful in borrowing equipment from a local electronics firm. The simplest arrangement is one in which the camera is mounted in a fixed position over the demonstration bench. The camera and lights are controlled by the instructor while he is presenting the demonstration. A more sophisticated arrangement may involve two cameras with a variety of lenses, an audio-visual console, and monitors all of which are mobile. In this case, of course, a camera crew and other technicians will be needed.

Numerous accounts in the literature of the field attest to the growing interest being shown in closed-circuit television by both high school and college teachers of industrial arts. Some of this interest is doubtless due to a more widespread knowledge of its use and effectiveness, but more likely is due to the declining costs of hardware because of advancements in production and technology. As the cost of equipment becomes within the reach of the average teacher, it is to be expected

---

[1] Harry A. Herbert, "Single Room Closed-Circuit Television in Industrial Education" (unpublished master's thesis, Stout State University, Menomonie, Wisconsin, 1964).

that more common use will result. Figure 15-2 shows a television lesson in a high school drafting class.

Fig. 15-2. Televised instruction in drafting. (Courtesy, Eberhard Thieme, Rochester City Schools, New York)

Some years ago one high school teacher provided valuable educational experiences with closed-circuit television by having the students in his electronics classes operate the equipment for the school. The hardware which the students learned to operate included two mobile cameras and monitors, a video console, and an audio console. In addition they used movie and slide projectors, tape recorder, radio tuner, and microphones by which video and sound signals were picked up and transmitted to any or all rooms in the school. A talk-back system permitted students in the television audience to ask questions and receive answers from the teacher. The electronics students who operated this equipment were rotated at regular intervals to ensure a well-rounded training experience.[2]

More recently a large city high school in California provided an occupational education program in which students studied electronics and gained practical experience by manning the school's closed-circuit television system.[3] Five hours weekly were spent in technical study of electronics in the newly-equipped classroom-studio and five hours were spent in practical work. Some of the practical activities

[2]Walter H. Brzezinski, "A Closed-Circuit TV System for Enriching Electronics," *Industrial Arts and Vocational Education,* vol. 50, no. 5 (May 1961), p. 26.
[3]Thompson, Dorothy "Applying the Media to a New Electronics Career Course," *School Shop,* vol. XXVII, no. 8 (April 1968), p. 86.

included television camera and videotape recorder operation, audio and video mixer controls, studio set construction and lighting, and program directing.

Certain uses and advantages of direct-wire television in connection with the laboratory demonstration were presented in some detail in Chapter 10; also several illustrations were provided (see Figs. 10-3, 10-4, and 10-5). It will be recalled from this discussion that in using closed-circuit television the class watches and listens while the teacher demonstrates or talks in the usual manner. At special times during the presentation the class may be directed to watch the television screen. Since the audio portion of the television system is not used ordinarily, the teacher continues to talk and explain the points in the lesson while the class watches the screen. Since television is not used continuously in this kind of presentation, the center of interest shifts at the discretion of the instructor to the television screen, to the lecture itself, or to some other instructional media. In other uses, of course, entire lessons or demonstrations can be watched on television. This latter use is valuable where potential danger exists from too-close observation of a process or an operation. In Fig. 15-3 a televised demonstration on tool sharpening is being sent to several classrooms simultaneously. Large groups will be able to see this demonstration better on television than if they were standing at the grinder and without exposure to any possible danger.

It is of paramount importance that the instructor carefully practice and rehearse the presentation in front of the camera lest the effectiveness of the medium be lost. It must be remembered, too, that a live demonstration, once televised, is lost forever and must be repeated each time just as the conventional demonstration. Videotape recording is, of course, the answer to this problem. By this means the demonstration is "canned" for showing whenever desired. Videotape recording will be discussed in the next section of this chapter.

Despite its advantages and student appeal, there are many instances in the teaching situation when the planning and rehearsal for a television presentation will consume more teacher time and effort than the results justify. This discounts, too, any consideration of the initial financial expenditure for equipment or maintenance. Thus closed-circuit television is not a panacea for all instructional problems nor is it the only way to teach. Like other educational media, closed-circuit television has its place and, if judiciously employed, can make unique and perhaps significant contributions to the learning process. Certainly it is an exciting instructional tool which holds promise of advancing the methodology of teaching.

3. Videotape Recording

As mentioned previously, the great drawback to live television presentations is that no record of performance is possible. Once televised, each scene is lost forever. A repeat showing or performance requires just that, a repeat performance. Through videotape recording, however, the performance is preserved on tape so that both sound and video may be replayed either instantly or years later. Thus, videotape recording is closed-circuit television with the addition of one component—a recorder which records both sight and sound on a magnetic tape. Videotape recorders are available either in black and white or in color. Figure 15-4 shows a videotape recorder.

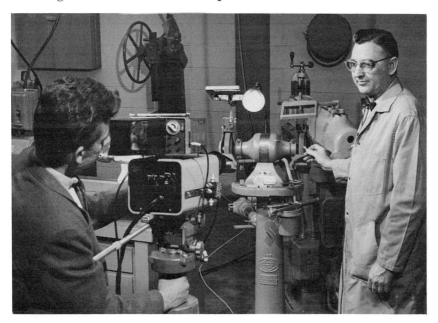

Fig. 15-3. This televised demonstration is being sent to several classrooms. (Courtesy, D. P. Barnard and H. A. Herbert, University of Wisconsin—Stout)

There are a number of interesting uses for videotape presentations in industrial arts. Teacher demonstrations and lessons for all grade levels can be taped and used any time with groups or individuals. There is some merit in presenting exactly the same demonstration techniques and information to each section of pupils. Such uniformity of instruction is easy to accomplish with videotapes. The use of a videotape library of key demonstrations and lectures will provide the teacher with extra time for individualized instruction. Videotapes are also useful in self-

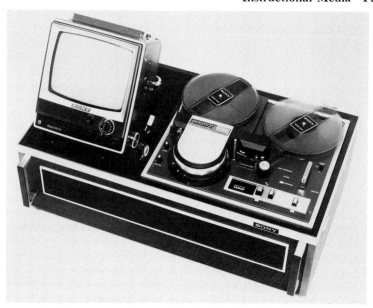

Fig. 15–4. A one-half inch tape Videocorder with built-in monitor. (Courtesy, Sony Corporation of America)

paced instruction, in remedial teaching, and in make-up lessons for absentees.

Videotapes are a natural vehicle for enriching the instruction in industrial arts. For example, where industrial visits are not feasible for any reason, such as hazardous conditions in the industry, travel considerations, or costs of the trip, videotape recordings made by the teacher can be used to good advantage. In this connection they are great time savers as well. Also, when resource persons cannot be scheduled easily, videotape recordings make excellent substitutes.

If well done, videotapes can be a valuable public relations tool for the industrial arts program. For instance, videotapes showing students at work in the laboratory may become an important part of an industrial arts display for use in or out of school. The use of videotapes can also be an interesting medium to illustrate a talk by the teacher who addresses a club or organization on his subject.

In teacher education videotapes have been used in teacher recruitment efforts. They are often used for videotaping the lessons of student teachers at the teaching site as well as in methods classes where students are learning how to give demonstration and information lessons (see Fig. 10–7). Videotape is ideal for this purpose because it can be stopped anywhere for scrutiny of a scene or it can be rewound at any point and re-run as often as desired. When prospective teachers hear and see themselves as others do, they are usually highly motivated to make an honest and accurate self-appraisal of themselves in order to improve future performances.

Videotape equipment is still considerably more expensive than aural recorders, but the great educational effectiveness in both seeing and hearing performances may well justify the additional cost. The growing demand for videotape recorders has resulted in improvements in design and production so that costs are now but a fraction of what they once were. Furthermore, the newer models are portable and so easy to use that even a child can operate them after a few minutes of instruction. Some cameras are equipped with several lenses, such as regular, wide angle, and zoom, a built-in microphone, and an optical view finder; others are battery powered and can be carried anywhere. The recording time of these models ranges from 20 minutes to one hour. Color reels presently have shorter running times than monochrome reels of equal size. The newer videotape recorders are exceedingly compact and lightweight. One battery-powered model weighs less than 20 pounds complete with recorder, batteries, camera with built-in microphone, and carrying case.

Unfortunately, for reasons of both an economic and a technical nature, videotape still lacks the flexibility of audio-tape recording insofar as editing is concerned—that is, editing videotapes is a complicated process. However, skill in initial recording can help to compensate for this shortcoming and rapid developments in technology doubtless will bring about greater improvements and economy in the near future. There is little doubt but that growing numbers of industrial arts teachers will soon be using videotape equipment routinely in their instruction.

4. Films

Next to the demonstration in terms of reality (see Chapter 14), and certainly in terms of student interest, is the motion-picture film. Sound films act upon the senses of sight and hearing simultaneously with the special added advantages of color and motion. Films offer more than motion alone; they provide sequential motion—that is, there is continuity in the action. Another advantage in using films is that little, if any, reading skill is involved, thus removing a serious handicap to the poor reader. Films arouse student interest and provide strong incentive to learn. By means of special photographic effects, such as slow and stop motion, animation, time-lapse photography, and extreme close-ups, motion-picture films open wide the doors to learning in a manner not heretofore possible.

Films make it possible to overcome some of the physical barriers to learning. In industrial arts, films are especially valuable in bringing to the laboratory certain aspects of industry which cannot readily be seen at first hand. For example, it would be impossible for most classes to visit a lumber camp to see how trees are made into lumber. Even if such a trip were feasible, it is doubtful if all of the various aspects of lumbering could be seen in a limited time. On the other hand, a film can bring the complete story in a vivid, colorful, and interesting manner to the group and at the time when this particular information is most needed.

Another use of films in industrial arts is the teaching of safe working practices. There is perhaps no other medium that can better impress upon the young learner the necessity for working safely than the motion-picture film.

In addition to teaching understandings, appreciations, and attitudes, films can be used effectively to teach skills. In some ways a movie is better than a live demonstration because everyone in the laboratory can have the best possible view of the process. A film permits each student to have a "front-row seat" at the demonstration. Through films the most skillful demonstrator can be brought to the class, using the latest equipment and showing the essential steps with an accompanying explanation. The use of special photographic effects referred to previously can build insights and understandings not possible with the conventional demonstration. Although only a limited number of films which exactly cover the desired industrial processes is available to industrial arts teachers, more films are certain to become available as the demand grows. In the hands of skillful teachers the use of films for teaching basic skills is sure to increase rapidly.

Improvements in movie projectors have made the task of showing films a great deal easier for the teacher than ever before. Some of the newer features of 16-mm projectors include a built-in screen which opens for table-top viewing by audiences of one to 20 persons, capability of operation in a lighted room, automatic film threading and focusing, and better sound fidelity. One model operates with 16-mm film having either optical or magnetic sound tracks. The use of a magnetic sound track permits the addition to silent films of teacher commentary, background music, or sound effects. Similarly, improvements have been made in movie screens. For example, a newly-developed concave aluminum screen has such a high reflection efficiency that the picture image is sixteen times brighter than the image reflected by a conventional matte screen.

Usually it is too expensive for industrial arts departments or even schools to maintain extensive film libraries; instead, films are rented from college libraries or commercial sources. In addition, films may be borrowed from certain industrial firms or trade associations. The *Educational Media Index*[4] is a major source of films, although considerably more instructional aids than films are listed. This index consists of 13 subject-area volumes and one index volume. Industrial Arts is included in Volume 9, "Industrial and Agricultural Education." Here is listed material that is suitable for use in grade 7 through college and adult levels. Each item is identified by title and type. The following are the kinds of educational media included in this reference set: slides, transparencies, phonotapes, flat pictures, phonodisks, videotapes, programed instructional materials, models, mock-ups, films, kinescopes, cross-media kits, maps, and charts.

Another major source of film is the *Educators Guide to Free Films*.[5] This guide is published annually. The prime advantage of this guide is that it is not merely cumulative, but as new films are added, older ones are deleted. Currently more than 200 films are listed under the subject heading, *Industrial Arts*. In addition, other headings, such as *Transportation, Science,* and *Safety*, list certain films which seem appropriate for use in industrial arts classes.

The United States Government lists thousands of films, filmstrips, audio and videotapes as well as the audio-visual materials available under the auspices of 66 foreign governments.[6]

Often lists of films appear in state bulletins, professional brochures, and curriculum guides. New titles and reviews are found in the current periodical literature of the fields of industrial arts and audio-visual instruction. Manufacturers and distributors of audio-visual equipment as well as educational film producers publish lists of films. Other potential sources of information on films and other audio-visual aids include (1) the school librarian, (2) the audio-visual director in the school system, (3) the town or city librarian, and (4) the local Chamber of Commerce.

For many years film research has been an area of great interest to teachers, educational psychologists, graduate students, and other researchers. One of the most comprehensive investigations ever made into the improvement of films was conducted at The Pennsylvania State University under the joint sponsorship of the Army and the Navy. Sixty-five of these research reports have been summarized in a single volume.[7] From these reports selected conclusions which seem of spe-

[4] *Educational Media Index* (New York, N.Y.: McGraw-Hill Book Company, 1964).
[5] *Educators Guide to Free Films* (Randolph, Wis.: Educators *Progress Service*), published annually.
[6] *Motion Pictures and Filmstrips of the United States Government*, National Audio-visual Center (Washington, D.C.: National Archives and Record Service, GSA).
[7] *Instructional Film Research Reports*, vol. II, Technical Report No. SDC 269–7–61, NAVEXOS P–1543 (Port Washington, Long Island, N.Y.: U.S. Naval Training Device Center, 1956).

cial interest and value to the industrial arts teacher are reported as follows:

1. Use films to reach instructional goals. Films should be shown with a definite purpose in mind. They should be integrated into the course and closely related to the objectives of the unit. For this reason films with specific content should be selected. Those with broad, superficial content aimed at a generalized audience are likely to be less effective than films with well-specified content aimed for a particular grade level of students.

2. Let the film do the teaching. Good films can be used as the sole means for teaching some kinds of factual material and performance skills.

3. Tell the class they are expected to learn from the films. The teacher should inform the pupils firmly that they are expected to learn from the film. If possible, tell them that they will be tested on what they have learned and then test them. This procedure will result in increased learning.

4. Advise the students to listen carefully to the sound track. The narrative often contains important information to be learned in an informational film.

5. Discourage note-taking during the viewing. Students should be advised not to take notes during the average film showing because it interferes with attention and hence learning.

6. Use the study guides accompanying the film. Students will learn more from the film if the printed study guides are used before and after the film viewing.

7. Use films often. The ability to learn from films improves with practice in learning from films—that is, students learn to learn from films. In other words, the more films one sees, the more he learns from other films.

8. Use films of interest to students. Films that are perceived by students to contain useful material will provide the greatest amount of learning.

9. Use films to teach simple motor skills. Motor skills that are at least as complex as operating a sound motion-picture projector can be taught by means of films alone. Increased learning of these skills will occur if the film shows common errors and how to avoid them. Short film loops which can be repeated continuously as many times as desired appear to be a good way of teaching difficult skills. An instructor can increase his effectiveness by using film loops to teach a skill to groups while he devotes his time to coaching individuals. Learning will increase if the viewer practices a skill while it is being presented on the screen, provided the film develops slowly enough, or provided periods of time are

allowed which permit the learner to practice without missing new material shown on the screen.

10. Use film loops in the practice area. One showing of a film dealing with a complex skill may be insufficient. The instructor should make arrangements to show a film in the practice area so that the student can easily refer to the film model as often as necessary. This can be accomplished by rear projection on daylight screens in the work area. Optimum viewing occurs within twelve screens widths and 30 degrees from the centerline.

11. Instruct students to practice skills mentally. Even though they do not have the equipment available, students can partially learn a skill by watching a film and imagining that they are performing that skill—that is, by going through the skill "mentally." Films can provide a model for guided "mental" practice.

12. Vary the length of the film sessions. Film-viewing sessions of informational material can extend to at least one hour without reduction in training effectiveness.

13. Test students after viewing a film. It should not be assumed that learning has occurred simply because a film, even a good film, has been shown. The teacher should evaluate the effects of the film by giving a test. This may be either written or oral in nature. If formal evaluation is not desired, a class discussion of the film should be conducted to reveal misconceptions and weaknesses in understanding.

14. Strive for the maximum amount of learning possible. One of the most effective means for increasing learning is to *repeat the showing* of an entire film or selected parts within it. Learning can be increased by *pretesting and posttesting* with student knowledge of the test results. Greater learning occurs when the *film is introduced* by the teacher who states the purpose and importance of the showing and tells the students exactly *what they are expected to learn*. At this time students should be told to *ask questions* about any part of the film which they did not fully understand and to *request a reshowing* if so desired.

Increased learning can be expected to occur if these and other research findings are applied when films are used in the instructional program. It is only through research that new knowledge is added to the field and teaching techniques are improved. The mark of a superior industrial arts teacher is to search out the latest research findings and apply them in everyday teaching.

8-mm educational films. The preceding discussion has been directed largely toward the laboratory use of conventional 16-mm sound films and equipment. However, increasing use is now being made of 8-mm film. The phenomenal growth in popularity of the small-size film is based on four recent developments: (1) the addition of a sound track,

(2) the new "Super 8" format, (3) the film cartridge or film loop, and (4) the introduction of the "single concept" theme.

Silent 8-mm films have been in use for many years, especially for home movies. But it was not until 1960 when sound became available through magnetic film striping and 1963 when the optical sound track system was developed that educators began to look seriously at the educational possibilities of this equipment. The battle still rages between manufacturers of optical and magnetic sound tracks for 8-mm film. It appears at this time, however, that the pendulum is swinging in favor of the magnetic sound track.

In 1965 "Super 8" film was introduced on the market. This new format provides a brighter, sharper screen image which is more than 50 percent larger in size than the standard 8-mm film. "Super 8" has now superceded the standard 8, but some projectors will handle either type of film.

The film cartridge or as it is more commonly known, the film loop, has given great impetus to the use of 8-mm film in the school because no threading or rewinding is required and the completely automatic projector can easily be operated even by preschool pupils. The film is an endless loop which is encased in a small plastic cartridge. In use the cartridge is simply pressed into a slot in the projector and the machine is turned on. Since the film is a loop, it will run continuously through the projector until the machine is turned off. The projector can be stopped at any point for close inspection of a single picture without burning the film. Although the equipment is ideal for self-paced instruction, it can be used as well with an entire class. The picture size can be varied from a small image on a piece of notepad paper to a "blow up" as large as 6 ft by 8 ft.

A variety of 8-mm projectors of both the silent and sound type are available today. Some models project the image onto a screen (see Fig. 15–5) while others are of the rear-projection type and resemble a small television set (see Fig. 15–6). This latter type can be operated in a fully-lighted room because the high intensity of the projected image covers a relatively small screen area. This feature lends 8-mm films for ideal use in the laboratory and in study carrells. A small, lightweight projector as shown in Fig. 15–5 can house a film loop with up to four minutes of running time. Larger models can hold a cartridge with 20 minutes of sound color film or 30 minutes of black and white sound film. Projectors are available with either a standard, a wide angle, or a zoom lens. A wide-angle lens provides the largest possible picture when the projection distance is limited. The zoom lens adjusts the size of the image to fit the screen without moving the projector. Some models have a dual brilliance control. "Normal" is used on short throws or in darkened rooms while "super" is used when the maximum brilliance is desired.

The 8-mm single-concept film is simply a cartridge-type film which features a single idea or thought where the use of motion is essential in creating the mental impression. "Boring a hole with an auger bit" and "setting type in a composing stick" are examples of typical titles of 8-mm single-concept films. These films vary in length from 30 seconds to four minutes. Industrial arts is rich in operations, procedures, and processes which lend themselves well to teaching by means of single-concept films.

In 1967 some 4,000 titles of commercially available 8-mm single-concept films were listed. In only a four-year span the number increased to well over 8,000 films available in Super 8 cartridges.[8] In the years immediately ahead this figure is expected to increase many fold. An astonishing number of 8-mm single-concept films suitable for use in industrial arts classes are presently available. In addition, there is evidence that more and more teachers in the field are producing their own film loops. Figure 15–7 shows an industrial arts student using an 8-mm projector. Note the unique teacher-made viewing box which is built into the wall of the laboratory.

Fig. 15–5. An 8-mm cartridge-type film projector suitable for use in an individual study carrell or with a large group of students. (Courtesy, Technicolor, Inc.)

The experimental comparison of single-concept film loops with other methods of instruction has invited the attention of a number of

[8]Richard Kahlenberg and Chloe Aaron "The Cartridges are Coming," *Cinema Journal*, vol. IX, no. 2 (Spring 1970), p. 7.

Fig. 15–6. This cartridge-type 8-mm sound projector looks like a television receiver. (Courtesy, Fairchild Camera and Instrument Corporation)

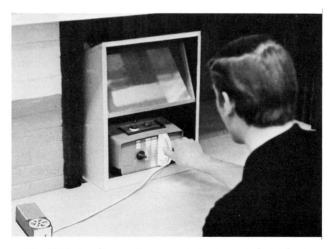

Fig. 15–7. Loading an 8-mm projector in an industrial arts laboratory. (Courtesy, Eberhard Thieme, Rochester City Schools, New York)

researchers. In a pilot study, Shemick and Sherk[9] investigated the teaching of a manipulative skill by means of 8-mm sound film loops. An experimental group of six ninth grade students received instruction in metal spinning solely by watching four single-concept film loops while an equal number of students received instruction via live demonstrations by the teacher (see Fig. 15–8). The results showed that the experimental group produced superior work by 13 percent even though this group averaged less shop experience, lower intelligence scores, and less hand strength. The experimental group requested less teacher assistance in spinning a metal bowl, but required more time to complete the work than did the live-demonstration group.

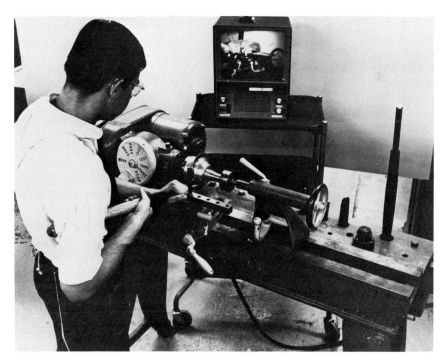

Fig. 15–8. Using 8-mm concept film loops to teach metal spinning. (Courtesy, John M. Shemick, The Pennsylvania State University)

The investigators reported that 8-mm film loops might be especially valuable in teaching manipulative operations to pupils who are not

[9]John M. Shemick and Dennis H. Sherk, "A Pilot Study in Evaluating the Use of Teacher-Made 8-mm Daylight Projected Film Loops for Individual Teaching of Motor-Sensory Skills," Central Fund for Research (University Park, Pa.: The Pennsylvania State University, 1965).

verbally oriented and to those less gifted academically. In the latter instance they noted an advantage in that learners can see the film loops again and again without fear of ridicule from their peer group. They concluded that "the concept film then is potentially an extremely useful technique especially if it is used intelligently and judiciously to meet individual pupil needs and teaching conditions.[10]

In another research study Sommer sought to answer this question: Does the use of silent single-concept loop films facilitate the acquisition of manipulative occupational skills for nonacademic students?[11]

Forty-two vocationally-bound eighth and ninth grade students who were low in academic achievement were divided into three treatment groups: (1) teacher only, (2) teacher plus films, and (3) films only. The acquisition of occupational skills was judged from performance test scores.

The findings showed that students acquired significantly more skill with teacher plus films than with either of the other two treatments. There was no significant difference noted in the acquisition of skill between teacher only and films only treatments.

In 1968 Kruppa[12] tested the relative effectiveness of three teaching techniques: (1) class demonstration only, (2) overview-film demonstrations followed by identical class demonstrations, and (3) overview-film demonstrations followed by optional student use of concept films with supplementary study guides. A major finding of this research was that a substantial savings in teacher time was noted in the use of the overview-film demonstrations plus concept films technique without any loss in the acquisition of technical knowledge or in student performance of constructed products. Additional details on this study will be found in Chapter 10.

Sources of 8-mm films and film loops will be found in *Selected References* at the end of this chapter.

5. Filmstrips

A filmstrip consists of a series of still pictures in sequential order on 35-mm film. A typical filmstrip consists of 25 to 50 individual frames or pictures. Filmstrips are made in color as well as in black and white. The

---

[10]John M. Shemick, "Learning Skills via 8-mm Concept Film Loops: Boon or Boondoggle?" *The Quarterly*, Western New York School Study Council, vol. XVII, no. 2 (January 1966), p. 8.

[11]Seymour A. Sommer, "The Use of Silent Single Concept Loop Films to Facilitate the Acquisition of Occupational Skills" (unpublished doctoral dissertation, Rutgers University, 1971).

[12]J. Russell Kruppa, "A Comparison of Three Teaching Techniques Using Instructional Films in Selected Units in Industrial Arts" (unpublished doctoral dissertation, The Pennsylvania State University, 1968).

Instructional Media—Part II 357

pictures are then projected onto a screen by means of a projector which makes it possible to study each picture and then advance the strip to show successive frames. Captions or explanatory notes may appear either on separate frames or be superimposed on the pictures which precludes the necessity for teacher explanation or comment. The captions are limited in most cases to about 10 to 15 words. Sometimes the teacher reads the captions aloud to the class and often adds supplementary information. Generally filmstrips are used with a class, but they can also be used by individual students either with a projector or with a hand viewer.

Sound filmstrips utilize a tape recorder or a phonograph record to supply background music or a running narration coordinated with the individual frames. In this case a bell-like signal is heard at the appropriate time which alerts the operator to advance the filmstrip manually to the next frame. Some filmstrips and projectors utilize an inaudible signal in the sound track which trips the change mechanism of the projector automatically.

New models of "suitcase style" projectors feature manual and remote controls, instant threading, and rear view screens especially suited for viewing in brightly lighted rooms. Some models permit a combination of filmstrip and slide projection, tachistoscopic instruction, and the projection of microscope slides.

The filmstrip is really a combination of certain features of both the motion picture and the slide. Filmstrips are relatively inexpensive, easy to store, and easy to use. Despite these advantages, filmstrips have two major weaknesses:

1. Filmstrips are inflexible in that the pictures must be shown in a predetermined order. If used with a recorded commentary, the frames must be kept in synchronization with the sound track or record. This precludes the possibility of supplementary instructional information being added by the teacher.

2. It is difficult to indicate motion with filmstrips, although some progress has been made in this direction. Reference has already been made to the extensive film research conducted at Penn State under the auspices of the Army and the Navy. In one of these studies Roshall[13] concluded that filmstrips are less effective than motion pictures for teaching skills in which motion is involved.

Many excellent filmstrips developed by the armed services for training are now available to industrial arts teachers. Filmstrips are available through sources similar to those for films. Major sources for filmstrips

---

[13]S. M. Roshall, "Effects of Learner Representation in Film-Mediated Perceptual-Motor Learning," Technical Report—SDC 269-7-5 (Port Washington, Long Island, N.Y.: U.S. Naval Training Device Center, 1949).

will be found in *Selected References* at the end of this chapter.

Some industrial arts teachers have experimented with making their own filmstrips. Although satisfactory results have been obtained in most cases, considerable skill is necessary to produce a strip that will approach the commercial product. Industrial arts teachers have had much greater success in producing color slides which will be discussed in the next section.

6. Slides

As a teaching device, slides are especially useful in industrial arts education. The ease with which these color transparencies can be made and their adaptability to many teaching situations are strong recommendations in their favor. For these reasons the use of teacher-made slides has increased phenomenally in recent years.

Two types of slides commonly used are the glass slide and the transparency. Although still available, glass slides have been superseded by the transparency, particularly the color transparency. Glass slides yield a larger, more brilliantly lighted image whose detail is considerably greater than a color transparency. Despite these advantages, glass slides are used only rarely today.

As its name implies, the transparency is a picture made directly on a transparent film and is designed especially for projection on a screen. Most slides of this type are made from pictures taken with the 35-mm camera. For black and white transparencies, the film is developed as a negative which is then printed on positive film. On the other hand, color transparencies are developed on the original film by a special reversal process. After separation from the strip, each transparency is mounted in a cardboard frame as a 2×2-in. slide.

If the slides are subject to frequent showing or will be used by students, it is desirable to have them glass mounted instead of cardboard-mounted to prevent wear and tear and to eliminate possible damage due to fingerprints and dust. It is possible to crop slides if a double-glass mount is used. In cases where slides are used frequently, but the teacher prefers not to use glass mounts, considerable protection can be given to the slides by inserting them in aluminum binders.

The production of a teacher-made series of 2×2-in. slides based on a single theme depicting sequence or continuity in action is not at all difficult. Such slide sets can be invaluable in teaching informative units as well as operations, processes, and safe work habits. Slide sequences are relatively inexpensive to produce and have the unique advantage of being custom-tailored to each instructional program and laboratory facility. It is often easier and more effective for the teacher to make his own set of slides than to adapt his instruction to fit the slides available from commerical sources. Most students are eager to assist in shop-made productions thus providing a ready supply of stage hands, assistants, and even "actors."

The basic equipment for making a set of color transparencies is a 35-mm camera. Only simple camera techniques need be used to produce slides which are educationally effective. Initially, titles can be made by writing on the matte surface of a "write-on-slide" or by photographing lettering on the chalkboard. Wallpaper and throw rugs also make interesting backgrounds for titles. White letters superimposed over a background scene are attractive and add finesse to titles, and this double exposure technique is easy to accomplish without additional equipment. The industrial arts teacher who is interested in making slides probably will want to visit the local photographic store in search of ideas for adding professional touches to his slides. He may wish to acquire one of the several types of inexpensive titling sets currently available. A wealth of how-to-do-it literature is available from photo stores, public libraries, and industrial arts periodicals.

It is often desirable to add sound effects, narration, explanatory notes, or background music to a set of slides. Sound makes a slide presentation much more interesting and effective. "Live" sound can be added during the showing by just talking or by means of a microphone and speaker. Better yet, the use of a tape recorder or audio cassette relieves the burden of the teacher by ensuring a uniform presentation each time the slides are shown. First, a script must be carefully written for the narrator; ad libbing should be avoided in all cases. Coordinating details must be planned also if other sound effects are used, such as a musical introduction or ending of the program. Records of background music made especially for use with amateur shows are available from local photographic dealers; in addition, classical records provide a good source of appropriate background music. Proper volume for musical introductions and "swells" of music between slides with proper fading to the next slide can be attained with a little practice.

If the synchronization between the tape and the slides is to be handled manually during the presentation, some sound signal (bell or bong) indicating time to change slides will have to be included in the script. If the slides are to be used with certain types of automatic projectors, it is possible by means of a special device to place a magnetic impulse on the tape which causes the projector to advance the next slide automatically each time the signal is detected.

A new system called sound-on-slide features a single recorder-projector which both records and plays from a sound track and also projects the slides. A conventional slide is mounted in a 4-in. square plastic mount on one side of which is a circular magnetic sound track. This track has a maximum recording time of 35 seconds. After the commentary is recorded on a slide, a button may be depressed on the machine which records an inaudible signal that will cause the next slide to be advanced thus providing automatic projection. Old sound may be erased and new sound inserted whenever desired. Since each slide has its own sound track, the slides may be rearranged in any order without affecting narration for that slide. Figure 15-9 shows a sound-on-slide recorder-projector which holds a tray of 36 slides.

Fig. 15–9. This compact unit both records and plays sound as well as projects 2 × 2-in. slides. (Courtesy, 3M Company)

Slides may be a little more difficult to handle than filmstrips because they get turned upside down or reversed and are easily disarranged out of sequence. However, their flexibility is also one of their chief advantages because it permits easy revision, updating, and selecting of specific slides to fit special teaching situations. Some slide projectors feature trays which double as storage containers for the slides as well as part of the slide-changing mechanism. With such devices the slides are held in proper position for showing and cannot become disarranged accidentally.

The newer type projectors greatly facilitate the use of slides in the laboratory by offering both manual and automatic slide changing, remote control including focusing, and zoom lenses for enlarged pictures.

Instructional Media—Part II   361

7. Overhead Projections

These are simply projections on a wall or screen of large transparencies via a special machine called the overhead projector (see Fig. 15-10). A transparency for overhead projection is a thin sheet of clear or colored acetate, 8½ × 10½-in., mounted in a cardboard frame whose inside dimensions or visible working field is 7½ × 9½-in. The sketch or words to be projected may be hand-drawn with grease pencil or felt-tip marking pen directly on the acetate, or the image may be photocopied by means of a special copying machine. In use the teacher merely places the transparency, sometimes called the projectual, on the lighted glass stage of the projector, thus projecting the image onto a conventional wall or ceiling-hung screen.

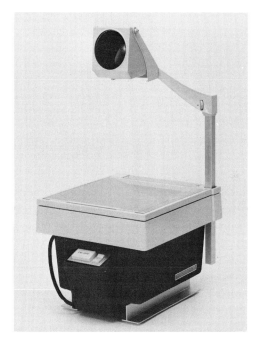

Fig. 15–10. An overhead projector.
(Courtesy, 3M Company)

The overhead projector is a unique visual teaching tool; some of its advantages include the following:

1. When using the overhead projector, the teacher faces the class while the instructional material is projected on a screen either in front of or slightly to one side of the group (see Fig. 15-11).

Fig. 15-11. One advantage in using the overhead projector is that the teacher faces the group. (Courtesy, John M. Shemick. Photo by Dominic Bencivenga, The Pennsylvania State University)

2. Class interest in the lesson can be maintained because transparencies need not be on the screen continuously; instead they are used only when the instructor wishes to illustrate a point. By turning the machine off and on, the teacher may shift the attention to the lecture, class discussion, or to some other instructional aid.

3. Projectuals can be seen well in a normally lighted room. Note in Fig. 15-11 the fluorescent lights are turned on in the laboratory.

4. Once made, projectuals are permanent and will neither fade nor discolor. This means that the teacher is spared many hours of preparation, redrawing, chalkboard work, and the like.

5. Projectuals can be made from original copy quickly and easily by a copying machine such as is illustrated in Fig. 15-12. Original copy can be prepared by writing, lettering, or drawing on a sheet of white paper. Excellent copy can be made quickly on a primary (large letter) typewriter. The copy is placed in contact with the transparency film and fed into the infrared copying machine which exposes and "dry develops" the transparency in a few seconds. Hand-drawn transparencies may be

made with grease pencils or felt-tip marking pens. Transparencies can also be made by photographing the copy and using the negative as a transparency. This technique has been used with electrical meter faces and with floor plans used in shop layout courses. Then there is the diazo technique for making transparencies.

Fig. 15-12. This copying machine makes a transparency in a matter of seconds. (Courtesy, 3M Company)

Overhead projection is an effective means of illustrating informative lessons in industrial arts and of teaching the knowing aspects behind manipulative skills—that is, in teaching step-by-step procedures in both shop and drawing work. The industrial arts field is rich in sequential operations, internal working constructions, processes, cycles, and procedures which lend themselves well to the use of the overhead projector. Relationships of components, working parts, cross sections, and exploded views which reveal interior details are only a few of the situations in which the unique contributions of overhead projection can be utilized. In addition, overhead projection is useful for flash-type drill, review, and testing.

Some of the techniques by which the industrial arts teacher can use the overhead projector in the laboratory to facilitate learning and to create high student interest include the following:

1. Pointing. It is unnecessary for the teacher to stand near the screen and use a long pointer to direct the students' attention to some detail. Instead, any opaque object placed over the transparency will appear in silhouette form on the screen and will focus attention on that portion of the picture. For example, a pencil, pen, cardboard arrow, or even a finger can be used as a pointer. To avoid movement and possible student distraction, the pointer should touch the transparency and the teacher's hand should rest on the projector or the glass stage. Note in Fig. 15-11 that this is exactly what the teacher is doing with a pencil as a pointer.

2. Revelation. This is a technique whereby the teacher covers part of the transparency and so controls what the class sees and reads. It is a device simply to keep the class together with attention focused on a certain portion of the transparency. A sheet of opaque paper makes a good mask. As the points in the teacher's explanation unfold, the mask can be shifted to uncover new material. A 6-point printer's slug (or other suitable weight) can be cemented along the top edge of a cardboard sheet (8½ × 11-in.) to make an excellent weighted mask. The weight permits revelation clear to the bottom line of the transparency without the mask falling off the projector stage. For other than line-by-line revelations, the use of masks which are hinged to the cardboard mount offer a useful technique.

3. Chalkboard Substitute. By writing with a grease pencil on a sheet of clear acetate, the overhead projector can be used in lieu of the chalkboard. The advantage is, of course, that once a transparency has been prepared it may be used for years whereas chalkboard material must be prepared anew each time. A partially completed transparency can be prepared in advance and later brought to completion with a grease pencil in front of the class. Erasures are made quickly and easily with a soft cloth or paper tissue. One company currently features a "lightboard." This consists of a sheet of clear acetate and an opaque film which are mounted together in a cardboard frame. Any blunt-pointed instrument can be used as a writing stylus to write on the top film thus producing a white-on-gray image on the screen. When the top film is lifted, the image disappears and the lightboard may be used over and over again.

Some overhead projectors are fitted with a scroll attachment on which long lengths of clear acetate can be rolled across the projector stage. The teacher can then either roll into place for projection prepared notes and sketches or use the surface as a chalkboard merely by rolling new acetate across the stage as needed. Figure 15-13 shows a projector with a scroll attachment.

4. Color. Overhead transparencies can be made more attractive by the use of color. In addition to the conventional black image on a white background, a wide variety of available film types makes possible other interesting combinations including black image on a colored background, white or colored image on a black background, and colored image on a white background. These transparency films are used with original copy and the infrared copying machine.

Instructional Media—Part II

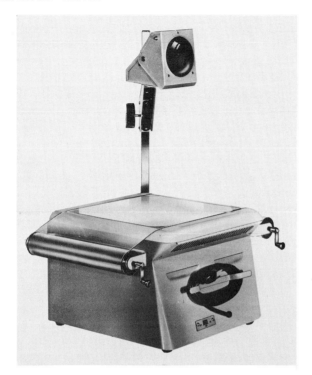

Fig. 15–13. This overhead projector is equipped with a scroll attachment. (Courtesy, Projection Optics Company, Inc.)

There are a number of ways by which color can be added to transparencies. Five simple techniques are: (1) use one of the film types indicated above, (2) color the desired area with a felt-tip marking pen, (3) apply transparent acetate inks with a small brush, (4) block in areas with self-adhesive colored film, or (5) apply transparent colored tape.

Color transparencies can also be made by the diazo process which is a dry-developing method involving ammonia vapor and azo dyes. This is a machine method of producing color transparencies and involves a special exposure machine and developing equipment. Careful attention to detail will produce professional looking results.

Also, by means of "lift films," color pictures suitable for overhead projection can be lifted directly from magazines which have been printed on clay-base paper.

5. Overlays. One of the most effective ways to use the overhead projector is with overlays. This technique is invaluable for presenting complex material which can be broken down or analyzed into smaller segments or steps. Basically, an overlay consists of several transparencies which are in register and hinged together so that when shown sequentially the images are superimposed upon each other, thus producing a progressive build-up of a complex whole.

To make an overlay, the first step is to make a basic transparency which is taped or stapled to the cardboard frame. Then the next segment of information to be presented is made into a transparency. This is placed in register over the basic transparency and hinged to one side of the mount or frame. Other transparencies depicting additional steps or informational sequences are made, registered, and hinged to other sides of the mounting.

In use, the transparency is placed on the projector stage with all overlays opened and hanging down over the sides and front of the projector. After presenting the basic transparency, the teacher then "flips over" the first overlay so that it rests on top of the basic transparency. As the lesson unfolds, other overlays are "flipped over" in turn. In this manner complex processes can be built up one step at a time.

Overlays are especially applicable in industrial arts because this field is so rich in complex material, procedures, and sequential operations that lend themselves well to analysis into simpler components or segments. Specific examples of the uses of overlays include exploded views of parts of tools, machines, and equipment; projections and constructions in mechanical drawing, sketching, and planning; procedures and operations for hand and machine tools; electrical diagrams; and concepts of fundamental processes of industry.

6. Transparent Objects. Certain tools, such as, rules, scales, protractors, lettering guides, and slide rules are available in clear plastic. When placed on the stage of the overhead projector, these objects will be projected onto the screen with their markings clearly visible. Clear plastic rules, triangles, and curves are also useful in drawing on a transparency while it is being projected. The use of transparent objects can add considerable interest to lessons on scale and angle reading, measurement, and slide rule work as well as simple drawing, sketching, and planning.

7. Opaque Objects. Not all objects used with the overhead projector need to be transparent. Any opaque object placed on the stage will produce a silhouette on the screen. The use of cardboard arrow as a pointer was referred to previously. In addition, cardboard and metal cut-outs can be used to demonstrate gears, wheels, and other shapes. A magnet and iron filings can be used to show magnetic lines of force on the overhead projector.

8. Animated Effects. It is possible to depict motion with the overhead projector by means of animated transparencies. These are clear plastic plates on which movable parts, some in color, are mounted. For example, two or more colored gears mounted on a plastic plate can be utilized to provide a working demonstration of gear ratios and synchronization. The device can be either manually operated or driven by a small low-speed motor attached to the drive gear with a flexible shaft.

In addition to showing motion, animated transparencies are also useful in teaching linear and angular measurement, relationships between moving parts, meter reading, vernier and micrometer reading, and in those instances where adding-to and taking-away representations are needed. Angular measurements, circular arrays, and geometric constructions are examples of animated transparencies which are available commercially.

Some electrical meter movements have transparent faces which when used with the overhead projector simplify the teaching of meter reading to a group. If the meter face can be separated physically from the meter movement, but still connected electrically, realistic teaching will ensue when the students observe the meter actually measuring varying loads which are introduced into the external circuit.

9. Simulated Motion. Apparent motion can be depicted in selected areas of an overhead transparency by using polarized film material in conjunction with a special polarizing spinner attachment affixed to the head of the overhead projector. This technique produces a "magic" effect because liquids, gases, electrons, or lines appear to move from one point to another. Apparent linear motion, rotational movements, vibrating motion, radiational effects, reversing actions, and turbulence can be simulated. Changes in light intensity can be effected thus causing areas of the transparency to flash on or off, to fade away, or even to change color.

Polarized film, or "motion material" as it is sometimes called, is available in the following forms: linear motion (from very slow to fast), reciprocating motion, vibrating motion, turbulence (from fine to heavy), radiational motion, reversing action, and on-and-off action. This material is in sheet form and can be easily cut and fitted to the desired area of the transparency. Adherence is assured because of the pressure-sensitive backing on the motion material.

When in operation the spinner unit on the projector rotates a polarized filter disk at speeds from 40–80 rpm, thus producing a blinking effect which in combination with the polarized motion material on the transparency creates the illusion of motion. Figure 15-14 shows a polarizing spinner attachment.

Useful transparencies are now commercially available for all major

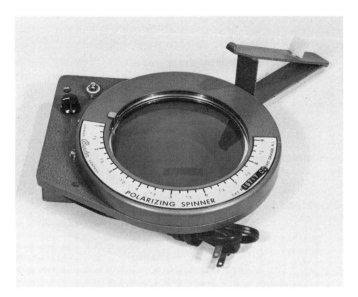

Fig. 15–14. When used with polarized motion material, this polarizing spinner attachment gives the illusion of motion on the overhead projector. (Courtesy, Charles Beseler Company)

areas of industrial arts education. At least one company produces printed copies of original drawings for the teacher to make into transparencies. However, one of the principal advantages of the overhead projector is that the teacher can quickly and easily make his own projectuals. With a copying machine a transparency can be made from original copy in a few seconds.

Compared with other teachers, the industrial arts instructor is in an enviable position in regard to making original copy involving drawing, sketching, and lettering by virtue of his professional training and experience. What may appear difficult to others is a relatively simple task for him. For this reason industrial arts teachers have been quick to accept and to utilize overhead projection in the laboratory.

8. Still Pictures

The category of still pictures encompasses a great variety of non-projected visual materials, such as flat pictures, charts, drawings, maps, sketches, diagrams, and graphs. The best use of flat pictures requires that they be studied carefully by the student; in fact, when used in this manner they are called *study prints*. For this reason the teacher should avoid using too many still pictures at one time. Simply decorating the laboratory with a large number of prints is, for the most part, a great waste of educational potential.

An effectual means of utilizing study prints in the industrial arts laboratory is in combination with a study assignment sheet (see Chapter 9). For example, a detailed picture of some industrial process is selected by the teacher who then prepares a set of questions based on the picture. The questions may involve (1) simple naming of tools, machines, and equipment appearing in the picture, (2) grasping overall importance of the industrial process depicted; (3) observing pertinent details, such as dress of workmen and safety devices in use; (4) connecting the picture to previous learning; and, (5) making inferences and generalizations. Questions related to time, distance, product texture, and relative size may also become important study guide questions. Sometimes two pictures are useful in promoting comparisons between industrial processes or between old and new methods, and the like. Although black and white pictures are highly satisfactory for use as study prints, some believe that color pictures may have greater student appeal and motivating effect.

Some industrial arts teachers use a series of flat pictures which depict the correct operational sequence or steps-to-follow in operating a certain machine tool safely. The pictures are mounted and hung alonside or over the machine so that the student may quickly make a final check before using the equipment.

Still pictures can, of course, be used to advantage in teacher lectures and presentations. If pictures are held up in front of the class, probably not all students will be able to see them well. If they are passed around the lecture area, class attention will be divided and confusion may ensue. The pictures can be displayed before the class and then placed on the bulletin board for further consideration by class members. Better yet, the pictures may be projected in an opaque projector during the class presentation. This technique provides simultaneous presentation and class discussion.

Still pictures are plentiful and inexpensive to obtain. Many industrial and commercial firms are willing to make flat pictures available upon request. Some magazines and textbooks are good sources of flat pictures. In addition, the teacher can bring back pictures from vacation trips and industrial visits.

If flat pictures are used to any extent in the industrial arts laboratory, it is advisable that careful attention be given to mounting them and to developing a filing system. Compared with unmounted prints, mounted pictures last longer, are easier to use in the opaque projector and on the bulletin board, and lend themselves better for use as study prints. Mountings also make pictures more attractive and generally facilitate filing and handling. Dry-mounting tissue is recommended because rubber cement eventually causes discoloration. A standard size for mounting boards should be selected to expedite handling and storing. A stiff mounting board, preferably of some neutral shade, is recommended. Part of the mounting involves protecting the surface of the picture. This can best be done by (1) plastic lamination, (2) brushing on a coat of flat lacquer, or (3) using a pressurized can of plastic spray.

The filing system should make it easy to locate any picture needed. Use of subheadings by course is a good method of filing still pictures. Either a regular or a legal-size filing cabinet makes an adequate storage facility. Provision should be provided in the filing system for expansion.

9. Opaque Projections

The opaque projector is a device which is much used as an aid to instruction. As the name implies, it is designed to reflect the images of opaque materials such as pictures, drawings, or even actual objects so that they may be viewed on a screen. Basically, an opaque projector is simply a means of reflecting light from any desired drawing or object, focusing it through a series of lenses, and projecting the resulting image on the screen. Figure 15–15 illustrates an opaque projector.

The opaque projector is especially useful in supplementing lectures or class discussions because such a large variety of materials may be used. A selected page from a textbook or reference book may be inserted in the projector and viewed for any length of time by the whole class without harm to the book. Unmounted pictures or sketches can be used with equal ease. Small articles or parts may also be accommodated. Another advantage of the opaque projector is that the reproduction on the screen is in natural color.

The major disadvantages of the opaque projector are that considerable heat is generated and, unless a fan is built into the projector, the paper is likely to scorch. Also, the intensity of the reflection is relatively low and so a fairly dark room is essential for a bright image. Another disadvantage of the opaque projector is related to its size and weight. At best the opaque projector is heavy and bulky compared with other types of projectors. There is an opaque projector on the market which is hand-held over a drawing or object resting on a table. This is a useful device, but unfortunately, it has a small field which precludes its use with whole pages of textbooks, drawings, or illustrations.

Instructional Media—Part II

Fig. 15-15. Flat pictures and small three-dimensional objects can be projected with the opaque projector. (Courtesy, Projection Optics Company, Inc.)

PHOTOGRAPHY FOR THE INDUSTRIAL ARTS TEACHER

The use of instructional media is of such importance in teaching the skills and knowledges of industrial arts that it behooves the teacher to become proficient not only in their use and application but in their production as well. The industrial arts teacher can greatly increase the effectiveness of his instruction by preparing audio-visual materials especially fitted to his instructional program. One of the best and easiest means of accomplishing this is through the medium of photography. Effort expended in the mastery of the principles of photography will be repaid many times over. Modern camera features, such as range finders, electric eye lenses, automatic film advance, easy film loading, and the latitude of the newer film emulsions have greatly simplified the matter of taking good pictures quickly and easily.

The industrial arts teacher will find many uses for photography in his daily teaching of both classes and individual students. Flat pictures, slides, or overhead projectuals can make a teacher's presentation more effective and interesting. Extreme close-ups are especially valuable in teaching certain industrial operations, such as the action of a cutting tool or the revelation of action which the learner could not otherwise see. When a close-up of some detail is projected on a screen and thus enlarged many times, the result is often nothing less than dramatic. Enlarged, close-up views of wood grain, foundry type, screw threads, leather lacing, and other small objects will be found to attract great student attention and to add considerable interest and color to the lesson.

Pictures taken through a microscope can reveal views of grain, structure, and other details that ordinarily would go unnoticed. Pictures of projects and students at work are useful in public relations programs. Industrial visits, on-the-job scenes, and individual or class research problems can be recorded on film for later use.

In addition to single flat pictures, the production of a series of 2×2-in. slides based on a single theme with sound added via a tape recorder or audio casette can do much to upgrade the instructional program. These are not difficult to produce and have been described briefly in a preceding section of this chapter.

The 35-mm camera is probably the best all-around camera for the industrial arts teacher to use in producing the variety of pictures previously described. This type of camera will produce color or black and white prints with enlargements up to 8×10-in. or even larger, depending upon the quality of the negative; 2×2-in. color or black and white slides; and black and white or color filmstrips.

It will be to the advantage of the industrial arts teacher to acquire a 35-mm camera and to become proficient in its use. Careful study of one or more of the books on basic photography and a little patient practice with the camera will produce satisfying results far beyond the cost and effort expended. The end result will greatly improve the teaching of industrial arts.

SUMMARY

Continuing the presentation of selected instructional media which began in Chapter 14, this chapter considered the following aids to instruction:

1. *Educational television* has opened a whole new field of educational possibilities. Enriched educational experiences are now attainable on a scale never before dreamed possible.

2. *Closed-circuit television* is being utilized by both high school and college instructors of industrial arts primarily to supplement demon-

strations. Videotape recordings have contributed much to the flexibility and effectiveness of closed-circuit television.

3. Motion picture *films* can be used effectively to teach understandings, appreciations, attitudes, and certain skills. Selected conclusions based on 65 film research studies were cited in the chapter. *8-mm single-concept film loops*, some with sound, are being used to teach concepts of industry as well as processes and operations.

4. *Filmstrips* with or without sound are useful in teaching some of the informative aspects of industrial arts. They may be used with an entire class or viewed by individual students with hand viewers.

5. The use of *slides*, especially teacher-made color transparencies, has increased greatly in recent years. In addition to being relatively inexpensive to produce, they can be custom-tailored to fit individual instructional needs. Sound can be added easily to slide sets by means of a tape recorder, an audio cassette, or a sound-on-slide projector.

6. *Overhead projections* of black and white or colored tranparencies are especially effective for illustrating informative lessons, constructions, and step-by-step procedures. Projectuals are quickly and easily made from original copy. The illusion of motion, animated effects, and the projection of transparent objects are techniques in overhead projection which contribute significantly to the learning process and to student interest.

7. *Still pictures* are easy to obtain from a variety of sources. When used for instruction, they must be studied carefully by individual students. For this reason the use of study assignment sheets in combination with still pictures is recommended. Still pictures should be mounted and filed for ready reference.

8. *Opaque projections* are projections in natural color via an opaque projector of any opaque material, such as pictures, drawings, and charts, as well as small objects which can be fitted into the projector.

In view of the many instructional aids which are based on photographic processes, the industrial arts teacher is urged to master the fundamentals of photography by becoming proficient in the use of the 35-mm camera.

DISCUSSION TOPICS AND ASSIGNMENTS

1. Make a study of the television programs available in your area and indicate which of them might be useful for educational purposes.

2. To what uses, other than supplementing the demonstration, do you think closed-circuit television can be adapted in the industrial arts laboratory?

3. Explain the idea behind 8-mm single-concept film loops. Cite several examples of possible titles.

4. How would you use film loops in the industrial arts laboratory? Cite research findings to support your statements.

5. From a teacher's point of view, what are the unique advantages of using overhead projectuals?

6. Besides those mentioned in the text, what other examples can you suggest of animated transparencies useful in teaching industrial arts?

7. Select an appropriate still picture related to some aspect of industry or technology and prepare a study assignment sheet to be used with it.

SELECTED REFERENCES

Abrams, F. Russell. "Techniques for Preparing Slide Film Lessons for Industrial Arts," *Industrial Arts and Vocational Education*, vol. 53, no. 9 (November 1964), pp. 26–27.

Amthor, William D. "How to Make Your Own Filmstrips," *School Shop*, vol. XXVI, no. 9 (May 1967), pp. 46–47.

Barnard, David P. "Change, Educational Technology, New Media and You," *Industrial Arts and Vocational Education*, vol. 58, no. 6 (June 1969), pp. 18–19.

Beatty, Donald T. "Enlarging the Use of the Overhead Projector," *School Shop*, vol. XXVII, no. 8 (April 1968), pp. 103–104.

Brzezinski, Walter H. "A Closed-Circuit TV System for Enriching Electronics," *Industrial Arts and Vocational Education*, vol. 50, no. 5 (May 1961), pp. 26–27.

Cipolletti, George. "Custom Filmstrips for Your Shop," *Industrial Arts and Vocational Education*, vol. 53, no. 1 (January 1964), pp. 25–27.

Doan, Cortland C. "How to Produce Sound-Slide Shows," *Industrial Arts and Vocational Education*, vol. 51, no. 9 (November 1962), pp. 26–27.

Dudley, Stanley A. and John Bieber. "Extending the Use of the Overhead Projector," *Industrial Arts and Vocational Education*, vol. 58, no. 6 (June 1969), pp. 30–32.

Dutton, Bernard. "A Guide for Teacher-Produced 8-mm Films," *Industrial Arts and Vocational Education*, vol. 53, no. 10 (December 1964), pp. 35–36.

Educational Media Index. New York, N.Y.: McGraw-Hill Book Company, Inc. 1964.

Educators Guide to Free Films. Randolph, Wis.: Educators Progress Service, Inc.

Educators Guide to Free Filmstrips. Randolph, Wis.: Educators Progress Service, Inc.

8-mm Film Directory. Complied by Educational Film Library Association. New York, N.Y.: Comprehensive Service Corporation, 1969.

Friedman, Nathan L. "Instant Playback in the Shop," *Industrial Arts and Vocational Education*, vol. 57, no. 1 (January 1968), pp. 34–35.

Hayes, Harold D. "Using Audio-Visual Materials in Industrial Education," *Industrial Arts and Vocational Education*, vol. 58, no. 6 (June 1969), pp. 20–23.

Herbert, Harry A. "Single Room Closed-Circuit Television in Industrial Education" (unpublished master's thesis) Stout State University, Menomonie, Wis., 1964.

Hess, Harry L. "Producing Closed Loop 8-mm Instructional Films," *Industrial Arts and Vocational Education*, vol. 58, no. 10 (December 1969), pp. 28–29.

Hocking, Charles. "How the Use of Media Can Help the Industrial Arts Instructor," *Industrial Arts and Vocational Education*, vol. 58, no. 6 (June 1969), pp. 24–26.

Index to 16-mm Educational Films. Complied by National Information Center for Educational Media at the University of Southern California. New York, N. Y.: R. R. Bowker Company, 1969.

Index to Overhead Transparencies. Complied by National Information Center for Educational Media at the University of Southern California. New York, N. Y.: R. R. Bowker Company, 1969.

Index to 8-mm Motion Cartridges. Complied by National Information Center for Educational Media at the University of Southern California. New York, N. Y.: R. R. Bowker Company, 1969.

Index to 35-mm Educational Filmstrips. Complied by National Information Center for Educational Media at the University of Southern California. New York, N. Y.: R. R. Bowker Company, 1969.

Instructional Film Research, Vol. II, Technical Report No. SDC 269-7-61, NAVEXOS P-1543. Port Washington, Long Island, N. Y.: U.S. Naval Training Device Center, 1956.

Kemp, Jerrold E. "A Better Way to Show the Way," *School Shop*, vol. XXX, no. 9 (May 1971), pp. 29–31.

Kenneke, Larry J. and Robert D. Rose. "Putting Motion in Your Overhead Transparencies," *Industrial Arts and Vocational Education*, vol. 60, no. 5 (May–June 1971), pp. 24–26.

Kruppa, J. Russell. "A Comparison of Three Teaching Techniques Using Instructional Films in Selected Units in Industrial Arts" (unpublished doctoral dissertation, The Pennsylvania State University, University Park, 1968).

Lauda, Donald P. and Robert D. Ryan. "Zoom-in on Industry and Instructors in Action," *School Shop*, vol. XXIX, no. 4 (December 1969), pp. 40–41.

Library of Congress Catalog: Motion Pictures and Filmstrips. Washington, D. C.: Library of Congress, Card Division, Bldg. 159, Navy Yard Annex (published annually).

Manning, Harold D. and William H. Turner. "Step-by-Step Photos Teach and Re-Teach," *School Shop*, vol. XXIX, no. 5 (January 1970), pp. 36–37.

Motion Pictures and Filmstrips of the United States Government. National Audiovisual Center, Washington, D.C.: National Archives and Record Service, General Services Administration.

Peterson, Charles H. "Shooting Super-Loops in Super-8," *School Shop*, vol. XXX, no. 6 (February 1971), pp. 62–64.

Schoenhals, Neil L. "The A-V Hardware is Here . . . But Where's the A-V Software?" *Industrial Arts and Vocational Education*, vol. 58, no. 6 (June 1969), pp. 28–29.

Schoenhals, Neil L., Frederick R. Brail, and David W. Hessler, "Before You Buy Audio-visual Hardware/Software," *Industrial Arts and Vocational Education*, vol. 60, no. 9 (December 1971), pp. 18, 20–21.

Schultz, Morton J. *The Teacher and Overhead Projection.* Englewood Cliffs, N. J.: Prentice-Hall, Inc., 1965.

Schwab, William C. "Videotape Demonstrations in the General Shop," *School Shop*, vol. XXIX, no. 7 (March 1970), pp. 54–55.

Seal, Michael R. "How to Make Single Concept Films Without Elaborate Equipment," *Industrial Arts and Vocational Education*, vol. 57, no. 4 (April 1968), pp. 40–41.

Shemick, John M. "Learning Skills via 8-mm Concept Film Loops: Boon or Boondoggle?" *The Quarterly*, Western New York School Study Council, vol. XVII, no. 2 (January 1966), pp. 6–8.

Sommer, Seymour A. "The Use of Silent Single Concept Loop Films to Facilitate the Acquisition of Occupational Skills" (unpublished doctoral dissertation, Rutgers University, New Brunswick, 1971).

The Audio-Visual Equipment Directory. Fairfax, Va.: National Audio-Visual Association, Inc. (published annually).

The Blue Book of Audio-Visual Materials. Published annually in August issue of *Educational Screen and Audio-Visual Guide*. Chicago, Ill.: Educational Screen & Audio-Visual Guide.

Thompson, Dorothy. "Applying the Media to a New Electronics Career Course," *School Shop*, vol. XVII, no. 8 (April 1968), pp. 86–87.

Warner, M. E. "How to Use the Overhead Projector to Publicize Industrial Arts," *Industrial Arts and Vocational Education*, vol. 52, no. 6 (June 1963), pp. 18–19.

Wolbert, Warren D. "Making Diazo Transparencies for the Overhead Projector," *Industrial Arts and Vocational Education*, vol. 54, no. 5 (May 1965), pp. 36–37.

Wooldridge, Robert E. "Teaching the Vernier Caliper," *Industrial Arts and Vocational Education*, vol. 60, no. 6 (September 1971), pp. 36–37.

Chapter 16

Community Resources

As used in this text, a community resource may be defined as any resource, either human or material, used to further the objectives of industrial arts and available in the community where the school is located. The industries, museums, natural products, and the people of a given community are examples of community resources. Thus the term *community resources* refers to the use of human talent and material resources of a locality for educational purposes. It is another example of realistic learning in which students obtain knowledge directly through first-hand experiences with people, places and things in their community.

Human and material resources within a community can play an influential role in vitalizing the offerings of a school. Community resources are among the most important materials of the curriculum because they involve in the learning process both the pupil and the environment in which he lives. Wittich and Schuller note:

> A program calling for *planned use of the community* enables us figuratively to go "through the walls of the classroom" and see for ourselves what people and their activities are like, and to catch a glimpse of what life may hold for us tomorrow.[1]

In most communities there is usually a wealth of material resources and human talent which offer innumerable opportunities for first-hand learning experiences by school pupils. Properly used, these resources will both enhance and enrich the curriculum as well as provide an inexpensive means to upgrade the quality of education. The use of local resources is applicable to any subject area and to any grade level from preschool through senior high school. However, the integration of such resources into a given curriculum calls for teacher effort, initiative, and ingenuity.

---

[1] Walter A. Wittich and Charles F. Schuller, *Audio-Visual Materials: Their Nature and Use* (New York, N.Y.: Harper & Row, Publishers, 1967), p. 227.

The use of community resources in the classroom is nothing new. Philosophers of old had advocated the use of direct experiences in a natural setting. However, the use of community resources today seems to have the potential to add new dimensions to the learning process and to make the school more meaningful to students of all ages. An extreme view on the use of community resources involves the concept that the community itself is the classroom or laboratory of learning. One innovative school has no site and no buildings; its classrooms are the city's institutions and businesses.[2] In this connection Williams notes:

> ... the possibility that in the last quarter of this century more and more of our children will be educated off school grounds, outside the hallowed classrooms, seems certain.[3]

As applied to industrial arts, the idea of using the community as a laboratory is an intriguing one; whether or not this will ever transpire remains for the future. But today, curriculum planners, teachers, parents, and students are giving the community an increasingly important role in education, especially in the areas of occupational information and career involvement. For this reason the industrial arts teacher should focus increased attention on the educational advantages of using community resources in his program.

ROADBLOCKS TO USING COMMUNITY RESOURCES

Unfortunately, industrial arts educators are not in the vanguard in the utilization of community resources. Except for the industrial visit, relatively little has been written by industrial arts teachers concerning the use of local human and material resources. This seems strange not only because of the growing importance of this topic, but also because it seems natural to link closely the industrial aspects of the community and the industrial arts activities in the schools.

Among the reasons why teachers do not utilize community resources more fully are the following:

1. Tradition

Many industrial arts teachers seem dedicated to the project method of teaching and regard any deviation as sacrilegious. They look skeptically upon the use of local resources and feel it would disrupt the

---

[2]"The Anywhere School," *School Management*, vol. 13, no. 12 (December 1969), p. 46.
[3]George L. Williams, "Beyond the Classroom—Life Experiences in the Field," *The Clearing House*, vol. 45, no. 2 (October 1970), p. 81.

learning process by keeping students from doing their "real" work in the laboratory. They seem unmindful of the dangers of isolation from the community and prefer the more comfortable laboratory approach to learning.

2. Lack of Leadership

Some teachers have neither the initiative nor the leadership qualities to plan, organize, and integrate community resources into their daily lessons or courses of study.

3. Lack of Knowledge

Some teachers do not use community resources simply because they have never heard of them. They are innocent of the benefits accruing from judicious utilization of community resources. In this respect their blindness continues undiminished because industrial arts textbooks and periodicals make little or no mention of community resources. As noted previously, a dearth of information on the use of community resources exists in current industrial arts literature.

4. Lack of Administrative Support

It is nearly impossible to utilize the resources of a community without some support by the school administration. Many busy, harried administrators have not the time, energy, or even the inclination, to assist and support their teachers, especially if extra funds are involved.

5. Community Attitude

In certain communities the point of view has evolved that education is the business of the school and it can best be conducted on school grounds. While these communities are willing to support education financially, they are unacquainted with their full role in the educative process. No doubt this attitude has developed, at least partially, because teachers and administrators have been remiss in their total responsibilities.

IMPORTANCE OF USING COMMUNITY RESOURCES

There are a number of good reasons why the industrial arts teacher should utilize community resources. Two of the more important of these are: (1) they increase educational efficiency, and (2) they promote closer relationships between the school and the community. A brief discussion will indicate how these results are brought about and why they are so important.

1. Increasing Educational Efficiency

Reference to the list of suggested behavior changes and the proposed learning activities for bringing them about (see Chapters 5 and 6), will show the strong emphasis that has been laid upon the utilization of community resources. Visits to industries and museums, reports on visits, and use of resource persons all presuppose cooperation with local individuals and establishments. Many lessons, especially those dealing with a first-hand knowledge of industry, its materials, and its personnel and organization, can be taught best through cooperation with local agencies.

Community resources make possible a unique form of occupational guidance in which students may observe a variety of professionals at work or hear them talk about their occupations. Several techniques for using local resources in partially fulfilling industrial arts guidance objectives include: (1) industrial visits, (2) inviting resource persons to the class to talk about their occupations, (3) playing tape-recorded telephone interview conversations between the teacher or a class representative and a resource person who may be unable to come to the school laboratory, (4) videotape presentations of resource persons in action, (5) career days in which students observe people working in their chosen careers, and (6) on-the-job or at-home interviews by students with persons in selected occupational fields. The factual information gained through such pupil experiences is seldom available in textbooks. Of more importance, perhaps, is the exposure of pupils to talent and excellence.

The timely use of community resources usually arouses much class interest and often kindles deep-seated motivation. These in turn contribute to improved student learning.

Educators have long recognized the value of involving both pupil activity and the local environment in the learning process. They stress the educational importance of familiarity with the immediate locale as an initial step in developing pupil understanding of his ever-expanding environment. The use of community resources in the curriculum enables students to participate in meaningful experiences in everyday living in their immediate world.

In today's urban society children often live literally in isolation from the workaday world. They have little or no real understanding of the people or the variety of trades, occupations, and industries that make up their own community. Students have been known to live within a few blocks of an industry without knowing anything about what was manufactured there until they visited the plant with an industrial arts class. Few students have knowledge of the industrial materials and information which are available in their own community. It is important that these be brought to their attention because, other things being equal, the more they know of the human and material resources of their immediate environment, the richer lives they can lead as citizens, and the more likely they will be to look for similar resources in any community in which they may later live. The industrial arts teacher has a real responsibility to bring to the attention of his students the many resources which surround them. Many worthwhile educational experiences lie within walking distance of the laboratory. A careful survey of the resources available in any community will usually bring to light many people, places, and things that can be used in the industrial arts program and that would be unobtainable through any other means.

Another way in which the use of the local environment tends to increase educational efficiency is that local resources are more likely to be available when needed. For example, if the industrial arts teacher is accustomed to purchasing metal from a local factory, it is less likely that this material will be out of stock in the laboratory than would be the case if that material had to be shipped from another city. Moreover, teachers who make a practice of utilizing community resources seem never to be low on supplies for, if some particular material is not available, another can always be found. Students are quick to recognize the resourcefulness of the teacher and they will frequently help to find and develop sources that would otherwise be overlooked.

2. Developing Closer Relationships between School and Community

Every contact which can be developed between the school and the community helps each to know the other better. Such contacts help to break down the undesirable viewpoint that learning is the business of the school only and that it can be best accomplished in the classroom. Collings stresses the need for good school-community relationships when he states:

> Educators increasingly recognize the principle that schools need contact with the realities of life; that curricula are most effective when closely related to the communities they serve; and that boys and girls learn best when dealing with direct, concrete experiences.[4]

[4]Miller R. Collings, *How to Utilize Community Resources* (Washington, D.C.: National Council for Social Studies, 1960), p. 1.

Community resources not only benefit the pupils but also provide the local citizens with new insights and understandings of *their* schools. Thus, when a group of industrial arts students visits a local industry, both the students and the management of the plant get to know each other better. This in turn is likely to lead to mutual understanding of each other's problems. Understanding often leads to active support. Or, if a class visits a local museum, those in charge immediately show interest in the school. If an expert in some field is invited to speak to an industrial arts class concerning his craft, he immediately takes a friendly attitude toward the school. Every tie which the school can make with the community is valuable, and using the resources of the community is one of the most effective methods of cementing such bonds.

TYPES OF COMMUNITY RESOURCES

The two general classifications of community resources are (1) those in which the pupils leave the laboratory to visit some human or material resource, and (2) those in which the resources themselves are brought to the classroom. It would be impossible to particularize upon the exact types of resources found in every community. However, most resources suitable for industrial arts purposes will fall into one or more of the following categories: (1) industries, (2) libraries, (3) business concerns, (4) resource persons, (5) museums and collections, and (6) materials.

1. Industries

Almost every community, regardless of its size, has some kind of industry. It is important that students know how industries are operated, what they manufacture or process, what raw materials are used, and what types of work the employees do. Trips to such industries will be interesting and valuable to industrial arts students (see Fig. 16-1). The educational value of the industrial visit has already been noted in Chapter 14. It will be recalled that an industrial tour is highly effective because of its realism. The senses of hearing, smelling, seeing, and sometimes touching and tasting may be brought into play through first-hand impressions and experiences during a visit to industry.

2. Libraries

The use of public libraries by industrial arts students should be encouraged. Frequently such libraries will have reference material on industries and industrial materials which cannot be found in the school

Fig. 16-1. The industries of a community offer educational resources which cannot be matched by any other means. (Courtesy, Ford Motor Company)

collection. Class or individual assignments in connection with informative lessons should be made to encourage acquaintanceship with these supplemental sources. The use of public or college libraries is almost a necessity for secondary school pupils engaged in research-oriented independent studies.

3. Business Concerns

Stores and other similar establishments often have much to offer the industrial arts student. In one instance a store in a small city invites the industrial arts teacher and his students to visit the store each year to examine and study trends and designs in furniture. Another department store encourages the teacher to bring his classes for the purpose of getting good project ideas. Other stores have been known to sponsor lectures on various types of merchandise, machinery and tool maintenance, and how to get the most out of specialized equipment. The service and help which can be rendered by these establishments are frequently overlooked by the industrial arts teacher.

4. Resource Persons

Many talented and knowledgeable people in a given community would be quite willing to share their expertise with school pupils if only they were contacted. Recently in one city system, 400 resource people were located who not only were eager to share their experiences, but were willing to take time from the regular working day to come to the school. The contributions that can be made to the industrial arts program by individuals in a community are often significant. More serious attention should be given to the educational potential of resource persons than seems to be the case in the typical industrial arts program. Resource people can make valuable contributions to the attainment of the objectives of industrial arts in a variety of ways including the following:

Class Demonstrations. Amost every town or city has a number of craftsmen who would be pleased by an invitation to demonstrate their skill to school pupils. One teacher invited a master molder to show his general metals class how a mold should be made. There was an immediate improvement in standards for this materials area. The same was true when an expert demonstrated screen process printing. A model builder, who showed and explained his hobby to a number of industrial arts classes, left an impression on many students which still remains with them in the form of home workshops and leisure-time activities.

Independent Studies. Resource persons play a key role in independent studies, for example, in the research and experimental approach to industrial arts. Figure 16–2 shows a junior high school student procuring data for his research topic from an expert in the field. In this case the telephone saves much valuable time for both the expert and the student.

Conferences and Seminars. The use of resource persons in conferences and seminars can greatly enrich units of instruction in industrial arts. For example, as part of a unit on industrial safety, a class may decide to hold a safety conference. After selecting a student-chairman for the conference, the class may invite resource people from the community to talk on industrial or shop safety. These may include personnel as (1) a safety engineer from a nearby plant; (2) a local representative from the National Association for the Prevention of Blindness, the National Safety Council, or the American Red Cross; (3) insurance or fire underwriter's representatives; or (4) members of local industrial safety councils. The sessions should be planned to involve the students, perhaps through a question-answer discussion period. Similar conferences

Community Resources

Fig. 16–2. In this industrial arts research laboratory, the telephone is an important communications tool for contacting out-of-school experts. (Courtesy, Alan P. Keeny, Montgomery Hills Junior High School, Silver Spring, Maryland)

or seminars involving students and resource people could be held on labor-management relations, industrial organization, personnel problems in industry, and the like.

Vocational Guidance. Students making career choices are often influenced by the lure of high wages, travel, or rapid promotion in glamorized announcements by the news media. More realistic career information can be obtained from persons already in that vocation. With this in mind, officials in one New Jersey school[5] established a "Vocational Resource Directory" which was a list of 270 local people whose 147 different jobs corresponded to those in the *Dictionary of Occupational Titles*. These school-minded citizens provided realistic career information to individual pupils who expressed interest in a given occupation. In addition to providing general information, these career specialists also advised and counseled the student as to training requirements, personal qualifications, working conditions, potential

---

[5]Benjamin Barbarosh, "Developing a Community Resource Directory," *The Vocational Guidance Quarterly*, vol. 14, no. 3 (Spring 1966), p. 179.

earnings, and future job outlook. This plan has proven to be a valuable supplement to the guidance services of the school.

One of the difficulties in utilizing resource people is coordinating their schedules with that of the class. The problem is compounded, of course, when multisections of pupils are involved. Two solutions seem apparent: (1) use of the telespeaker and (2) use of videotape recording.

The telespeaker is useful for an entire class to listen or participate in an interview with a resource person by telephone. This piece of equipment features a microphone by which individual members of a group can talk to the resource person whose responses are then returned through a loudspeaker for the entire group to hear. Of course, the whole interview can be recorded on a tape recorder for later class discussion or for use with other classes. Such speakerphone equipment is useful in making both local and long-distance calls to experts and authorities wherever they may be. The costs of utilizing long-distance speakerphone interviews are usually far less than a personal visit by an individual or a class. It is understood, of course, that phone interviews should be planned carefully in advance by the student or the class concerned. Key questions should be carefully written out to guide the student or class and to help keep the interview focused on the problem under consideration.

Videotapes of resource people in action offer fine substitutes for live presentations. Despite the fact that personal contact is absent, at least one study reports that student achievement does not suffer. Earp and Rink compared the factual and conceptual knowledge gained by sixth-graders through a live presentation and a 30-minute videotape presentation by the same resource person.[6] First a live presentation was made by the resource person. Notes were made of student questions during the discussion period. The presentation was then revised by the resource person to include those points and videotaped. Statistically no significant differences were found between the live and the videotaped presentations, so the investigators concluded that student achievement was equal regardless of mode of presentation. They recommended, however, that videotape presentations be preceded by a live presentation or two to ensure that the material will be directly related to the interest and purpose of the class.

It must be remembered that a visit by a resource person to the school is really an important public relations activity. The teacher, the

[6]N. Wesley Earp and Ortho Rink, "Resource Persons Provide Effective Learning Experiences Via Videotape," *Audiovisual Instruction*, vol. 15, no. 10 (December 1970), p. 55.

class, and the school itself will be judged by the resource visitor in terms of the reception he receives.

5. Museums and Collections

Many communities have private or public museums which are open to individual students or groups. Valuable personal collections may also be available within convenient distance from the school. These serve as excellent bases for the study of materials, projects, and discussion topics. Teachers should not only urge students to visit and learn to know museums, but also should plan class visits if these are in harmony with the course of study.

In some communities where museum facilities are not available, the industrial arts teacher may be able to interest his classes in starting a school museum of industrial materials. Such a project has almost unlimited possibilities. The educational returns from this activity accrue not only to those who collect and prepare the materials, but to all classes that follow. The lack of a suitable room need not deter the teacher from initiating the development of a school museum, since a start can be made with little more than a cabinet or book case. As the collection grows, it is usually possible to win support from the administration for more adequate housing.

6. Materials

Frequent reference has been made in this chapter to various types of materials that may be procured from sources within the community. It is not meant to imply that all the materials used in an industrial arts class can or should be obtained locally. Neither is it possible to make a complete list of all the supplies which could be obtained in any given community. These will vary greatly from one locality to another. However, a general classification of the types of materials will, perhaps, direct the thinking and activities of the teacher to possible sources.

Printed materials. Supplementary aids from industrial firms, such as booklets, pamphlets, films, and other printed materials are often useful in enriching the instruction. They may reinforce previous learning as well as add variety and interest to the learning activities. Obviously, the use of those materials overladen with sales propaganda should be avoided.

Metals and Ores. Some schools are located in communities where there are metalworking industries. Most of these will have large amounts of scrap metal which are no longer suitable for their purposes, but which can be used effectively in the school shop. Such scrap can usually be purchased at junk prices, or in some cases it may be given to the school. This scrap may include sheet metal, tin plate, sheet copper, brass, aluminum, or nickel silver. Scrap aluminum, lead, brass, and bronze are also frequently obtainable for metals casting purposes. If the school is located near mines or smelting and refining industries, samples of various ores may be procured for the shop museum or to illustrate class reports and discussions.

Woods. Many schools have successfully utilized the products of nearby woodlands. Projects of a highly artistic and useful nature may be designed and made from the limbs, bark, and trunks of small trees. The handicraft societies of New England and of some southern states have developed the use of such community resources to a high degree. Important also to the industrial arts teacher are the scrap and waste materials of various types of woodworking industries. Short lengths of lumber which are ideal for use in the school laboratory frequently can be purchased at a fraction of the cost of full-length material.

Textiles and Leather. Textile working industries frequently have odd-size material which can be purchased at a relatively low price and which is useful in making rugs, mats, etc. Plants engaged in leatherwork, such as the manufacture of shoes, are an excellent source of leather which can be used in making a variety of small projects. Sometimes leather materials can be procured directly from a local tannery.

Clays. Clays suitable for ceramics work are to be found in widely distributed beds over a great part of the country. Considerable educational value lies in digging, drying, and preparing clay for use. The time and effort spent in trying out various clays which may be available are certainly repaid if even one good deposit is found.

Stones and Minerals. There is some interest in lapidary work in the industrial arts laboratory. In almost every community there are a variety of stones and minerals which can be shaped and polished for the making of rings, pins, and costume jewelry. The collecting and polishing of local stones has been found to be a most fascinating activity by those schools in which it has been introduced and encouraged. Petrified wood is also available in some localities.

LOCATING COMMUNITY RESOURCES

A first step in the utilization of community resources requires that the teacher know what is available, where it is located, and how it may be acquired or used. Through normal community contacts many sources will come naturally to the teacher's attention. It is unlikely, however, that even a major portion of the possibilities will be known even through the most extensive of such contacts.

One thing is certain, however. If community resources are to be integrated into the curriculum—and surely they merit inclusion—then careful plans must be made, beginning with a community survey of all resources which seem potentially useful.

Kapstein[7] proposes a "Community Resource Roster Plan" in which the local school administrator accepts the responsibility for compiling a master list of creative and talented persons from the locality who are willing to donate their services to the school. Each resource person is asked to give 12 hours of service per year by addressing small groups of students on his specialty. Interested teachers merely select appropriate resource persons from the master list; the administrator's office then makes all necessary arrangements with the individual concerned.

Industrial Arts Department		Pleasantville, Pa.
	Community Resources Survey	
	I – Resource People	
NAME:	George H. Brook	
ADDRESS:	183 N. High Street	PHONE: 364-7706
POSSIBLE CONTRIBUTIONS:	Gives talk, illustrated with color slides, and demonstrates screen process printing.	
BACKGROUND:	Owns and operates Screen Process Printing Inc. which he founded in 1940. Artist and craftsman par excellence.	
REMARKS:	Brings own demonstration tools and equipment.	
	A busy business man – very willing to speak to I.A. classes, but must be contacted at least one week in advance.	

(Reverse side of card)

DATE	CLASS	TEACHER'S EVALUATION	CLASS REACTION
10/15/72	9th grade	Highly satisfactory. Understands this age level. Fine demonstration – even "printed" on water!	Excellent

Fig. 16–3. Suggested form for surveying a community for resource people.

[7]Sherwin J. Kapstein, "Implementing Community Resources for Education," *The American School Board Journal*, vol. 152, no. 3 (March 1966), p. 24.

Industrial Arts Department	Pleasantville, Pa.

Community Resources Survey

II – Industrial Visit

FIRM:	Ford Paper Company, Inc.
ADDRESS:	408 Glen Road
CONTACT:	Richard A. Smith, Public Relations Manager
PHONE:	364-7609
POSSIBLE CONTRIBUTIONS:	Shows manufacture of paper from wood chips to finished product.
PUPIL HAZARDS:	None, but pupils must stay with guide.
REMARKS:	Guided tour takes 45 minutes. Can be scheduled on one day's notice.

(Reverse side of card)

DATE	CLASS	TEACHER'S EVALUATION	CLASS REACTION
11/20/72	9A, 9B	Good trip. Could not hear guide in chipping room. Students were given samples of products.	Enjoyed trip: wanted to stay longer.

Fig. 16–4. Suggested form for surveying a community for prospective industrial visits.

Some teachers prefer to conduct their own community surveys so that they will gain personal familiarity with the potential existing in their locality for their particular subject matter discipline. A good place for the teacher to begin is with the parents of his pupils. A simple inquiry either mailed or carried home by pupils may produce unbelievable results. The teacher should keep the educational needs in mind and frame his questions accordingly. Similar requests can be sent to civic, service, and other organizations in the community as well as to interesting people whose names and activities have been mentioned in the news media. Figure 16–3 illustrates a sample card for use by teachers in surveying and maintaining records of resource people in a given locality. Figure 16–4 shows a card which may be useful in recording pertinent data for making an industrial visit. These cards are best filed in a card index which should be kept up-to-date. New cards should be added when new resources become available and old cards withdrawn when a resource proves less useful than anticipated or is no longer available.

If students can be alerted to the desirability of locating and using the resources of the community, they can frequently suggest many items which would otherwise go unnoticed. Instances have been known where the teacher and the members of his classes have undertaken extensive surveys of their community and have developed complete lists of the resources which might be used by the school. The activities involved in the survey have, in themselves, proved to have high educational values.

OPPORTUNITIES FOR COMMUNITY SERVICE

There is another aspect in the use of community resources which certainly is as important as any of those that have been discussed. *Opportunities for community service,* if properly used and organized, may afford educational possibilities which can seldom be realized within the four walls of the school. A number of schools have been organized on the basis of community betterment. Under such circumstances the community itself becomes the basis for the curriculum and the problems which are undertaken by students represent real situations in which each has a personal interest. A further result of this kind of organization is that the school becomes the social and cultural center for adults as well as the children. There is considerable reason to believe that the community school movement will spread and become important in many localities.

In other schools certain classes engage in community study projects which sometimes lead to improved conditions in the locality. Reforestration efforts, clean-up campaigns, pollution control programs, ecology studies, and the establishment of play areas are examples of typical civic projects by which school pupils acting as future citizens can help make their community a better place in which to live.

Industrial arts education can play an important role in community service. As shown in Fig. 16-5 some classes repair and refinish toys for underprivileged children at Christmas time. At other times classes may engage in group projects of civic value, for example, a float for a parade or the construction of playground equipment.

Instruction in various home mechanics jobs as well as those related to avocational pursuits can render a real service in raising the living

Fig. 16–5. Repairing Christmas toys for underprivileged children is a fine service which an industrial arts program can render to the community. (Courtesy, California State Department of Education)

standard in many homes. In fact, even where the total school is not organized on the basis of community service, there is much that the alert industrial arts teacher might do to utilize the opportunities for service which the average community offers and in so doing, enrich and vitalize his program. It must be remembered, also, that community service is an invaluable public relations medium.

SUMMARY

Community resources refers to the use of human talent and material resources of a locality for educational purposes. Typical community resources include people, industries, libraries, business concerns, museums, and such raw materials as metals, woods, textiles, etc. Such resources are among the most important of all curriculum materials. Curriculum enrichment, pupil motivation, and a general upgrading of the quality of education can be expected to occur with proper use of appropriate local resources. Community resources not only benefit pupils, but in addition, provide citizens with new understandings of their schools. Thus, community resources can be regarded as a valuable public relations medium. The initial step in using local resources is to discover what is available by means of a community survey. Forms for this purpose were presented in the chapter. In localities where a community school program is organized, industrial arts will find many opportunities to use the needs of the community as a source of desirable activities.

DISCUSSION TOPICS AND ASSIGNMENTS

1. Devise a form which would fit your idea of needs and purposes in making a community survey of resource persons.
2. Select some community with which you are familiar and make a complete list of the resources that might be used by industrial arts classes.
3. Design a project which could be made entirely from materials that might be found in a selected community.
4. Make a list of the community services that a typical industrial arts program might provide.
5. Outline your ideas for using the community as an industrial arts laboratory.
6. Outline plans for a conference on labor-management relations as part of a unit of instruction.

SELECTED REFERENCES

Barbarosh, Benjamin. "Developing a Community Resource Directory," *The Vocational Guidance Quarterly*, vol. 14, no. 3 (Spring 1966), p. 179.
Burts, Eleanor. "The World in a Neighborhood," *Childhood Education*, vol. 42, no. 5 (January 1966), pp. 302-303.
Collings, Miller R. *How to Utilize Community Resources.* Washington, D.C.: National Council for Social Studies, 1960.
Earp, N. Wesley and Ortho Rink. "Resource Persons Provide Effective Learning Experiences Via Videotape," *Audiovisual Instruction*, vol. 15, no. 10 (December 1970), p. 55.
Kapstein, Sherwin J. "Implementing Community Resources for Education," *The American School Board Journal*, vol. 152, no. 3 (March 1966), p. 24.
Learner, A. L. "Work-Experience Field Trip," *School Shop*, vol. XXV, no. 4 (December 1965), p. 32.
Morton, Berry E. and Richard L. Burns. "Telecture in Industrial Education," *Industrial Arts and Vocational Education*, vol. 55, no. 6 (June 1966) p. 40.
Pasley, J. Gordon. "The Local Community as a Social Studies Lab," *School and Community*, vol. 52, (April 1966), pp. 28-29.
Staley, Frederick A. "Community Resources—The Forgotten World of Knowledge," *Educational Screen and Audiovisual Guide*, vol. 44, no. 10 (October 1965), p. 27.
Stringer, Mildred B. "Call on Your Community," *The Texas Outlook*, vol. 49, no. 10 (October 1965), p. 40.

"The Anywhere School," *School Management,* vol. 13, no. 12 (December 1970), pp. 46-55.

Williams, George L. "Beyond the Classroom—Life Experiences in the Field," *The Clearing House,* vol. 45, no. 2 (October 1970), pp. 81-85.

Wittich, Walter A. and Charles F. Schuller. *Audio-Visual Materials: Their Nature and Use.* New York, N. Y.: Harper & Row, Publishers, 1967.

Zak, Allen P. "Vocational Guidance in the Elementary School," *Chicago Schools Journal,* vol. XLVI, no. 7 (April 1965), pp. 308-313.

Chapter 17

Evaluation in Industrial Arts

Evaluation is the process by which one estimates carefully the value of something. The term is used in this chapter to indicate the procedure by which the results of instruction in industrial arts are appraised. Evaluation also indicates that something more than mere measurement is required. The distinction between *evaluation* and *measurement* is drawn by the California Test Bureau as follows:

> The emphasis in measurement is upon single aspects of subject matter achievement or specific skills and abilities; emphasis in evaluation is upon broad personality changes and major objectives on the educational program.[1]

The Bureau further notes that the term *evaluation*

> ... involves the identification and formulation of a comprehensive set of major objectives of a curriculum, their definition in terms of pupil behavior, and the selection or construction of valid, reliable, and practical instruments for appraising specified phases of pupil behavior. ...[2]

TRADITIONAL PRACTICE

There has been little by way of standard practice in arriving at grades for industrial arts students. In general, final marks have been determined either by an overall subjective appraisal by the teacher or by grades assigned to finished projects or to informational tests. Sometimes a combination of two or more of these bases is used. It is evident, however, that grades have been determined largely on two types of evidence: (1) development of skills as indicated by the finished project,

[1] *A Glossary of Measurement Terms* (Monterey, Calif.: California Test Bureau), p. 7.
[2] *Ibid.*

and (2) acquisition of information as shown by ability to pass a test. *Unless one is to assume that growth in these two areas is a satisfactory index of desired achievements in all phases of industrial arts, then some more comprehensive basis for evaluation must be found.*

There are several "blind spots" which present practices in evaluation seem to have ignored. Until very recently there was almost no attempt on the part of teachers to measure the growth of students in their ability to think and to solve problems—and yet this has been identified as the central purpose of American education. However, in 1970 one researcher, Cillizza, focused attention on this problem by developing a test of critical thinking ability for grades seven and eight.[3] The instrument was validated by a jury of experts and when tested experimentally was found to yield a reliability coefficient of .90+. The instrument was deemed an adequate measure of critical thinking skills for the population tested.

Other areas have been equally disregarded by teacher evaluation, including (1) the extent to which a student develops consumer perception, (2) his ability to judge and appreciate design, (3) his growth toward the development of a hobby or recreational interest, and (4) the development of desirable social relationships. If growth in these areas are important outcomes of industrial arts, then certainly some effort should be made to evaluate them.

EVALUATION IN TERMS OF OBJECTIVES

The thesis has been advanced earlier that the selection of learning activities should be in terms of the behavior changes which it is desired to bring about. If this point of view is accepted, then it is evident that evaluation should be on the same basis. The extent to which a course has been successful should be in direct proportion to the advancement of students toward the achievement of the desired behavior changes, and likewise to the progress of any individual in growth toward these ends.

It is difficult to make an accurate appraisal of growth toward many of the desired objectives. This difficulty is due partly to the fact that some behavior changes are rather subjective, but a more important reason is that teachers and researchers have not sought to find or develop suitable evaluative techniques.

[3]Joseph E. Cillizza, "The Construction and Evaluation of a Test of Critical Thinking Ability (Grades 7–8)," (unpublished doctoral dissertation, Boston University, 1970).

TECHNIQUE OF EVALUATION

The steps in the development of evaluation techniques are much the same as those in preparing an instructional program. In general, they are:

1. Statement of Objectives

Clearly, one cannot evaluate until he knows what his final objectives are. Since the aims of a course should represent the outcomes which are desired, one can evaluate on no other terms than the extent to which they are being achieved. A first step in evaluation becomes, therefore, a clear concept and statement of such objectives. In Chapter 4 the objectives of industrial arts were stated as derived from an analysis of the goals of general education.

2. Analysis of Objectives in Terms of Behavior Changes

Since objectives are usually stated broadly and accordingly are likely to defy measurement, a second step toward evaluation is to break down and analyze such objectives in terms of observable behavior changes. In Chapter 5 the objectives of industrial arts were analyzed in terms of the expected behavior changes from students. If the teacher has already used this technique for determining his course content, the same set of behavior changes will be used as a basis for measurement.

3. Isolating Situations Where Behavior Changes May Be Noted

The only method for measuring behavior changes is to determine whether they have occurred. For example, if one of the behavior changes which is desired is that students should develop a home workshop, then the proof of achievement is clearly in whether such shops have actually been set up. This could be determined by teacher visits to homes, or somewhat less accurately by asking students to describe their home workshops. Parental conferences or casual conversation during open house may also provide data. At least, these represent possibilities for making an evaluation on this behavior change.

Each desired behavior change should be reviewed, and the teacher should ask himself: *Under what circumstances or by what method can I obtain objective evidence on growth toward this change?* From the results of such self-questioning, the teacher will be able to develop an evaluation program which will be much more effective than those generally in use today with industrial arts classes.

4. Recording Results

After situations have been isolated which will give evidences of behavior changes, the next step is to determine some means whereby a record can be made of such changes. The exact form that such a record will take depends almost entirely on the type of behavior change being measured. The following forms of evaluation are offered merely to indicate the range of possibilities.
  1. Direct questions to student
  2. Conferences with parents
  3. Direct observation of pupil behavior
  4. Anecdotal records
  5. Check lists, score cards, and rating sheets (to be completed by teacher, student or officer in pupil personnel organization)
  6. Teacher-made tests (paper-pencil and performance types)
  7. Standardized tests

AN EXAMPLE OF BEHAVIORAL CHANGE EVALUATION

At this point a specific example of how a teacher might set up an evaluative program would, perhaps, be helpful. It will be recalled that in Chart I (Chapter 4) the relationship of industrial arts to general education was depicted and the objectives of industrial arts were derived. It will be recalled, also, that Chart III in Chapter 5 and Chart IV (Chapter 6) were continuations of Chart I and showed the expected behavior changes from students as well as the student and teacher activities for achieving these desired behavior changes. Now, Chart X carries these charts to logical conclusion by suggesting evaluation procedures for the student behavior changes.

As shown in Chart X, nine expected behavior changes from students are listed for Objective 8, *to develop safe working practices.* The teacher may well ask himself: Under what circumstances and to what extent could the growth of students toward these types of behavior be evaluated? The following suggestions are indicative of how this problem might be approached:

1. Behavior Change 1—They Will Exhibit a High Degree of Safety Consciousness. Chart X suggests that *direct observation* by the teacher is one method for evaluating this beahavior change. While safety consciousness itself cannot be seen, tangible evidence of its presence in terms of pupil actions and words may be noted by directly observing the overt behavior of the pupil. Or, a *rating form* could be devised which would define safety consciousness as the summation or composite score for the other seven behavior changes listed for Objective 8.

2. Behavior Change 2—They Will Observe All Safety Regulations and Rules. In Chart X it is again suggested that *direct observation* by the teacher will provide data for evaluating this behavior change. If the teacher were truly interested in gathering data on this point, he might carry a small pocket notebook with a page for each pupil. It would then be relatively easy to jot down observed infractions of safety rules or commendatory notes. Another good way to test pupil observance of safety rules is through *performance tests*. While knowing the rules is no guarantee that they will be observed by pupils, the teacher may use written *safety tests* to evaluate knowledge of safety regulations.

3. Behavior Change 3—They Will Wear Safe Clothing and Clothing Suited to the Work Being Done. Clearly, one method of evaluation would be *teacher observation*. Or, a check list could be utilized, say, at the beginning of the shop period. This check could be made by the safety engineer in the pupil personnel organization.

4. Behavior Change 4—They Will Use Protective Devices Such as Safety Glasses and Gloves When Needed. This change is easy to observe and record; again, the pocket notebook could be utilized effectively. In addition to *observation by the teacher*, the superintendent, foreman, or safety engineer could make *safety checks and reports*. These could be either "spot checks" or continuous checks.

5. Behavior Change 5—They Will Make Certain That All Guards Are in Place before Starting a Machine. Techniques similar to those suggested for 3 (above) could be employed. When a shop machine is turned on, the noise will attract the attention of an alert teacher as a magnet attracts iron. He will swiftly and almost automatically make an appraisal of the situation. Questions like these will race through his mind: Is this student cleared to operate the machine? Are the guards in place? Is this student performing the operation safely? It would be a simple matter to record data like these.

6. Behavior Change 6—They Will Call Attention to Unsafe Tools or Machines. A *direct check* on this behavior change could be made by the teacher by simply keeping a record of the number of instances and the students who reported unsafe tools or machines. Here again evaluation and recording are accomplished easily.

7. Behavior Change 7—They Will Warn Others Who May Be Working in an Unsafe Manner. The technique (direct check) suggested for 6 (above) could be utilized. This method could be abetted by encouraging students to report to the teacher when they give or receive a warning from a member of their peer group.

CHART X
Suggested Evaluation Procedures

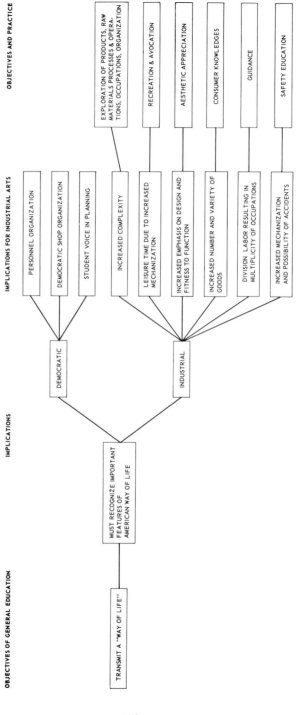

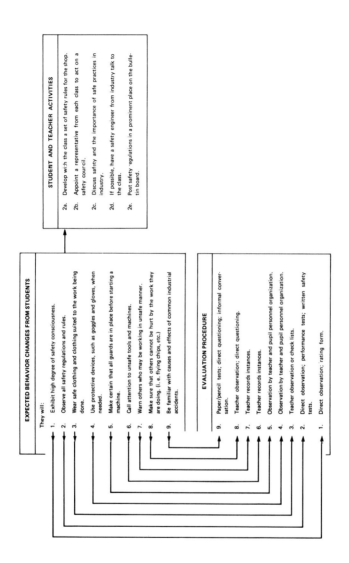

8. Behavior Change 8—They Will Make Sure That Others Cannot Be Hurt by the Work They Are Doing (i.e., flying chips, etc.). *Observation* by the teacher as well as *checks by officers of the personnel organization* could be employed. In addition, *direct questioning* of students at work may also provide data on this point.

9. Behavior Change 9—They Will Be Familiar with Causes and Effects of Common Industrial Accidents. Scores or grades on *paper-pencil tests* could be used here. Or, some indication of behavior change may be obtained by *direct questioning* or *informal conversation* with each pupil.

In examining these suggestions for evaluation, one is impressed that the devices and methods used are far removed from standard practices in measurement. It is evident that many and diverse techniques must be utilized if one is truly interested in getting information on growth in behavior.

It should also be clear that no teacher could hope to secure direct evidence in regard to all of the behavior changes in which he may be interested. At best the teacher might pick out a few items which appear to him to be unusually significant, and assume that growth in these indicates the general direction of development.

EXAMPLES OF EVALUATIVE TEST ITEMS

It should not be assumed from the discussion so far that there is no place in industrial arts for various kinds of test questions or that the results of tests cannot be of significant help in arriving at a true evaluation of pupil growth and development.

As in other areas of education, evaluation in industrial arts should assess insights and understandings, interests, attitudes, and appreciations, as well as skills of the pupil. Bloom et al. refer to these as the cognitive, the affective, and the psychomotor domains.[4] Subsequent sections of this chapter will deal with examples of test items to evaluate phases of the cognitive and the psychomotor domains.

1. The Cognitive Domain

This is the "knowing" domain involving knowledge or perception. It refers to those mental operations by which we become aware of objects by thought or perception. As the *Taxonomy of Educational Objectives* is presently structured, it contains six major divisions ar-

---

[4]Benjamin S. Bloom (ed.), *Taxonomy of Educational Objectives, Handbook I: Cognitive Domain* (New York, N.Y.: David McKay Company, Inc., 1956).

ranged in hierarchical order. These classes are: knowledge, comprehension, application, analysis, synthesis, and evaluation.[5]

The following examples illustrate how the taxonomy may be applied to industrial arts education. Careful reflection and study on the part of the teacher will suggest additional test items for these and other objectives of industrial arts whose behavioral changes lie within the cognitive domain.

It must be kept in mind that successful completion of a test item or two related to a behavioral change is no assurance that the change has actually been realized by the pupil. Such items usually sample only one aspect of the behavior change desired. A number of items repeated over stated intervals of time may be required to ascertain the direction of pupil growth and development.

The examples contained in this section were prepared by students enrolled in graduate classes in "Evaluation in Industrial Arts" at The Pennsylvania State University. Recognition for permission to use these items is extended to the following: M. James Bensen, items 3, 4, 7, 9, 12; Gerald E. Brown, 10; Curtis G. Hepler, 6; Carl J. Hoffman, 5, 8; and John F. Nehring, Jr., 1, 2, 11.

Objective 1—To explore industry and American industrial and technological civilization in terms of its evolution, organization, raw materials, processes and operations, products, and occupations.

Behavior Change 2: They will recognize the scientific principles underlying industrial operations and processes.

EXAMPLE 1: When a metal is heated to a bright red color and then oxidized quickly to transform it to an iron oxide, you are:
___ A. casehardening the metal.
___ B. carbuerizing the metal.
___ C. spot welding the metal.
_x_ D. cutting the metal.
___ E. arc welding the metal.

Behavior Change 6: They will be able to define common industrial terms.

EXAMPLE 2: Which of the following statements concerning the properties of metals is true?
_o_ A. Ductility and elasticity are synonymous.
_+_ B. A tough metal can withstand sudden force without breaking.
_o_ C. Brittle metals are the strongest metals.
_o_ D. Tensile strength refers to the ability of the metal to resist compression.

[5] *Ibid.*, p. 18

Behavior Change 7: They will be familiar with industrial organizations and relate them to the personnel organization of the industrial arts laboratory. Their cooperation in the pupil personnel system will increase.

EXAMPLE 3: Construct a flow chart communicating to the personnel in a typical large industry the organizational structure of a particular corporation.

Start with the Board of Directors and carry through to the man on the assembly line or in a similar function. Include auxiliary functions such as training programs, research and development, and sales—showing their relationship to the total structure.

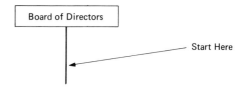

Behavior Change 8: They will be familiar with the sources of some of the raw materials of industry and will be able to discuss their transportation, processing, and industrial uses.

EXAMPLE 4: In the United States the greatest amount of iron ore for our steel industry has been supplied by the
- ____ A. Vermilion Range (Minnesota)
- ____ B. Iron Springs District (Utah)
- ____ C. Birmingham District (Alabama)
- _x_ D. Mesabi Range (Minnesota)
- ____ E. Iron Mountain Range (Missouri)

Behavior Change 15: They will read about the evolution of labor unions and will be able to interpret current problems of capital, management, and labor.

EXAMPLE 5: A strike that disrupted air transportation throughout the country was in the headlines recently. The mechanics who service the commercial jet airliners felt they were underpaid in proportion to the vital services they perform. They produced figures showing that automobile mechanics were making more money than airplane mechanics.

The airline management offered a 6 percent pay raise. The union wanted twice that amount. The productivity of the airlines indicated that they could afford the higher pay raise. Both groups were under pressure from the government to settle the dispute without an inflation-

Evaluation in Industrial Arts

ary price hike. As the strike continued, the government appeared willing to relax its anti-inflationary guidelines in order to reach a settlement.

Directions: Examine the conclusions given below. Assuming that the paragraphs above give a fair statement of the problem, underline the conclusion which you think is justified.

Conclusions:
A. The mechanics were entitled to their demands.
B. The mechanics should have accepted the 6 percent pay raise.
C. More information is needed to decide the issue.

Directions: For each of the following statements encircle the "A" if it explains why your conclusion is logical; encircle the "B" if it does not explain why your conclusion is logical; or encircle the "C" if you are unable to decide about the statement.

Statements:

A (B) C   1. The airline management knows how much they can afford to pay their employees.
A B (C)   2. Government officials should arbitrate the dispute and impose a binding decision.
A (B) C   3. The mechanics were trying for more money than they deserve.
A B (C)   4. The airlines should be nationalized to ensure efficient air transportation in this country.
(A) B C   5. Employees should receive a pay raise in proportion to the productivity of the industry.

Objective 3—To increase an appreciation for good craftsmanship and design both in the products of modern industry and in artifacts from the material cultures of the past.

Behavior Change 1: They will know the names of outstanding craftsmen and designers of the past and the present and will be able to recognize the work for which they are known.

EXAMPLE 6: Which of the following designed a home in Western Pennsylvania called "Falling Water?"
___ A. Donald H. Drummond
___ B. Herman H. York
_x_ C. Frank Lloyd Wright
___ D. Raymond Lowery
___ E. Buckminster Fuller

Objective 4—To increase consumer knowledges to a point where students can select, buy, use, and maintain the products of industry intelligently.

Behavior Change 2: They will understand the common processes by which materials are shaped, formed, and assembled in industry.

EXAMPLE 7: The concept that describes the process where two materials are fastened together by a third bonding agent is

   ___ A Cohesion
   ___ B. Mechanical linkage
   ___ C. Seam
    x   D. Adhesion

Objective 5—To provide information about, and insofar as possible, experiences in the basic processes of many industries in order that students may be more competent to choose a future vocation.

Behavior Change 8: They will be familiar with the health risks and dangers of various jobs and occupations in several industries.

EXAMPLE 8: Construct a line graph using the following data:

DISABLING WORK INJURIES BY INDUSTRY IN STATE OF TRANSYLVANIA

Division	1960	1965	1970
Agriculture	17,121	16,104	16,000
Mineral Extraction	1,966	2,165	2,043
Construction	23,457	24,754	28,209
Manufacturing	39,330	41,744	45,060
Transportation, Communication, Utilities	13,424	14,834	15,254
Service	14,906	15,556	17,468
Government	17,880	18,891	21,597

Objective 6—To develop critical thinking as related to the materials, tools, and processes of industry and to encourage creative expression in terms of industrial materials.

Behavior Change 6: They will be able to formulate and test hypotheses for given problem situations.

EXAMPLE 9: *Introduction*—Providing fresh, clean air for the working personnel in many industries is an acute problem faced by management. Dust, fumes, smoke, and steam are some of the causes for polluted air. The lack of oxygen in seemingly clean air is also a health hazard in industries where special equipment demands massive amounts of air in their operation. *The Problem*—A student wished to test the hypothesis that the rate of respiration of a blast furnace operator is dependent upon the amount of available oxygen that he has to breathe under working conditions.

The student fashioned a face mask which allowed him to rebreathe the same air supply, assuming that the amount of oxygen in the air would progressively decrease.

The chief error in this procedure is that:

  _x_ A. The effect of increased carbon dioxide concentration has not been considered.
____ B. A student-made mask could not be sufficiently tight for this experiment.
____ C. The temperature of the air supply will continually increase.
____ D. The pressure exerted by the air supply will continually decrease.
____ E. The amount of oxygen in the sample will remain relatively constant.

Objective 8—To develop safe working practices.

Behavior Change 2: They will be able to cite and will observe all safety regulations and rules.

EXAMPLE 10: When working in the industrial arts laboratory safety glasses should be worn:

____ A. only when there is dangerous work being done.
  _x_ B. at all times, no matter what type of work is in progress.
____ C. whenever you are operating machinery.
____ D. at all times, except when you are working with hand tools.

Behavior Change 7: They will promote safe practices in the laboratory and will be familiar with emergency procedures.

EXAMPLE 11: You are pouring molten aluminum into a flask and some spills on the floor. The first thing you should do is:

____ A. call the instructor at once.
  _x_ B. throw sand on the molten metal.
____ C. throw a bucket of water on the metal.
____ D. get a fire extinguisher.
____ E. throw a fire blanket over the molten metal.

Objective 9—To develop a degree of skill in a number of basic industrial processes.

Behavior Change 2: They will be able to use common hand tools and machines safely and effectively.

EXAMPLE 12: Which of the following drawings illustrates an *improper* use of the claw hammer?

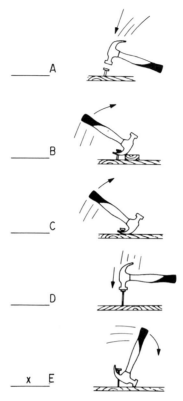

Some of the behavior changes which are normally expected of industrial arts students are that they (1) will read about industries, (2) will talk with workers about their occupations, (3) will study industrial terminology, and (4) will visit industries. All these activities are directed toward the end that the students should become oriented to industrial civilization and should explore its various aspects. One outcome which might be expected from such behavior would be a growth in the vocabulary related to industry. A check on the extent of a student's vocabulary at both the beginning and end of a given course would, therefore, be some indication of growth in desired behavior changes. The following matching items in the area of electricity illustrate a type of vocabulary examination that might prove of value.

<table>
<tr><th>Definitions</th><th>Terms</th></tr>
<tr><td>1. The electrical unit of power</td><td>A. Ammeter</td></tr>
<tr><td>2. A device for changing alternating current to pulsating direct current</td><td>B. Ampere
C. Anode</td></tr>
<tr><td>3. The positive terminal of an electric source.</td><td>D. Armature
E. Cathode</td></tr>
</table>

4. The unit of current flow through a conductor
5. The opposition of a substance to the passage of an electric current
6. A compound which is subject to decomposition by the passage of an electric current through it
7. A device for generating an electric current through chemical action
8. A device for detecting very small electric currents
9. A piece of soft iron or steel used to connect the poles of a magnet
10. A material that does not readily conduct a current of electricity
11. The practical unit of quantity in electricity
12. The practical unit of electrical capacity
13. A device for converting mechanical to electrical energy
14. Chemical decomposition through the action of an electric current
15. A device for measuring the difference in potential in electric currents

F. Commutator
G. Coulomb
H. Dry cell
I. Dynamo
J. Electrolysis
K. Electrolyte
L. Electromotive force
M. Electroplating
N. Farad
O. Galvanometer
P. Insulator
Q. Magneto
R. Motor
S. Ohm
T. Rectifier
U. Resistance
V. Transmitter
W. Volt
X. Voltmeter
Y. Watt

2. The Psychomotor Domain

The development of a degree of manipulative skill is an objective of most industrial arts courses (see Objective 9, Chapter 4). Performance tests have been used to a limited extent to measure skills. It is questionable, however, if the time required to construct, administer, and grade these tests is worthy of the teacher's effort, considering the limited time available in industrial arts and the fact that the completed project stands as a record of skill performance. For these reasons, it has been common practice to measure skill development by grading the completed project. That this method leaves much to be desired can be demonstrated easily by having any group of teachers give individual grades to several projects. Experience indicates that a single project may be graded from excellent to failure, depending on the standards of the teacher who makes the evaluation.

It must be remembered that the teacher is more concerned with pupil behavioral changes than merely skill development. Although the completed project itself stands as a record of pupil skill development, it is but a by-product of instruction and not the end sought. In other words, *the teacher should be more concerned with what the project did to the boy than what the boy did to the project.*

The following check list is an attempt to rate the development of skills on a more objective basis than the mere examination of a finished project. Use of this check sheet several times a term for each student will give a more accurate estimate of the student's growth in skills than can be obtained by the project grading method.

CHECK LIST ON THE USE AND CARE OF TOOLS

Measurement in the Use and Care of Tools

Name of Worker .... Course ....
Instructor ....

Classifications	Composite Ratings		
	1st 6 wk	2nd 6 wk	Final

DEGREE OF SKILL
1. Clumsy and awkward; seems "lost" ....
2. Awkward but tries ....
3. Handles tools with some coordination and rhythm ......
4. Handles tools with considerable skill ....
5. Handles tools expertly ....

SELECTION OF TOOLS and/or EQUIPMENT
1. Selects tools which cannot be used; lacks understanding of tool functions ....
2. Selects tools which can be used but are awkward ......
3. Selects tools which are suitable ....
4. Expresses a desire for better tools but doesn't know what they are ....
5. Selects the most suitable and best available tools; may invent tools ....

EXTENT OF INSTRUCTION NEEDED
1. Needs repeated instructions ....
2. Needs repeated instructions; tries to follow them ....
3. Needs additional instructions only on some jobs ....
4. Needs only initial instructions ....
5. Self-propelling ....

| | Composite Ratings | | |
Classifications	1st 6 wk	2nd 6 wk	Final

SPEED IN THE USE OF TOOLS and/or EQUIPMENT
1. Uses tools so slowly that it hinders progress on the job ..
2. Uses tools with expected speed only when pressure is applied ....
3. Uses tools with expected speed without pressure ....
4. Uses tools as rapidly as job requires ....
5. Uses tools more rapidly than is expected ....

EXTENT OF CARING FOR TOOLS and/or EQUIPMENT
1. Damages tools through careless or improper use ....
2. Uses dull or dirty tools ....
3. Sharpens or cleans tools on some occasions ....
4. Takes care of tools conscientiously ....
5. Is meticulous in the care of tools and equipment ....

DEGREE OF ACCOMPLISHMENT
1. Tries to get by with doing as little work as possible .....
2. Works hard but gets very little done ....
3. Does the rquired work when forced to do so ....
4. Does the required work without being pushed ....
5. Does more than is required or expected of him ....

DEGREE OF SAFETY
1. Exposes himself and others to dangerous situations through unsafe use of tools ....
2. Is careful about his own safety, but not that of others, when using tools ....
3. Is careful about others' safety, but not his own, when using tools ....
4. Uses tools in a safe manner most of the time ....
5. Uses tools in a safe manner at all times ....

STANDARDIZED TESTING

In addition to the techniques for evaluating behavior changes described previously in this chapter, the teacher may also consider using standardized tests. The pros and cons of standardized testing in industrial arts have been a subject of discussion for a number of years. The major academic deterrents to the development of a standardized test in the field have been the great variance existing among programs across the nation, especially in objectives, behavioral changes, and subject content. Micheels states it this way:

> One definite ... weakness is the lack of a commonly accepted body of subject matter content in industrial arts ... The teaching content often varies from school to school and state to state. This brings up the problem of trying to identify what kinds of behavior the students are expected to exhibit at the end of a course or a training program. This makes it difficult to develop instruments for general use.[6]

Before relying on the results of any standardized test the teacher should make certain that the test is valid for his pupils. Validity is a specific characteristic of a test, that is, a test may be valid for one group of students and not for another one even if both groups are taught by the same teacher. The teacher is in the best position of all to judge whether or not a given test is valid for his students.

Presently the latest standardized tests in industrial arts education are the *Cooperative Industrial Arts Tests*.[7] Five tests comprise the series: *General Industrial Arts, Drawing, Electricity/Electronics, Metals,* and *Woods*. The tests are available in two parallel forms. Each test consists of 50 four-distractor multiple choice items to be completed within a 35-minute time limit.

The tests represent a joint effort among the American Industrial Arts Association, the American Vocational Association, and the Educational Testing Service. Test development committees were nominated by the two professional associations. Test items were developed after the preparation of detailed specifications which reflected the most important outcomes of instruction in terms of knowledge, understandings, and skills that students might be expected to acquire. The Educational Testing Service provided editing, review services, pre-testing, item analysis, final assembly of items, and standardization.

While these tests are not intended to evaluate all the behavioral changes for all the objectives in industrial arts, they may provide the teacher with useful information on pupil achievement at certain levels of the cognitive domain (knowledge, comprehension, and application) in the materials areas indicated by the test titles.

[6] H. H. London, W. J. Micheels, C. B. Porter, R. C. Bohn, and R. M. Reese, "Standardization in Tests and Measurements," *Industrial Arts and Vocational Education,* vol. 55, no. 1 (January 1966), p. 23.

[7] *Cooperative Industrial Arts Tests,* Educational Testing Service, Princeton, New Jersey, 1969.

EVALUATION AND GRADING

The teacher who has been accustomed to marking students on the basis of projects and tests on information, is likely to wonder how such information and data as indicated in this chapter could be translated into a percentage or letter grade to place on a report card. This problem is pertinent because most schools require some kind of composite grade for each student at regular intervals.

An estimate made by the teacher on the basis of what evidence he has been able to obtain concerning changes in behavior, translated into a single figure or letter, will probably be as satisfactory and defensible as one arrived at by testing and the marking of projects. Some schools, however, have attempted to make their reporting of student progress more significant by a comprehensive summary of growth in desirable directions. The form that follows is an adaptation of a card recommended by Wrinkle and Gilchrist[8] which appears to have certain advantages:

EVALUATION OF STUDENT ACHIEVEMENT AND PROGRESS IN INDUSTRIAL ARTS

Name of Student   Course or Activity

Evaluations are made in terms of what might be expected normally of a student of similar age and grade placement.

O — Outstanding
S — Satisfactory
N — Needs to make improvement
PN — Has made unusual progress but needs to make further improvement
U — Unsatisfactory
IE — Insufficient evidence

Evaluation by

Student	Teacher	Objectives
_____	_____	1. He *explores industry,* reads about processes and products, visits industries, can define common industrial terms.
_____	_____	2. He *develops recreational and avocational activities,* selects and develops a hobby, spends spare time in the laboratory, makes equipment for his hobby, develops a home workshop.

[8]William L. Wrinkle and Robert S. Gilchrist, *Secondary Education for American Democracy.* (New York, N.Y.: Rinehart and Company, Inc., 1942), p. 427.

_____ _____ 3. He *develops an appreciation of good craftsmanship and design,* applies good design in projects, redesigns projects to improve design, avoids overdecoration.

_____ _____ 4. He *selects, buys, uses, and maintains the products of industry,* looks for constructional features in purchases, knows and looks for trade names, maintains and correctly uses equipment.

_____ _____ 5. He *studies occupations,* knows entrance requirements, working conditions, and wages for many trades and occupations, considering tentative choice of vocation.

_____ _____ 6. He *engages in critical thinking* with tools, materials, and processes of industry, increasingly attempts to solve his own problems.

_____ _____ 7. He *evinces creativity,* designs and makes new projects, thinks through correct procedures, experiments with new ways of solving constructional problems.

_____ _____ 8. He *develops desirable social relationships,* takes active part in personnel organization, cooperates with others on group projects, participates in club activities, gives help and advice willingly, works willingly with individuals who differ in race, color, and creed.

_____ _____ 9. He *develops safe working habits,* uses tools and machines safely, gives attention to the safety of others, reports unsafe equipment and practices.

_____   _____
(student signature)   (teacher signature)
_____, 19___   _____. 19___

(*Note:* Teacher's comments, particularly on N, PN, and U evaluations, are written on the opposite side.)

DISCUSS PROGRESS WITH STUDENTS

It is important for all students to be familiar with the objectives toward which they are working. This information should be made available to them both in terms of general industrial arts objectives and in expected student behaviors. It is equally important for the students to know how their behavior progress will be measured and reported. If score cards or rating sheets are to be used, these should be made available for student examination. A student should never be in doubt as to his grade or class standing in an industrial arts course at any given time. The teacher should practice an "open" rather than a "closed" grading system. Student progress should be a topic of discussion in the frequent conferences held between the teacher and the pupil.

SUMMARY

The grading of industrial arts students in the past has been a highly subjective process. Measurement has been restricted largely to the teacher's estimate of a grade for projects and to scores on informational tests. A more defensible evaluation procedure will include (1) a clear statement of objectives, (2) analysis of objectives in terms of behavior changes, (3) isolation of testing situations, and (4) a method for recording results. Sample test items based on this approach to evaluation were included in the chapter and a method for determining final grades in industrial arts was presented.

Standardized testing has had limited use in industrial arts education because of the great variance in objectives, behavioral changes, and subject content in programs across the nation. Before relying on the results of any standardized test the teacher should make sure the test is valid for his pupils.

DISCUSSION TOPICS AND ASSIGNMENTS

1. In parallel columns contrast evaluation with measurement from the standpoint of scope, method, purposes, and outcomes.
2. How can growth in consumer literacy be evaluated? Develop a suitable technique or instrument.
3. Propose a method for evaluating growth in desirable recreational activities.

4. Growth in desirable social relationships is generally accepted as an outcome of industrial arts. Propose a method for evaluating such growth.

5. Elaborate upon the pros and cons of standardized testing in industrial arts.

6. Devise a rating form to assess safety consciousness (see Objective 8, Behavior Change 1).

7. Select one of the objectives for industrial arts and prepare objective-type test items designed to measure pupil growth or attainment of the several behavior changes listed (see Chapter 4).

SELECTED REFERENCES

Bloom, Benjamin S. (ed.) *Taxonomy of Educational Objectives, Handbook I—Cognitive Domain.* New York, N.Y.: David McKay Company, Inc., 1956.

Cillizza, Joseph E. "The Construction and Evaluation of a Test of Critical Thinking Ability (grades 7–8)" (unpublished doctoral dissertation, Boston University, Boston, Mass., 1970).

Cooperative Industrial Arts Test (handbook and specimen sets) Princeton, N.J.: Educational Testing Service, 1969.

Koble, Ronald and Robert Thrower. "Research Related to Evaluation in Industrial Arts," Chapter 3 in *Status of Research in Industrial Arts.* 15th Yearbook of the American Council on Industrial Arts Teacher Education. Washington, D.C.: American Industrial Arts Association, 1966.

Krathwol, David R., Benjamin S. Bloom, and Bertram B. Masia. *Taxonomy of Educational Objectives, Handbook II—Affective Domain.* New York, N.Y.: David McKay Company, Inc., 1964.

London, H. H., W. J. Micheels, C. B. Porter, R. C. Bohn, and R. M. Reese. "Standardization in Tests and Measurements," *Industrial Arts and Vocational Education,* vol. 55, no. 1 (January 1966), pp. 23–25, 70.

Micheels, William J. and M. Ray Karnes. *Measuring Educational Achievement.* New York, N.Y.: McGraw-Hill Book Company, Inc., 1950.

Nelson, Lloyd P. and William T. Sargent (co-editors). *Evaluation Guidelines for Contemporary Industrial Arts Programs,* 16th Yearbook of the American Council on Industrial Arts Teacher Education. Washington, D.C.: American Industrial Arts Association, 1967.

Pendered, Norman C. "Test Items That Teach," *Industrial Arts and Vocational Education,* vol. 45, no. 1 (January 1956), pp. 4–6.

Steeb, Ralph and Marshall Hurst. "Evaluating Your Industrial Arts," *Industrial Arts and Vocational Education,* vol. 57, no. 2 (February 1968), pp. 20-29.

Wrinkle, William L. and Robert S. Gilchrist. *Secondary Education for American Democracy.* New York, N.Y.: Rinehart and Company, Inc., 1942.

Chapter 18

Industrial Arts and Public Relations

The public has an interest and a stake in the schools. Supported as they are by public funds, the schools cannot logically nor expediently withdraw and exist apart from the community and community interests. It is imperative that a free interchange of information and ideas take place between the school and the public. The activities designed to promote such an interchange are commonly known as the public relations program of the school. The principal purposes for such activities are to (1) generate good will, cooperation, and mutual understanding between the community and its schools; (2) gain financial support of the tax-paying citizenry; (3) minimize possible misunderstandings relative to educational programs and other school activities; and (4) continuously inform the general public about educational matters.

It may be well at this point to differentiate between *publicity* and *public relations* because these terms are not synonymous. Publicity is the dissemination of information about a person, place, or organization. It is a form of propaganda because its purpose is usually to influence people or to cause them to act in a certain way. An advertisement is an example of publicity. On the other hand, public relations are intended solely to improve the relationship between the school and the community by promoting understanding and good will. Public relations involve little showmanship and are quite unlike newspaper advertisements or radio and television commercials. A public relations program is not a "one-shot campaign," but rather is a never-ending process. In contrast to publicity, a public relations program may be likened to a two-way street which permits communication and promotes harmony between the school and the community. Publicity is a direct, planned, and often forceful means to demand attention. While public relations may involve some direct measures, the term also includes the many indirect, unplanned contacts with individuals in the community where no immediate gain, response, or support is sought. Prakken puts it this way:

> Mere "publicity" is not a public-relations program. . . . What we are after is public understanding. Politicians and show-business characters are happy if they get publicity, whether it is good or bad. What they are seeking is "notice" and not "understanding." Understanding only comes through an honest and straightforward presentation of the facts. A good public-relations program takes the public into its confidence and shares shortcomings as well as successes.[1]

If a school has a sound public relations program which has been working well over the years, it will have little or no need for publicity.

THE IMPORTANCE OF PUBLIC RELATIONS

Certain programs in the school curriculum, such as science, home economics, commercial subjects, vocational education, industrial arts, and others, may involve slightly higher per-pupil costs than the so-called conventional classroom teaching. This is not so true now as it was years ago because of student response systems, individualized instruction, computer-assisted instruction, closed-circuit television, teaching machines, and the host of other educational media which permeate today's academic-type instruction. But, there are still some who favor lowering educational costs and taxes by pointing to a high-cost instructional area and suggest that it could be eliminated without harm to the total school program. More often than not, such contentions have arisen from those who know little of the objectives or real worth of the subjects and, in their ignorance, have labeled them as "educational frills." For this reason it is necessary occasionally to defend the educational values of some subjects to uninformed lay groups and even to school boards. In most cases these misunderstandings have come to the surface because the administration and teachers have failed to communicate the educational values involved. In this connection Prakken notes:

> . . . Uninformed people will not come to you for information; partially-informed persons usually have some interesting misconceptions. Explanations to the public concerning what their taxes are providing, educationally, for the young people of the community will help. "Keeping everlasting at it" may insure partial success.

[1] Lawrence W. Prakken, "Let There Be Light!" *School Shop*, vo. XIX, no. 8 (April 1960), p. 2.

Constant, continuous effort to tell *all* to *everybody* is essential, since make-up of the public changes, people forget, and industrial educational changes.[2]

It is not to be expected that all areas of instruction will cost the same. Reasonable people will look beyond the immediate costs to the educational benefits to be derived. Once these have been explained, understanding and support usually follow.

The industrial arts teacher should stand ready to justify the costs of his program in terms of its educational values and outcomes. To those who are familiar with all that an industrial arts program can accomplish in preparing youth for successful citizenship, it is obvious that the field is in a position which can be defended easily. However, it is of paramount importance that the public be well informed concerning all phases of the industrial arts program. Not only must the teacher be thoroughly convinced of the value of his subject area, but he must be an articulate salesman who can convey his convictions to the public. More than this, he must set an example—keeping in mind that the public notices and appreciates a truly professional teacher. He must recognize the serious purpose behind public relations efforts and realize that they are as essential to his program as a good shop layout, adequate supplies, equipment, and instructional materials. In fact, his budget may well be related to his success in public relations.

Some of the more important reasons why industrial arts needs an active public relations program are: (1) A clear understanding of industrial arts will help win and maintain a high level of support from the community. (2) A good public relations program will ensure greater support for industrial arts from the school administration. (3) A sound program of public relations will foster understandings and appreciations of the industrial arts department by other teachers in the school. (4) A successful program of educating the public about industrial arts will tend to improve relationships between the entire school system and the community.

1. Community Support

Individuals and groups tend to be suspicious of things they know little about. Tierney reports that "... only one out of every six parents have had any firsthand experience with industrial arts courses . . ."[3]

---

[2]Lawrence W. Prakken, "Out of the Past—From School Shop Files," *School Shop*, vol. XXIX, no. 8 (April 1970), p. 184.

[3]William F. Tierney, "Public Relations and the Industrial Arts Teacher," *Industrial Arts and Vocational Education*, vol. 52, no. 7 (September 1963), p. 26.

Thus, if the industrial arts teacher has an excellent program but fails to impress the people of the community with what he is doing, he will probably find them apathetic to the importance or needs of his department. If, on the other hand, he takes every ethical means to acquaint them thoroughly with what he is accomplishing, he will more than likely find solid backing for any reasonable requests that he may make.

The importance of a sound public relations program for industrial arts education has been demonstrated countless times. In many schools where the teacher has not seen fit to sell his program to the community, industrial arts has not reached its full potential. But in those schools where the citizenry has been educated to the achievements and values of industrial arts, strong community support has followed requests for supplies, new equipment, additional staff, and expanded shop areas.

2. Support by School Administration

The industrial arts teacher must fully realize that his, or any other, educational program is doomed to failure unless he earns the support of his local administration. If the administration exhibits only lukewarm enthusiasm for industrial arts, it may well be due to an unfortunate experience or simply to a lack of understanding. But when an administrator discovers that his industrial arts teacher has addressed the local Rotary or Kiwanis Club concerning his work, or reads an account about the industrial arts department in the local newspaper, or sees an exhibit of excellent work in a downtown store window or on television, his whole attitude toward both the teacher and the department is likely to change to one of pride and interest. Many cases are known where an intelligent approach to a public relations program has changed the attitude of the administrator from one of noncooperation or indifference to one of such active interest that the industrial arts laboratory is the first place to be shown to school visitors. Usually support by the administration is not achieved overnight nor is it the result of a single incident. Rather it is earned through teacher effort over a period of time.

It is important for the teacher to keep his department head, his principal, and his superintendent fully informed at all times of significant activities in his laboratory. A simple way to do this is to send regular reports through channels. Information may be included on recent field trips, guest speakers, individual and group projects, community service activities, placement of graduates, results of tests and comparisons with other norms, innovative practices, and other successes or accomplishments of students. While a face-to-face report enables the teacher to provide full details orally, there is no written record and it is sometimes difficult for one level of administration to pass this information along to a higher level. A written report takes teacher time, but administrators can read it at their convenience and it does establish credit where credit is due.

3. Support by School Faculty

While it is important that the general public and the school administration be fully informed concerning the industrial arts program, it is equally important that other teachers in the school be as well informed. Too often in the past the industrial arts teacher has neglected or overlooked this audience. To make matters worse, he has been prone to assume that his colleagues were familiar with and appreciated the virtues of his field. Much of the blame for any lack of understanding of industrial arts which may exist within a given school can be traced to a lack of effort on the part of the industrial arts teacher himself. Even today some industrial arts teachers still practice "isolationism," that is, they fail to socialize formally and informally with other members of the faculty; they prefer to "talk shop" among themselves. Fortunately, however, the situation generally is much better than it was some years ago and continues to improve with each passing year and with each crop of new industrial arts teachers. Basically, the problem is largely one of communication. Wherever the industrial arts teacher has opened lines of communication with other teachers in his school, the results in the form of better understanding and appreciation of industrial arts have more than repaid the effort exerted.

The industrial arts teacher should eagerly seize every opportunity to inform the faculty of the purpose, methods, and results of his program. As shown in Fig. 18-1, this teacher is using the overhead projector to good advantage in informing faculty members about industrial arts. In addition, the teacher should seek actively to correlate the subject matter of industrial arts with other school subjects such as English, art, mathematics, and science. When feasible, interdisciplinary assignments may be made and cooperative projects and joint enterprises attempted. Finally, the industrial arts teacher should avail himself of the countless opportunities of professional contact in both formal faculty meetings and casual, social conversation to further the cause of industrial arts education. The support of the industrial arts program by the faculty is a coveted prize and should be sought by the teacher.

4. Improved School and Community Relations

A program of public relations by the industrial arts teacher has a further function which is not directly concerned with his own depart-

Industrial Arts and Public Relations

Fig. 18–1. Informing the school faculty about the industrial arts program. (Courtesy, 3M Company)

ment. The more favorably a school is known to the community, the better will be the overall relationships between them. Since *some* of the results of industrial arts work are more objective and are of greater primary interest to citizens in general, a fine opportunity exists to keep the general public informed through this department. People who see exhibits of projects or pictures of work done in industrial arts classes are likely to assume that it is representative of other good work in the school. When students go home enthusiastic about their industrial arts classes, general approval of the whole school program by the parents is apt to follow. Good will generated in this manner reflects not only to the benefit of the industrial arts department, but indirectly to the school as well.

DEVELOPING A PUBLIC RELATIONS PROGRAM

Public relations programs may be divided into two types: the *unplanned* and the *planned*. Some teachers may feel that informal or unplanned programs are not true public relations programs, but this is far from the truth because unplanned programs are often more effective than those which involve elaborate preparations. Whether the teacher wishes it or not, the manner in which he conducts his classes will mold public opinion concerning him and his school. In fact, everything which a teacher and his class do both in and out of school contribute to the general public opinion of the teacher, the students, the program, and the school itself. As mentioned in Chapter 14, when students and teacher leave a school to make an industrial visit or other field trip, they are actually engaged in public relations work. Likewise, good community relations are promoted when a local resource person is invited to the school laboratory to talk to the students about his specialty (see Chapter 16).

One of the most potent forces available to the teacher for good or bad is the opinion and reports of his students. Thus the place to begin a sound public relations program, planned or unplanned, is with the students themselves. Convince them of the worth of industrial arts and the battle is half won. Day after day from every student in the class a report will go into some home in the town or school district. A school principal in Illinois states:

> ... the only truly significant, positive factor in good school public relations is a "satisfied" student, a student who goes home each night feeling good about his education, his teachers, and his experiences generally, and expressing these feelings to his parents.[4]

Supplementing these oral reports, objective evidence in the form of projects will occasionally be shown to parents and friends. This evidence will often do more than anything else to determine the professional standing of the teacher and his subject in the community.

There are certain things which the teacher can do to affect, directly or indirectly, the general character of the reports and attitudes of his students. Among these are the following:

1. Maintain a laboratory which is so well arranged and attractive that students will enjoy working in it.

2. Plan the instruction carefully so that *all* students will have a good chance for success.

3. Be friendly with all students. Learn their problems and be worthy of their confidences. This can be done without becoming so familiar as to lose their respect.

4. Set standards as high as the ability of the student will permit. Students are not apt to respect and admire the teacher who accepts less than their best.

5. Be firm and fair in matters of discipline and organization. It is not the "easy" teacher who is most admired.

[4] Arthur H. Rice, "Public Relations Problems Impede Community Support," *Nation's Schools*, vol. 86, no. 2 (August 1970), p. 12.

6. Develop an organization which functions smoothly. Let the class feel that it is their plan.

7. Watch for special interests and aptitudes and help students develop them.

8. Try to visit the homes of students and become acquainted with the parents. Keep in mind that the key to successful public relations lies in the personal relationships among students, parents, teachers, and others in the school-community.

9. Above all, remember that you are teaching children and not a course of study. Effecting changes in the behavior of students is much more important than having them master a certain number of tool skills.

The industrial arts teacher who operates a program based on these and similar concepts need have no misgivings concerning that part of his public relations program which stems from his students.

The teacher cannot depend entirely on unplanned aspects of a public relations program. It must be supplemented and fortified by an active and well-structured series of activities designed to bring his work directly to the attention of the public. *Before the industrial arts teacher need give any serious thought to initiating a planned program of public relations, it is requisite that he have a program worth publicizing.* This means that he must have the goals of education clearly in mind and understand the unique contributions which industrial arts can make. He must know what behavioral changes he seeks to bring about in his pupils and how best these may be achieved in light of the changing needs of society and the pupils themselves. He must understand adolescents and like to work with them as he guides their learning activities in the laboratory. He must strive ever to become a master teacher.

Many of the methods commonly employed in public relations programs have been proved over a long period of time, while others are relatively new. Ten common means which the alert teacher should consider in informing the community of his educational program include: (1) exhibits and project fairs, (2) talks to service organizations, (3) open house, (4) newspapers, (5) school assemblies, (6) school publications, (7) display cases, (8) radio and television programs, (9) adult classes, and (10) community service. Each of these will be discussed briefly in the following sections.

1. Exhibits and Project Fairs

There has been serious criticism of the industrial arts exhibit on the grounds that it places unwarranted emphasis on the project, that it leads to the teacher doing the students' work, that the work of only the better students is shown, and that competition produces undesirable effects among students and among teachers. Undoubtedly the exhibit has been abused and, in isolated cases, it is probably guilty of the criticism leveled at it. However, if it is properly handled and students' interests and welfare are adequately safeguarded, it provides unusual appeal as a public relations tool.

It has been repeatedly emphasized that the industrial arts project is merely a means and is not an end in itself. Nevertheless, projects do result from industrial arts work. They are tangible evidence of what the students have done. They represent one of the products for which the community has spent its money. It is not strange, therefore, that parents and others interested in the school like to see and examine the projects which have been made. It is reported that in 1970 more than 30,000 people from all walks of life visited New York City's annual industrial arts exhibition.[5]

Industrial arts exhibits can be improved if the projects are not featured as "things" but rather as solutions to problems facing students. The "how" and the "why" of a project are often more important than the end result itself.

Exhibits and project fairs tend to fall into two types: (1) those held in the school where the public is invited to come and see them; and (2) those held in a place, such as a downtown building or store window, where people will naturally notice and examine them (see Fig. 18–2). In either case considerable preparation and care are required to make the exhibit of maximum value as a public relations device. The following suggestions may apply to either or both types:

1. Plan the exhibit well in advance of the opening date. Annual project fairs are usually a never-ending cycle; that is, by the time one year's display is over, it is time to begin to plan for the next one.

2. Give as extensive and complete publicity as possible. Use newspaper articles, assembly announcements, notices to be taken home by students, posters, and possibly radio and television facilities.

3. Make the exhibit as attractive as possible. Use table covers, colored paper, streamers, and any other devices to "set off" the projects. The introduction of motion by use of a revolving stand helps to attract attention. The use of the overhead projector with polarized transparencies creates striking and colorful effects. Cooperation with the art department and the home economics department in the arrangement of the display is frequently helpful.

4. Consider the use of tape recorders and audio cassettes to add sound to the display. A sound-on-slide presentation or the use of an

---

[5]"New York City's Annual Industrial Arts Exhibition," *Rockwell Power Tool Instructor*, vol. 18, no. 2 (1970), p. 3.

Industrial Arts and Public Relations

Fig. 18–2. The Western Pennsylvania Industrial Arts Fair provides parents and friends a fine opportunity to examine some of the construction activities of the school. (Courtesy, J. Philip Young)

8-mm film loop showing students at work in the laboratory can add much interest to the display.

5. Use printed or hand-lettered posters and cards to explain the theme of the exhibit and to credit students who constructed the projects on display.

6. Make the exhibit as representative of your program as possible. Avoid giving the impression to the public that the only results of the program are individually-constructed objects which emphasize skill and craftsmanship. Include in the display examples of independent research-oriented reports and studies, materials testing experiments, class efforts in line production, group projects, community service projects, and the like.

7. Avoid the temptation to exhibit only the best work of a few selected students. If ample space is available, every student may well have at least one project represented. If space is limited, it is better not to crowd the exhibit, but rather to display fewer projects and to change the display more frequently.

8. If practicable, have members of the industrial arts classes on hand to explain details and to answer questions from visitors.

9. Take every precaution to protect students' work while it is on display.

10. Avoid leaving any single exhibit on display too long. A week is probably the maximum effective time for which a display may be left unchanged.

11. Make a careful plan for advertising, setting up, maintaining, and taking down the exhibit. Assign definite responsibilities in each of these categories and check continually to see that they are carried out. If properly handled, the exhibit may be made a valuable learning experience for the students.

2. Talks to Service Organizations

Because the values, objectives, and methods of industrial arts are less widely known and understood than is the case with some other school subjects, every opportunity should be accepted to present this information to service organizations, such as the P.T.A., Rotary, Kiwanis, Business Men's and Women's Clubs, etc. The teacher need not be a polished speaker. If he is obviously interested in his work and sincerely believes in its contributions, he will be able to tell an interesting story. Any group will listen gladly to a straightforward discussion of the objectives of industrial arts and an explanation of how they are achieved. If examples of the attainment of individual students are interspersed, they will help to personalize the talk and make it objective.

When preparing such a talk, the teacher should endeavor to procure some form of supplemental visual aid such as that illustrated in Fig. 18-1. Some projects made by members of the class will be helpful. Slides or pictures of students at work are excellent. A motion picture taken in the school laboratory would be one of the finest public relations tools that could be devised. The use of visual aids in connection with a talk has the further advantage of providing a reason for appearing before a group. If one has pictures or slides to show, this in itself is sufficient reason for being invited. Visual material of this type also tends to give direction and coherence to the talk.

Good speech material can be drawn from both the professional aspects of industrial arts education and from new developments in materials areas. Concerning this point, Hackett suggests:

> Rather than explaining how valuable industrial arts is, we should teach them a portion of one of our better course units such as: "New Materi-

als of Industry," "Civilization Through Tools," "Automation," "The Furniture Industry," "Technical Occupations," or "Consumer Economics." Such a discussion serves to illustrate the subject matter content of industrial arts.[6]

3. Open House

There is a growing tendency to encourage parents and other citizens to visit the schools regularly and to become better acquainted with their programs. It has become common practice for a school to hold open house each year during American Education Week. Other schools have found it convenient to hold open house on other occasions. The industrial arts teacher should welcome such opportunities to educate the lay public concerning his program. The shop teacher has an added advantage which is not so evident in some other classes, that is, the students are engaged in activities which are understood, at least to some degree, by the parents.

In order to make the most of the opportunity offered by an open house, the teacher should plan every detail with care. Among the things to be considered are the following:

1. Make the program of activities as diversified as possible. It is not particularly interesting to see a whole class working on the same project.

2. Select projects or activities which are interesting and, to a degree, spectacular. This does not mean that the work should be unreal or staged. It does mean a judicious selection from the many activities which might be presented.

3. Use as many students as can work safely in the laboratory, in view of the maximum number of visitors expected. Every parent would like to see his boy or girl at work. Figure 18–3 shows a student at work during open house activities.

4. Eliminate any activities which might have any possible danger. Even a slight accident during an open house would more than undo any good will that might be built up.

5. Rehearse students in the jobs that they are to do. Make sure that each knows exactly what is expected of him.

6. Try to anticipate some of the more common questions which will be asked and make certain that students know the answers.

7. If feasible, have a supply of simple, pupil-made souvenirs for

[6]Donald F. Hackett, "Building Momentum Through Public Relations," *The Journal of Industrial Arts Education*, vol. XXIV, no. 1 (September–October 1964), p. 13.

Fig. 18–3. This student is busily engaged in demonstrating screen process printing during open house at the school. (Courtesy, Ward S. Yorks, Red Lion High School, Red Lion, Pennsylvania)

visitors to the school shop. A small note pad suitably imprinted by the graphic arts laboratory makes an ideal souvenir.

Of course, the teacher should be on hand during open house and make it clearly evident that he is glad to meet the parents and friends of his students.

4. Newspapers

An effective medium for keeping the public informed of educational matters in the community is the local newspaper. Unfortunately, a recent national study reveals that schools are not taking advantage of this means of informing the public. After analyzing fifty consecutive issues of twelve different newspapers in eleven states, Gordon concludes:

> The data of the study show that it is highly doubtful that the citizens were being adequately informed on the activities, programs, and goals of the schools by the local newspaper. School news constituted less than 2 percent of all the news. With the elimination of news of school sports and other extra-curricular activities, the figure was less than three-fourths of one percent![7]

[7] Richard J. Gordon, "School News in the Local Newspaper," *Journal of Educational Research*, vol. 61, no. 9 (May–June 1968), p. 403.

Generally, most newspapers are willing to print articles about school activities and especially if photographs are available. Parents form one of the largest reading groups. They are interested in what the schools are doing for their children. Today more than ever parents want to know what is being taught, why it is taught, and how it is taught.

The task of preparing interesting copy for a news medium can be facilitated if the teacher studies the basics of report writing which are outlined in English composition or journalism texts. He may decide to prepare the article exactly as it is to appear in print, or he may furnish the facts for a reporter to rework. In either case recommendations for content, sequence, and format are worthy of the teacher's perusal.

In general, the two types of news stories are (1) the straight news item and (2) the feature article. A straight news item is one that must be printed at once or it loses its news value. Announcements of an open house or an educational television program are examples of straight news stories. On the other hand, feature articles are those in which time of printing is relatively unimportant, that is, they can be printed this week, next month, or some other time. For example, not less than once each year the industrial arts teacher should write a feature article in which the objectives, purposes, and expected outcomes of the industrial arts courses are discussed. Editors and readers alike enjoy photographs, so a few pictures showing various aspects of the program will help ensure that the article is printed and read.

Articles concerning special activities, recent industrial visits or field trips, and unusual projects are always of interest. The addition of a new materials area to the laboratory, a new piece of equipment, an innovative approach, an unusual audio-visual aid, all have news possibilities. Students should be encouraged to write about their work. Human-interest stories supported by photographs usually have wide reader appeal. Sometimes a reporter can be worked into the student personnel organization. In some cases a cooperative effort can be carried on with the English department which may result in a suitable article.

If the newspaper has a regular reporter assigned to cover school news, the industrial arts teacher should make certain that his department is well represented in the column. Occasionally the reporter, or even the editor himself, should be invited to the laboratory to see new equipment or to observe some unusual student activity. Professional news people such as these will often uncover news stories whose existence the teacher never suspected.

In addition to news items of local importance, the teacher should also take the responsibility to keep his public informed of significant industrial arts developments at regional, state, and national levels. Trends, innovative programs, and problems are good topics for newspaper coverage. While outside news can stand alone, tie-ins with the local program should be made whenever possible. In this way readers will relate their local program with others. To illustrate this point, a recent national survey showed that the public is greatly in favor of national testing and comparisons of the educational achievement of students. The industrial arts teacher could prepare a timely news account in which the latest standardized tests in industrial arts are described (see Chapter 17). He could then compare the scores of his students with those of national norms—be they higher or lower!—and offer a rational explanation to the public.

5. School Assemblies

Another excellent opportunity to "sell" the industrial arts department, especially within the school, is the assembly. In many schools each department is expected to present one program during the year. The professionally-awake industrial arts teacher will not lose this fine opportunity to publicize his department. Among the types of assembly programs that have been given are:

1. A demonstration of industrial processes. One successful program was built around metal spinning and the use of the potter's wheel. While some students performed the operations, others described what was being done. The program was enthusiastically received by the assembly.

2. Illustrated talks by the industrial arts faculty. In this assembly program the faculty, attired in authentic colonial dress, presented short talks on industry in colonial times and described some of the more important crafts. These presentations were supplemented by slides and transparencies.

3. A talk by one or more students, illustrated with color slides taken in the industrial arts laboratory or on an industrial visit.

4. A panel discussion by students concerning the values of industrial arts. This type of program should prove of value not only to the audience but to those who participate as well. "Occupational Choices" might make a good theme for such a program.

5. The use of a commercially-made industrial motion picture as an industrial arts assembly program probably should be a last resort. It gives the least opportunity for informing the audience about the industrial arts program in the school.

Industrial Arts and Public Relations

6. School Publications

In considering newspapers as a public relations medium, the teacher should not overlook school publications, such as newspapers, quarterlies, yearbooks, leaflets, booklets, and the like. Many of these publications are carried home by students and often read by parents and others interested in school activities. In addition, some publications are sent to school officials and board members as well as to local newspapers and radio or television stations.

The school newspaper is an effective tool of communications and often students will be glad to write articles on industrial arts activities. The position of public relations reporter for the school paper could well be incorporated into the student personnel organization. Pictures either to support an article, or used alone, make interesting news. The industrial arts laboratory is a rich hunting ground for the news photographer, be he amateur or professional, because students are busily engaged in a variety of interesting activities.

It goes without saying, of course, that industrial arts programs offering graphic arts activities are in a superb position to produce a variety of printed materials useful for public relations purposes. Interesting, colorful leaflets, booklets, and folders which tell the story of industrial arts can be prepared by students as part of their manipulative activities (see Fig. 18–4). It may be possible in some instances to involve the English and the art departments in a cooperative project. However, it must be remembered that public relations material should be produced in the graphic arts area only if it is of educational value. Where instruction is concerned, the public relations aspects are but secondary by-products.

7. Display Cases in Corridors and Laboratories

Visitors at a school are impressed by what they see in the corridors and rooms. Students have been known to select courses on the basis of visible evidence of the work being done. In the senior high school, displays are one of the better ways to encourage students to elect industrial arts courses. It is common sense, therefore, for the industrial arts teacher to tell about his program through the use of corridor, entrance, and shop showcases.

Bulletin-board space in hallways, the library, or the classroom may also be used to advantage. For example, a display of industrial processes, history of tools, or styles of furniture could be developed with help from

Fig. 18–4. This student is inserting a plate in the offset duplicator in preparation for producing an industrial arts bulletin. (Courtesy, James O. Reynolds, Dayton Public Schools, Dayton, Ohio)

the librarian. Special display boards at temporary locations in the school are effective in attracting interest and attention because of their "newness." Mobile display boards consisting of a folding easel-type stand and a panel of pegboard can be constructed easily and quickly. As mentioned previously in Chapter 14, pegboard makes a good display board because projects, industrial samples, tools, and models can be easily mounted with hooks, wires, or pegs.

In planning a new building, the matter of adequate display cases is something that should be carefully considered. If the teacher finds himself in a school where these provisions were not made, he should improvise by securing permission to place ready-made glass cabinets in the corridors and by building display facilities within the shop. The most attractive projects should be placed in these cases and they should be changed as frequently as new material becomes available. Printed or lettered cards describing the projects, together with the students' names should be a part of the display.

8. Radio and Television Programs

The development of radio and television has brought to the schools powerful new tools for promoting public relations. As yet, the schools have not learned how to make the best use of these media.

Most of the techniques described for use with school assembly programs could be adapted for radio or television programs. Groups of students could discuss the purposes and results of industrial arts courses. The teacher could give a short paper about the local department. An announcer and the teacher could utilize the interview technique to discuss activities in the school shop. Or the teacher could perform actual demonstrations of operations and processes which might be of special interest to the homeowners of the television audience. A script could be prepared for a short play which describes some of the activities in a typical industrial arts shop. Here again the preparation of such a script might be worked out in cooperation with the English department.

Radio and television as effective public relations devices are just beginning to be appreciated by educators, but they are tools of communication that should become increasingly important as their full value is understood.

In Chapter 16, educational television was discussed briefly as an audio-visual aid to instruction. It is also a powerful public relations tool which reaches into the homes of thousands of persons over a wide geographical area. In addition to formal courses, educational television programs can carry news and information about school activities and programs to many people who would otherwise not be reached through the usual school publicity media. Figure 18–5 shows two industrial arts teachers at work during a weekly television series promoting craftwork and hobbies in the home workshop.

9. Adult Classes

The growing interest in classes for adults has provided another means for bringing citizens into the schools and familiarizing them with programs in public education. The industrial arts teacher in particular has an unusual opportunity to interest the adults of the community in his work. Almost every industrial arts laboratory has the necessary equipment and facilities for teaching one or more activities at the adult level. Evidence seems to indicate that classes of a craft or vocational nature are among those in greatest demand. Other adult classes that have proved popular are woodworking, wood carving, furniture refinishing, upholstering, model making, art-metal work, metal spinning, ceramics, weaving, household mechanics, and blueprint reading. Fig-

Fig. 18–5. Scene from a weekly television series on industrial arts. (Courtesy, Alvin H. Souerwine, York Public Schools, Pennsylvania)

ure 18–6 shows adults engaged in furniture construction and refinishing in an evening program.

It is understood that most industrial arts teachers have a full-time responsibility with their day classes. However, where time can possibly be spared, the results of working with the adults of the community in evening classes one or two nights per week will pay large dividends in terms of public confidence and respect. In certain instances teachers have been able to procure new laboratories, additional rooms, or more and better equipment through pressure brought by adults who had discovered the need for these new facilities because of their attendance at adult classes.

One teacher earned parental approval of his program by featuring a father-son night in the industrial arts laboratory.[8] In addition to gaining parental understanding and support for the industrial arts program, the evenings together provided fine opportunities to strengthen bonds between fathers and sons. In Fig. 18–7 a father and son are shown working together in the industrial arts laboratory after school hours.

[8]Michael R. Morton, "OK, Pop, Here's What You Do," *The Journal of Industrial Arts Education*, vol. XXVIII, no. 4 (March–April 1969), p. 27.

Industrial Arts and Public Relations 437

Fig. 18-6. These adults are happily engaged in worthwhile construction activities in the industrial arts laboratory and at the same time are serving as a valuable public relations medium for the school. (Courtesy, William J. Wilkinson, Nether Providence High School, Wallingford, Pennsylvania)

Fig. 18-7. A father-son combination at work one evening in the school's industrial arts laboratory. (Courtesy, William J. Wilkinson, Nether Providence High School, Wallingford, Pennsylvania)

10. Community Service

An indirect but highly effective means of inviting public attention to the industrial arts program is through community service. One way for the industrial arts teacher to do this is through adult classes which were mentioned in the preceding section.

Another way to perform public good is to involve industrial arts students in toy-repair programs at Christmas time. Such effort usually involves major and minor repairs, adjustments, and refinishing all of which can serve as practical applications of the skills and knowledge gained in the industrial arts laboratory. If it is not educationally desirable to use class time, the work can be accomplished by students working during free periods, after school hours, or on Saturdays. This kind of community service can become a fine cooperative effort involving several agencies interested in improving community life. For example, charitable agencies may collect and distribute the toys to underprivileged children, the industrial arts staff and students may make the necessary repairs, and the school board may underwrite certain costs by making the industrial arts laboratory available and by furnishing the materials and supplies.

Another method for securing a supply of playthings for children is through line-production projects in the school shop. This type of project can serve both as a valuable educational experience for industrial arts students as well as provide a source of toys and games for distribution to underprivileged children at home or abroad, or to children in hospitals, during the Christmas season or at some other appropriate time. Figure 18–8 shows an industrial arts class mass producing attractively bound notepads for senior citizens in a nursing home.

KEEP THE GAP CLOSED

The industrial arts teacher must make every effort to see that no gap exists between the public and his educational program. He must bear in mind that a good public relations program is a two-way street. Not only must accurate and up-to-date information about the program be disseminated, but the opinions of the public schould be considered. Taxpayers expect the best education possible for students in the community. They have a right to visit schools, to ask questions, and to receive truthful answers. This is part of the school's role in educational accountability to which the industrial arts teacher can make a significant contribution. Moreover, the industrial arts teacher should seek

Fig. 18–8. As a community service these junior high school students are hard at work producing notepads in an attractive cover for distribution to elderly persons in a nearby nursing home. (Courtesy, Ward S. Yorks, Red Lion High School, Red Lion, Pennsylvania)

feedback from other teachers and students. Today's students want to be heard. Such interaction can lead to mutual understanding of each other's role in the educational process. Finally, the teacher can tell if he is being successful in his public relations efforts if the community at large is aware that the industrial arts program is contributing to quality education in the school.

BUILDING THE INDUSTRIAL ARTS IMAGE

The foregoing discussion has centered largely on the public relations efforts which the individual teacher can make. In addition to this, agencies such as local, regional, and state professional associations, state departments of education, and teacher education institutions can do much to maintain good public relations and enhance the image of industrial arts education. These groups, working together or independently, may sponsor large-scale activities similar to those outlined previously for the teacher. Other professional undertakings of a public relations nature may include: (1) persuading the governor to declare an Industrial Arts Week throughout a given state, (2) sponsoring industrial arts clubs at the secondary school level, (3) awarding scholarships to outstanding students, and (4) providing college loans to deserving students. One state department of education was successful in spreading the virtues of industrial arts education on 71 cost-free billboard advertisements.[9]

SUMMARY

Public relations and *publicity* are not synonymous terms. Publicity is a form of propaganda designed to influence people or to cause them to act in a certain manner. Public relations are those activities engaged in by the school to promote a free interchange of information and ideas between the school and the public. Helping to promote sound public relations is an important duty of every industrial arts teacher. A strong public relations program for industrial arts is essential because (1) it will help ensure greater understanding by teachers and school administrators; (2) it will help win and maintain high-level support from the community; and (3) it will help to improve relationships between the entire school system and the community.

Before developing a public relations program the industrial arts teacher must be certain that he has a program worth publicizing. The two types of public relations programs are the *unplanned* and the *planned*. Unplanned public relations are those activities carried on without purposeful intent to inform the public. Examples include the way in which the teacher conducts himself, his relationships with his students, the conduct of pupils on a field trip, and so on. Planned public relations programs involve (1) exhibits and project fairs, (2) talks to service organizations, (3) open house, (4) newspapers, (5) school assemblies, (6) school publications, (7) display cases, (8) radio and television media, (9) adult classes, and (10) community services.

The teacher must make special effort to keep the gap closed between the public and the industrial arts program. He must keep in mind that a good public relations program is like a two-way street. Not only should he keep the public informed, but he should also solicit and consider the opinions of the public, the school staff, and the students.

Local, regional, and state professional associations, state departments of education, and teacher education institutions can make significant public relations contributions and enhance the image of industrial arts on a broader scale than individual teacher efforts.

[9]John O. Murphy, Jr. "Spreading the Good Word Via Billboard," *School Shop*, vol. XXX, no. 2 (October 1970), p. 78.

DISCUSSION TOPICS AND ASSIGNMENTS

1. Assume that you are teaching industrial arts in a small city. Prepare a newspaper account dealing with the contributions which your subject makes to the overall school program.
2. Prepare an outline for a talk to be given before a local group on either the objectives and outcomes of industrial arts or some new subject matter content in the field.
3. Plan a short radio or television script on some phase of industrial arts.
4. Outline an assembly program to acquaint students, the faculty, and the administration with the work of the industrial arts department.
5. Enumerate several community services which the industrial arts department could perform.
6. Prepare an 8-mm film loop or a slide-tape presentation for use with a display of projects by industrial arts students.
7. Conduct a survey to discover what lay persons in the community think of industrial arts education.

SELECTED REFERENCES

Cardinell, Jr., C. E. "Does the Open House Affect Our Public Relations?" *Industrial Arts and Vocational Education*, vol. 53, no. 2 (February 1964), pp. 56, 59.

Dapper, Gloria. *Public Relations for Educators*. New York, N.Y.: The Macmillan Company, 1964.

Dennison, Bobby. "Improving Industrial Arts Through Public Relations," *Industrial Arts and Vocational Education*, vol. 57, no. 6 (June 1968), pp. 40–41.

Feirer, John L. "Our Public Image," *Industrial Arts and Vocational Education*, vol. 55, no. 10 (December 1966), p. 11.

Figurski, Arthur J. "Introducing the Public to Industrial Arts," *School Shop*, vol. XXXI, no. 1 (September 1971), pp. 44–45.

Good, James E. and Donald MacCracken. "Total Program Awareness—Showing the Community the IA Process," *School Shop*, vol. XXX, no 9 (May 1971), pp. 32–33.

Gordon, Richard J. "School News in the Local Newspaper," *Journal of Educational Research*, vol. 61, no. 9 (May–June 1968), pp. 401–404.

Hackett, Donald F. "Building Momentum Through Public Relations," *The Journal of Industrial Arts Education*, vol. XXIV, no. 1 (September–October 1964), p. 13.

Hoenes, Ronald L. "Build Good Will With Your Camera," *Industrial Arts and Vocational Education*, vol. 59, no. 1 (January 1970), pp. 30–31.

Husak, A. Jack. "Try a Spring I-A Open House," *Industrial Arts and Vocational Education*, vol. 51, no. 5 (May 1962), p. 22.
"Industrial Arts and the School Paper," *School Shop*, vol. XXI, no. 7 (March 1962), p. 32.
Miller, Rex. "Methods of Public Relations," *The Journal of Industrial Arts Education*, vol. XXV, no. 4 (March–April 1966), pp. 30, 68–69.
Morton, Michael R. "OK, Pop, Here's What You Do," *The Journal of Industrial Arts Education*, vol. XXVIII, no. 4 (March–April 1969), pp. 26–27.
Murphy, Jr., John O. "Spreading the Good Word Via Billboard," *School Shop*, vol. XXX, no. 2 (October 1970), p. 78.
"New York City's Annual Industrial Arts Exhibition," *Rockwell Power Tool Instructor*, vol. 18, no. 2 (1970), pp. 3–7.
Nichols, George V. "The Status of Public Relations," *The Journal of Industrial Arts Education*, vol. XXVIII, no. 3 (January–February 1969), pp. 19–20.
Porter, Sam R. "Contests and Industrial Education," *School Shop*, vol. XXV, no. 6 (February 1966), pp. 44, 46, 48, 50–51.
Prakken, Lawrence W. "Let There Be Light!" *School Shop*, vol. XIX, no. 8 (April 1960), p. 2.
_____. "Out of the Past—From School Shop Files," *School Shop*, vol. XXIX, no. 8 (April 1970), p. 184.
Rice, Arthur H. "Public Relations Problems Impede Community Support," *Nation's Schools*, vol. 86, no. 2 (August 1970), p. 12.
Tierney, William F. "Public Relations and the Industrial Arts Teacher," *Industrial Arts and Vocational Education*, vol. 52, no. 7 (September 1963), pp. 26–28.
Warner, M. E. "How to Use the Overhead Projector to Publicize Industrial Arts," *Industrial Arts and Vocational Education*, vol. 52, no. 6 (June 1963), pp. 18–19.
Wright, Lawrence S. "Public Relations: Whose Responsibility?" *The Industrial Arts Teacher*, vol. XXI, no. 1 (September–October 1962), p. 11.
_____. "The Image We Build," *The Industrial Arts Teacher*, vol. XXII, no. 3 (January–February 1963), p. 12.

Chapter **19**

Laboratory Planning and Layout

A major responsibility of any teacher is to seek to arrange the learning environment in such a manner as to facilitate learning and to promote optimum results. Thus, every industrial arts teacher should strive continuously toward the most efficient arrangement of tools and equipment in the laboratory. While some teachers may never be called upon to design a new industrial arts facility or to refurbish an older one, each must face the problem of providing the best possible physical arrangement of the setting in which the learning is to take place.

Sooner or later, however, many industrial arts teachers are faced with the problem of planning or reorganizing an industrial arts laboratory. This may range all the way from working with the architect in deciding the size, shape, and general layout for an entirely new laboratory, to rearranging a room which has seen many years or service as a school shop, or refurbishing a room designed for an entirely different purpose. In each situation the planning and layout problems will differ. In the case of a new laboratory some control over the size, shape, and other essential features is possible. However, when refitting an old room, the general structural details must be accepted and worked into the rearrangement.

Because of the nearly limitless variations which will be present in planning or reorganizing a laboratory, no hard and fast rules can be established. This chapter will attempt, however, to set down some of the important factors to be considered. Some of the suggestions will apply only where a new laboratory is being planned while others will be equally applicable to both old and new layouts.

EDUCATIONAL SPECIFICATIONS

Before considering the design for a new laboratory facility, the industrial arts teacher should first assist members of the faculty, the school administration, and others in preparing the educational specifications for the proposed building. In a general way educational specifications tell the architect what kind of building is needed, what kinds of people will use it, and what kinds of activities will go on within its walls. The specifications portion of this term is somewhat misleading because educational specifications are neither so specific nor so precise as are building specifications for a home or factory. Hopefully, educational specifications are full of ideas, of possibilities, and suggestions for the architect. They are merely the means by which the school faculty and administration as well as the students and community leaders communicate with the architect. The educational specifications for a new school building should contain at least the following:

1. Goals of quality education.

2. Information about the community, school district, and the area where the school will be constructed.

3. The kind or type of school desired, its initial and utlimate pupil capacity, and other purposes for which the building will be used.

4. Descriptions of the pupil and teacher activities to take place in the several parts of the school facility. Activities involving the community should be included here.

5. Kinds and numbers of personnel (students, faculty, and other staff) for which provisions must be made.

6. General space requirements for the kinds and types of equipment to be used.

7. Desirable relationships among the parts or areas of the facility.

8. Instructional subjects to be offered and estimates of class sizes, numbers of sections, class schedules, and the like.

9. Relationships of facility to grounds, surroundings buildings, community, and region.

10. Transportation system planned.

11. Communications systems desired.

12. Utilities and other services needed.

PRELIMINARY PLANNING FOR THE LABORATORY

In this section the following topics will be discussed: (1) role of the industrial arts teacher, (2) steps in laboratory planning, (3) locating

Laboratory Planning and Layout

industrial arts facilities, (4) determining the number of shops required, and (5) minimum space requirements for the laboratories.

1. Role of the Industrial Arts Teacher

A well-trained and experienced industrial arts teacher can be of invaluable assistance to the school architect in planning laboratory facilities. His role can be an important one whether the layouts are for new or refurbished rooms.

In the first place the experienced industrial arts teacher is a professional who understands the educational objectives involved and is able to make plans for laboratories in which pupils can achieve these objectives. As a specialist in the field, he is thoroughly familiar with the requirements of the laboratories in terms of overall aspects and details. Further, by virtue of previous professional training, the industrial arts teacher understands the fundamentals of architectural drafting and can communicate his ideas not only on paper, but by means of three-dimensional models as well. In short, the industrial arts teacher is undoubtedly the best qualified person on the staff to render advice and counsel to the school architect on matters of industrial arts planning. This embraces all phases of laboratory planning including educational objectives, program, and learning activities as well as the selection and arrangement of tools, machines, supplies, and equipment.

Fortunately, the unique contributions which the industrial arts teacher can make to the school building program are usually recognized. More often than not the architect or the school administrator will indicate the area designated for industrial arts education and will charge the teacher with the responsibility for developing preliminary sketches and detailed plans for the laboratory as well as specifications for new tools, machines, and equipment.

2. Steps in Laboratory Planning

Before starting to develop a floor plan for any industrial arts facility, it is essential that the teacher identify the educational objectives of the program and know what learning activities are anticipated. A mistake frequently made is to plan and equip a laboratory and then build the instructional program to fit it. A clear concept of the total program and the learning activities to be conducted is requisite before planning begins.

While it is impossible to establish inflexible rules to which everyone would agree or which would apply in every situation, it has been found that the following steps represent a satisfactory approach to the planning of most industrial arts laboratories.

1. Outline with the school administrator and the architect the educational objectives of the program and the general requirements of the laboratories in terms of space requirements, site in relation to the main plant, and the like.

2. Determine the materials areas in which the learning activities will be conducted on the basis of the objectives of the courses to be taught.

3. Determine the number of laboratories and teachers required to accommodate the number of students involved. Details on how this is accomplished will be given later in the chapter.

4. Become familiar with the latest trends and designs in laboratory planning, current building codes, lighting standards, etc. This may be accomplished by reviewing the literature of the field and by trips to new installations with the architect and school administrator.

5. Decide what auxiliary rooms and special facilities will be needed.

6. Prepare a preliminary scale drawing of the room to be used, showing all windows, doors, and other structural details which will affect the placement of equipment. A scale of ¼ inch = 1 foot is generally satisfactory.

7. Lay out auxiliary rooms, if these are to be utilized. Insofar as possible, group auxiliary rooms together to avoid "cutting up" the room or making it irregularly shaped. Sometimes auxiliary rooms may be placed across one end of the room or along one side. In other cases architectural requirements make it necessary to locate auxiliary rooms so as to fill up recesses in a room that otherwise would be irregular in shape. Note the arrangement of auxiliary rooms in Fig. 19–1.

8. Prepare scale cutouts for all major pieces of equipment, machines, benches, or cabinets. These are to be used to determine the most effective placement of equipment. The cutouts are placed on the scale drawing and moved about until the most efficient use of the space is attained. Bits of plastic adhesive, pins, or loops of masking tape will hold the cutouts in place and at the same time permit them to be moved about easily. Or, small, flat magnetic strips can be cemented underneath the cardboard cutouts for use on a sheet of tinplate on whose top surface is cemented a drawing of the floor plan. Consult the periodical literature for sources of preprinted, scaled templates of machines and equipment. Consider the use of three-dimensional models. Some 3-D models are available commercially; others can be teacher-made by using scaled wooden blocks, cardboard, and other materials. Show the proposed layout to experienced colleagues for the purpose of improving it.

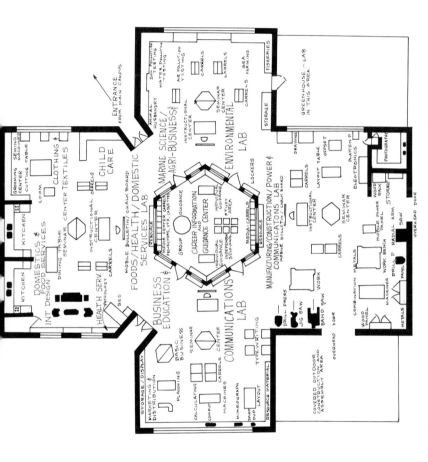

Fig. 19-1. A scaled drawing for Career Exploration laboratories. (Courtesy, Ralph Johnston, East Carolina University)

9. Prepare a final scaled drawing after the best possible arrangement of equipment has been made (see Figs. 19-1, 19-11, and 19-12).

10. Prepare the following for use with the plan: Details concerning *utilities* for water, gas, compressed air, low-pressure steam, and electrical service; *specifications* for tools, machines, and equipment to be ordered; and *sketches* for any features to be built into the room, such as, cabinets, tool panels, lockers, hoods, or exhaust systems.

11. Assist the architect in revising the proposed plan and in answering questions as problems develop during final planning.

3. Locating Industrial Arts Facilities

Several factors should be considered in selecting the site for industrial arts facilities in relation to the total educational plant. In the first place, the laboratories must be easily accessible since they will be used by both students and adult classes. A one-story wing which is an integral part of the total plant probably represents a good solution. Such a wing also lends itself well to future expansion of industrial arts activities. The point is that the industrial arts laboratories should not be isolated from the rest of the school nor located in basement areas. Although some learning activities are noisy at times, for example, construction or machine woodworking, it does not follow that all activities are noisy at all times. The use of acoustical materials on shop walls and ceilings will help prevent disturbance to adjacent classes and also materially improve the auditory comfort of the industrial arts classes. Figure 19-2 shows how an industrial arts facility can be located to good advantage in relation to the rest of the educational plant.

4. Determining the Number of Shops Required

The enrollment of the school is the most influential factor in determining the extent of industrial arts offerings in that school. Despite the philosophies and recommendations by professionals in the field concerning broad course offerings, the hard fact remains that only enough teachers can be retained in any school as is necessary to teach the number of students involved. For this reason the smaller school is handicapped and cannot hope to have as extensive offerings in industrial arts, or any other discipline for that matter, as can schools in larger or consolidated districts.

The following example will serve to illustrate one method for determing how many laboratories and teachers will be needed for a given school size:

1. Suppose that the school enrollment is 660 students and there are 35 class periods per week.

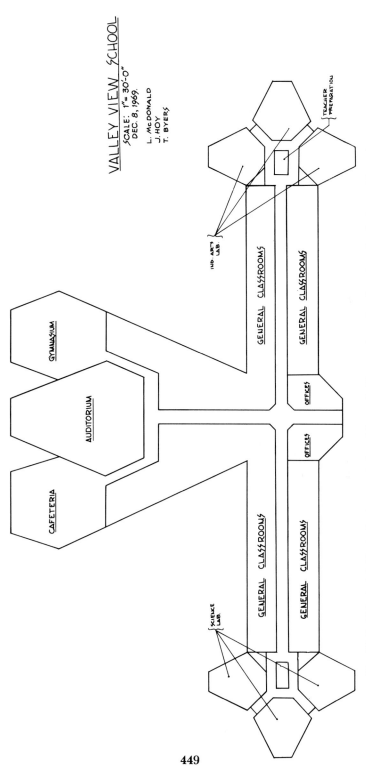

Fig. 19–2. This sketch illustrates how an industrial arts facility can be related to the school plant without being isolated and still provide for future expansion. (Courtesy, Tom Byers, James Hoy, and Larry McDonald, The Pennsylvania State University)

2. If approximately one-half of the school enrollment are boys and if industrial arts is mandatory for them, then the anticipated enrollment is 330 boys.

3. To this enrollment figure must be added the number of girls who will take industrial arts. For girls, industrial arts may be either required or on an elective basis. Assume in this example that 120 girls elect industrial arts. The total expected student enrollment is then 450 students.

4. For reasons of safety in the shop, 15 students are deemed to constitute one section, but this may vary according to local conditions. Thus, in this example there will be 22 sections for boys and 8 sections for girls or a total of 30 sections.

5. At this point it is necessary to convert the shop sections into teaching periods per week. To do this, multiply the number of sections by the number of meeting periods per week. For example, so many sections may meet five periods per week, so many sections may meet three periods per week, and so on. For purposes of easy illustration, suppose that the 30 shop sections meet for two periods per week or a total of 60 teaching periods per week.

6. The teaching load in this school is 30 periods per week, which gives each teacher one free period daily for class preparation and shop maintenance. Thus, in this hypothetical situation, two teachers and two laboratories will be needed. If instead of two meeting periods per week, industrial arts is offered for three periods per week, as it is in many schools, then an additional teacher will be needed.

The services of an industrial arts consultant for the elementary grades should not be overlooked in determining the number of teachers needed for a school district. If the services of a full-time consultant may not be necessary, the chief school administrator may find it convenient to assign an industrial arts teacher on a part-time basis to work with elementary teachers and to devote the remainder of his time to teaching at the secondary level. In situations where the number of industrial arts teachers needed at the secondary level does not come out as a whole number, the remainder of the teaching load for that teacher may well be taken up in supervising industrial arts activities at the elementary school level.

5. Minimum Space Requirements

The size of an industrial arts laboratory will vary with the type of organization—that is, whether it is a comprehensive general shop, a general unit shop, or a unit shop. In most cases the comprehensive

Laboratory Planning and Layout

general shop will require more space than either of the others, while the unit shop probably will require the least space.

Laboratory size will vary also with the subject areas being taught. For example, woodworking or metalworking equipment requires more floor space than industrial crafts or technical drawing. Thus it is unsound to attempt to standardize the sizes of school laboratories, either by type of organization or by subject areas being taught.

Instead, minimum space requirements seem best stated in terms of square feet of work area per pupil. Minimum work areas have been established in some states and range from 75 to 150 sq ft per pupil.

There is considerable basis for the argument that some types of industrial arts laboratories should have a larger area per pupil than others. For example, if unusually large pieces of equipment are placed permanently in the shop, the floor space which they occupy must be considered. This is frequently the case with power mechanics or transportation laboratories. Figure 19-3 shows that provision must also be made for the construction of large projects in the construction laboratory. Under such circumstances the square foot allowance per student may approach the maximum.

Fig. 19-3. A large working area with storage space is needed for group projects in this construction laboratory. Imagine the space required if this class had four or five such projects under construction simultaneously and if the teacher had five other classes meeting daily doing this kind of manipulative activity. (Courtesy, Julian Cleveland, E. B. Aycock Junior High School, Greenville, North Carolina. Photo by Walter F. Deal, III)

If an average figure must be selected, then perhaps 100 sq ft per pupil might be generally satisfactory. For a class of 20 pupils, this would yield a minimum floor area of approximately 2,000 sq ft. The per-pupil-allowance figures do not represent the actual working space in the laboratory allowed for each pupil; rather they represent the total area to be devoted to industrial arts and are inclusive of pupil work space and auxiliary rooms as well as machine and equipment space. The figures are exclusive, however, of dead-storage space, out-of-shop locker areas, and external washroom facilities.

An absolute minimum for a junior high school industrial arts laboratory should be 1,800 sq ft; junior-senior and senior high schools must have at least 2,200 sq ft.

FACTORS TO BE CONSIDERED IN PLANNING

There are many important items to be considered when planning an industrial arts laboratory. If not handled properly on the drawing board, some of these can create awkward teaching situations which will impede the teacher so long as the room is used for a shop. In other words, the design of a laboratory can create safety hazards, problems of class control, storage difficulties, and other problems. Thus, with a few strokes of a drawing pencil the shop planner can either create teaching problems or "design them out" of the laboratory.

The following factors are important and should be considered for any laboratory being planned.

1. Shape of the Shop

A rectangular shape is probably best for the industrial arts laboratory because it lends itself well to the arrangement of machines and equipment. However, not all rectangles make the best shops, because a square room can be better than a rectangle that is too long and too narrow. Proportionally, a rectangle in the ratio of 1:2 is ideal, but ratios of 1:1½ or even 1:3 are satisfactory. The width of a shop should not be less than 30 ft; widths ranging from 30–35 ft are satisfactory. Irregularly-shaped rooms, such as U, L, or T-shapes, should be avoided. The difficulties which may arise when trying to teach a class in a room with hidden areas are evident. Some school architects utilize hexagonal and pentagonal shapes as well as long, sweeping curves in designing classrooms, laboratories, and other facilities. Figure 19–12 shows a proposed laboratory in a hexagonal shape. Note also the shapes of the laboratories depicted in Fig. 19–2.

Laboratory Planning and Layout 453

2. Auxiliary Rooms

These are small rooms which are partitioned off from the main shop room or located adjacent to it. Auxiliary rooms are used for planning, classroom instruction, supply and storage, teacher's office, tool rooms, lavatories, darkrooms, and finishing rooms (see Figs. 19-1 and 19-11). Some industrial arts laboratories will have a number of auxiliary rooms; others will have few or none. The planner must consider the space available, the type of materials areas, and the needs of the teacher. Whenever possible, glass windows are used in doors and walls between the auxiliary rooms and the main shop area. These "vision panels" are designed to facilitate supervision of the room while the teacher is in the main laboratory.

3. Instructional Area

Because of the special importance of this topic and its close relationship to the planning center, it is dicussed more fully in Chapter 20.

4. Work Stations

It is essential that the laboratory be equipped with sufficient work stations for the largest class. In fact, it is desirable to have more work stations than the number of students enrolled in the largest class. A pupil work station or an individual work station may be defined as:

> ... a location at which a pupil may be engaged in the shop during the entire class period. Examples of pupil work stations are: a drawing board, lathe, printing press, approximately 30 linear inches of bench space, potters wheel, and a vise. Equipment items such as drill presses, power saws, buffers, grinders, squaring shears, etc., do not comprise work stations.[1]

5. Storage Areas

A variety of storage problems faces the shop planner. There is need for storing machine accessories; common hand tools; special or seldom-used tools; portable machines; eye safety face shields and glasses for both students and visitors; supplies and parts as well as materials of all kinds, forms, and sizes. Printed instructional materials, laboratory records, library books and periodicals, film loops and audio-visual equip-

---

[1] *Industrial Arts Shop Planning* (Albany, N.Y.: Bureau of Industrial Arts Education, 1956), p. 6.

ment will need to be stored. Student projects will also need to be stored during construction. These may be stored on either a class or an individual basis which may involve open racks, lockers, or cabinets. Elementary classes find it convenient to store their projects on a group basis in large cabinets (see Fig. 19–4). Adult classes can contribute significantly to the laboratory storage problem. Many of these projects are sizable, for example, Fig. 19–5 shows a chair being clamped for regluing. A class of 15 to 20 adults can tax the storage facilities of the typical laboratory with refinishing projects alone.

It is not feasible here to discuss ways and means for solving all of these storage problems. Much valuable information on storage facilities is available in the current literature of the field. The point to be made here is that storage facilities of the type mentioned above must not be overlooked nor taken for granted. Unless special attention is given to providing adequate storage space, the teacher will find his daily tasks that much more difficult, and perhaps, frustrating.

6. Planning Center

Details on planning centers are not presented here but are given special consideration in the following chapter.

7. Assembly Areas

A clear area for the assembly of large projects is often overlooked in planning an industrial arts facility. Some laboratories will not need an assembly area, but if large projects are to be undertaken, either now or in the future, some space should be planned for this purpose. In the construction laboratory an assembly area centrally located is preferable (see Fig. 19–3). Some of the large group projects, such as the dioramas depicting American industry, the sports biplane, the automobile, and the gyro glider (see Chapter 8) will require sizable assembly areas.

8. Teacher's Desk Area

An open area near the entrance to the laboratory makes a desirable location for the teacher's desk. Such a site favors the reception of visitors, the arrival of hand-carried bulletins from the office, and the control of the master electrical panel. Or the desk area may be located in the instructional-planning center (see Chapter 20). In addition to a locked desk, the area should be equipped with file cabinets, a shelf or bookcase for professional books, a place for conferring with pupils and-/or parents, and a locker for personal effects. Some teachers also like to

Laboratory Planning and Layout 455

Fig. 19-4. A double-door, full-length cabinet provides easy access to good storage space for the projects of these elementary school pupils. (Courtesy, California State Department of Education)

Fig. 19-5. Shop planners frequently neglect to make provisions for storing adult projects in the laboratory. (Courtesy, William J. Wilkinson, Nether Providence High School, Wallingford, Pa.)

have a work table on which duplicating equipment can be placed and used.

In certain new school plants, office space for teachers is provided in a central facility sometimes referred to as the instructional-preparation area. Carrels for faculty as well as typewriters and duplicating equipment are available. Usually a number of these facilities are located throughout the school to meet the needs of faculty members. Note teacher-preparation area in Fig. 19–2.

9. Utilities

The availability and placement of utilities is of the greatest importance in planning the industrial arts laboratory. If a new layout is being developed, it is usually possible to place the utilities in almost any desired location. But if an old shop is being renovated or a room designed for other purposes is being converted to an industrial arts facility, the location of utilities frequently influences the placement of areas and equipment. Restrictive compromises are sometimes necessary under such conditions.

Some of the service utilities will not need to be specified in detail by the teacher who assists in new laboratory planning. Undoubtedly the architect will provide standards for heating, ventilation, and general illumination. However, the teacher should make certain that the minimums specified will be satisfactory for industrial arts purposes. For example, if the laboratory is to be used for adult evening classes, then the heating and ventilating systems should be designed and controlled so that it will not be necessary to provide these services for the entire wing in which the facility may be located.

The following suggestions concerning utilities are based on the assumption that a new facility is being planned and that they may be located as desired.

1. Electrical service. In general, electric service should be estimated generously to produce an installation capable of handling future needs rather than a modest one capable of handling only present loads.

All power switches should be centralized in a locked master control panel, preferably near the laboratory entrance. Each wall of the shop should be equipped with a safety push button which, if an emergency arises, can be used to shut off all power in the shop.

The minimum foot candles of light recommended for general work in school laboratories is 70 ft-c and 150 ft-c for close work.[2]

[2] *Footcandles in Modern Lighting*, TP-128 (Cleveland, Ohio: General Electric Corporation, 1964), p. 15.

Laboratory Planning and Layout

Overhead wiring versus the underfloor conduit system for supplying 110–220-volt service for power machines should be considered. Some of the advantages and disadvantages of each plan can be noted by careful examination of Figs. 19–6 and 19–7.

Duplex outlets providing 110-volt service should be spaced at 10–15 ft intervals around the walls of the laboratory. These should be at work-level height from the floor. Switches with pilot lights should be specified for such equipment as gluepots, kilns, etc.

2. Water facilities. These should include both hot and cold water at a sink which should be located as near as possible to the students' exit door. Water may be needed elsewhere too, for example, in finishing rooms, darkrooms, etc. A drinking fountain should be provided in the laboratory.

3. Other facilities. Gas may be needed for the foundry unit and in the furnaces for forging, soldering, and heat treating. Compressed air may be needed for the metals bench, for the foundry and power mechanics areas, in the finishing area for spraying, and for an air hose for general cleaning. Low pressure steam may be needed in the woods technology area for molding or bending or in the power mechanics laboratory.

10. Exhaust Systems

Depending upon the materials areas involved, certain types of exhaust systems may be necessary. Dust, or acrid, toxic fumes must be exhausted from finishing rooms, soldering and electroplating areas, forging units, automotive areas, and woodworking areas. It is obvious that these obnoxious elements should be removed from the laboratory for health and safety reasons.

In general, the two kinds of exhaust systems are for (1) dust removal and (2) fumes removal. From a safety point of view, systems to remove both dust and fumes cannot be combined into a single exhaust duct. Consider the devastation that might result if an exhaust system for a forge unit were combined with one for a finishing room!

Wherever possible, some provision should be made for dust and scrap collection from woodworking machines. Where a general dust collection system is impractical, machines with individual dust collectors should be considered.

11. Soundproofing

Provision should be made for soundproofing the industrial arts laboratory. Acoustical plaster on both the walls and the ceiling is recom-

Fig. 19–6. An example of overhead wiring. (Courtesy, West High School, Davenport, Iowa. Photo by the Do All Company)

Fig. 19–7. An example of underfloor wiring. (Courtesy, Clausing Corporation)

mended. If complete treatment is impracticable, the soundproofing of the ceiling with acoustical tile or plaster will aid significantly in reducing the noise carried to other parts of the building.

12. Display Areas

At least one display cabinet should be provided for each laboratory. If possible, this should be lighted and built into the wall of a corridor through which many students will pass. Supplementary display space within the shop should also be planned.

13. Floors

Flooring should provide a long-wearing surface with a finish that will minimize the danger of slipping. Some resiliency is desirable in order to prevent undue fatigue from walking. The flooring should offer a pleasing appearance and require a minimum of upkeep and maintenance. In today's schools the use of carpeting is becoming more widespread. The industrial arts teacher may well consider carpeting for certain areas of the laboratory, such as the instructional-planning center, technical drafting rooms, and other areas where it is deemed appropriate.

Flooring materials will vary among laboratories and even within the same shoproom to meet the needs of the learning activities. In Fig. 19-8 the materials rated "1" are "first choice" for the shops and areas indicated; those rated "S" are considered "satisfactory."

GENERAL LAYOUT POINTERS

It is unlikely that every layout pointer can be incorporated into each plan for an industrial arts laboratory. Rather, each layout is the result of a balanced judgment based on careful consideration of the principles of planning in terms of the immediate situation. Following are a number of selected layout pointers that have proven useful over the years:

1. Plan the laboratory by areas. Work out the floor plan by materials areas rather than develop the shop as a whole. While there should be unity among areas, this can be achieved by consideration of their placement in relation to one another, even though each is planned as a unit. The foundry area shown in Fig. 19-9 is well-designed in itself, but it is also an integral part of the whole general metals area.

2. Locate related educational activities together. For example, woodworking and plastics are compatible. Activities which are incom-

RECOMMENDED FLOORING FOR INDUSTRIAL ART SHOPS

Shop	Maple	Vert. Grain Douglas Fir	End-Grain Wood Block	Concrete	Linoleum	Asphalt Tile
General Shop	1	S	S			
Wood, Cabinet						
Carpentry	1	S	S			
General Metal						
Sheet Metal	S	S	S	1		
Electric, Radio	1	S			S	S
Crafts	1	S			S	S
Graphic Arts						
Printing	1	S			S	S
Mechanical Drawing						
Drafting	1	S			S	S
Auto Mechanic						
Body & Fender				1		
Aviation				1		
Machine Shop			S	1		
Welding				1		
Foundry				1*		
Forging				1		
General Classrooms						
Offices	1				S	S

* An earthen floor is preferred.

Fig. 19–8. Recommended flooring for industrial arts laboratories. (Courtesy, Industrial Arts Division, Pennsylvania Department of Education)

patible or which may interfere with each other, such as woodworking and graphic arts, should be separated. Locate educational activities in the laboratory to best advantage of each. To be specific, locate the library or planning center in a relatively quiet area.

3. Have all floor space in the laboratory on one level. Avoid overhead or mezzanine work areas or storage spaces. If overhead storage space must be utilized, use a stairway—never a wall-hung ladder.

4. Lay out the laboratory so that all areas are visible to the teacher. Do not locate high tool panels or other tall equipment in central areas and thus obstruct vision throughout the shop. Use glass in partitions between the main shop area and auxiliary rooms. Close up recessed areas underneath stairways and recesses within the shop.

5. Orient the main window area to north light. If possible, have natural lighting from two sides. North and east light are preferred.

6. Provide two exit doors for fire protection. These doors must open outward and should be equipped with "panic" hardware. Provide a place where students can store books as they enter the shop. If possible,

Laboratory Planning and Layout

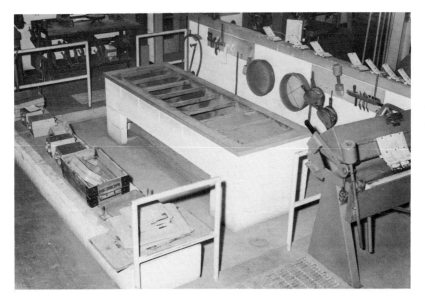

Fig. 19–9. A well-planned foundry area. (Courtesy, Joint School District, Downington, Pennsylvania. Photo by William J. Wilkinson)

have students enter by one exit and leave by another to minimize traffic problems.

7. Avoid too many common areas between rooms such as planning areas or storage rooms which are to be shared by teachers.

8. Consider the vista. In planning the laboratory it is well to consider the impression it will make on visitors or new students on their first entrance. Some interesting activity should be placed where it will immediately "catch the eye." Aisles should lead the eye to other centers of interest.

9. Provide traffic lanes. Allow ample aisles for passage throughout the laboratory. Avoid the placement of equipment so that it will restrict traffic flow. An aisle should be at least 4 ft. wide. Note the wide aisles shown in Fig. 19–10.

10. Locate forge and foundry units as well as gas welding and electric welding in low traffic areas. A wide emergency door leading directly to the outside is desirable.

11. Locate tool panels on or near natural traffic routes. It is desirable to have short travel distances from tool panels to work stations where the tools will be used. Details on tool panels, tool arrangements, and toolholders will be presented in Chapter 21.

12. Plan open spaces near entrances, exits, washing stations, and tool panels to minimize congestion.

Fig. 19-10. Note the wide traffic aisles and ample working space around this woodworking machinery. (Courtesy, Oliver Machinery Company)

13. Locate machines to provide for normal flow of materials from stock areas to finishing areas. Plan lines of work or sequential machine operations to minimize excess travel.

14. Plan sufficient working space around each machine for safe operation (see Fig. 19-10). Don't overcrowd the laboratory with machinery and don't crowd the machinery. Fasten equipment and machines to the floor securely. Paint safety zones and traffic lanes on the floor or mark them with pressure-sensitive tape. Use nonskid abrasive paint or nonslip adhesive tape where machine operators stand to prevent slips and falls.

15. Use explosive proof outlets in finishing rooms and explosive proof lamps under hoods and in areas where there is excessive heat.

PLANNING FOR THE FUTURE

Unfortunately, most schools must be built in accordance with present, not future, needs. Growth and development often necessitate

Laboratory Planning and Layout

changes in the use of rooms and areas. For this reason the principles of flexibility and expansibility should be incorporated into any new school building construction. This is especially applicable in the case of industrial arts, because as the school grows in size, so does the industrial arts department.

It is important, therefore, that industrial arts layouts be developed with flexibility and expansibility in mind. Some suggestions for achieving these are as follows:

> Ample site area should be left undeveloped where building additions are logical.
>
> The fenestration pattern should be continuous along the entire wall rather than grouped especially for each shop.
>
> Corridors should be carried through to outside walls wherever extensions are possible. Stairs should be placed in separate enclosures off the corridors rather than in corridor ends.
>
> Partitions between shops should be nonbearing curtain walls as free as possible from mechanical and utility installations.
>
> Partitions between shops should be so constructed that they can be removed to convert two shops into one, three shops into two, or other space arrangements as conditions in later years warrant.
>
> Heating and lighting services should be engineered so that controls serve relatively small areas within the shops.
>
> Conduit and other utility supply services should be based on a liberal rather than restricted estimate of future needs.
>
> Cabinets, lockers, shelves, and work benches should be standardized so far as possible.
>
> Shelving in cabinets, lockers, and other shelf areas should be of the adjustable type except where safety requires rigid shelving.[3]

TYPICAL LABORATORY LAYOUTS

Limitations in space, available funds, and equipment make it impossible in most cases to include every desirable feature in each industrial arts laboratory. However, in order to illustrate the suggestions contained in this chapter, floor plans for proposed laboratories which seem to be in keeping with present recommended practices are shown in Figs. 19-1, 19-11, and 19-12.

[3] *Industrial Arts in Pennsylvania*, Bulletin 331 (Harrisburg, Pa.: Commonwealth of Pennsylvania, 1951), pp. 99–100.

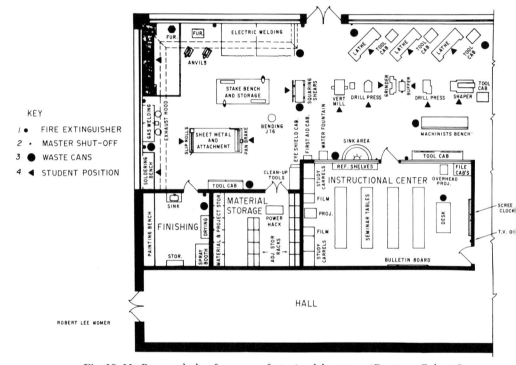

Fig. 19-11. Proposed plan for a manufacturing laboratory. (Courtesy, Robert Lee Womer, The Pennsylvania State University)

SUMMARY

Every industrial arts teacher may some day face the task of planning a new industrial arts facility or redesigning an old one. If a new school plant is being planned, the industrial arts teacher should assist the school administration and other faculty in preparing the educational specifications.

In this chapter eleven steps in industrial arts laboratory planning

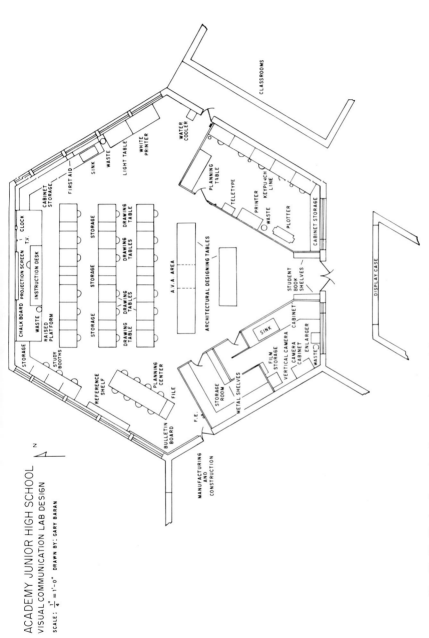

Fig. 19-12. Design for a visual communications laboratory. (Courtesy, Gary Baran, The Pennsylvania State University)

were presented together with recommendations for (1) locating the site, (2) determining the number of laboratories and teachers needed, and (3) establishing minimum space requirements.

In addition, the following factors were considered: shape of the laboratory, auxiliary rooms, work stations, and special areas, such as instructional, storage, assembly, display, and teacher's desk area. Service utilities, exhaust systems, and soundproofing were also discussed. A number of suggestions or layout pointers were presented. Recommendations for building flexibility and expansibility in new layouts were offered. The chapter concluded with illustrations of proposed laboratories which incorporated many of the suggestions and recommendations outlined previously.

DISCUSSION TOPICS AND ASSIGNMENTS

1. How many industrial arts teachers will be needed for a junior high school with a total enrollment of 1,500 students? Assume that boys in grades 7, 8, and 9 and girls in grade 9 only are required to take industrial arts for two periods per week. Assume, too, that there are 35 periods per school week.

2. Why do state-recommended minimums often become local maximums?

3. List several additions to those given in this chapter for achieving flexibility and expansibility in industrial arts laboratories.

4. Select any industrial arts facility with which you are familiar. Show by a sketch how the layout could be replanned for greater efficiency.

SELECTED REFERENCES

Baillargeon, Jarvis H. "A Blueprint for Flexibility in I-A Shop Design," *School Shop*, vol. XXVIII, no. 10 (June 1969), pp. 28–29.

Baker, G. E. "Remodeling a School Shop," *Industrial Arts and Vocational Education*, vol. 59, no. 3 (March 1970), pp. 50–53.

Basilone, James. "Modern Elementary Industrial Arts Room," *Industrial Arts and Vocational Education*, vol. 55, no. 3 (March 1966), pp. 68–70.

Boyd, Robert A. "Light: Its Effect on Teaching and Learning," *School Shop*, vol. XXII, no. 8 (April 1963), pp. 39–40.

Brown, Alan D. and George M. Haney. "Tailoring a Junior High for New Instructional Tactics," *School Shop*, vol. XXVIII, no. 8 (April 1969), pp. 100–103.

Campbell, Edward A. "The Learning Atmosphere," *Journal of Industrial Arts*

Laboratory Planning and Layout

Education, vol. XXV, no. 3 (January–February 1966), pp. 45–47.

———. "Let's Design for Efficiency," *Industrial Arts and Vocational Education*, vol. 55, no. 3 (March 1966), pp. 44–45.

Dean, C. Thomas and Irvin T. Lathrop. "Planning Facilities with 3-D Models," *Industrial Arts and Vocational Education*, vol. 53, no. 3 (March 1964), pp. 33–35.

Dumas, George W., Frederick P. Richard, and William L. Gagnon. "Industrial Arts at John F. Kennedy Junior High," *Industrial Arts and Vocational Education*, vol. 59, no. 3 (March 1970), pp. 68–70.

Educational Change and Architectural Consequences—A Report from Educational Facilities Laboratories. New York, N.Y.: Educational Facilities Laboratories, Inc., 1968.

Finsterbach, Fred C. "The Multiple Dimensions of School-Shop Planning," *School Shop*, vol. XXVIII, no. 9 (May 1969), pp. 38–41.

Fricker, Robert W. "Dual-Purpose Prototype for Suburbia," *School Shop*, vol. XXVIII, no. 8 (April 1969), pp. 104–106.

Joyner, David I. "How to Interpret Innovation for Facility Design," *Industrial Arts and Vocational Education*, vol. 60, no. 3 (March 1971), pp. 21–23.

Kuhl, Stanley A. "Dust-Collecting Systems for the Shop," *School Shop*, vol. XXV, no. 9 (May 1966), p. 35.

Larson, Milton E. "A Checklist of Facilities-Planning Musts," *School Shop*, vol. XXVIII, no. 8 (April 1969), pp. 98–99.

Melo, Louie. "Planning a Materials Laboratory," *Industrial Arts and Vocational Education*, vol. 57, no. 3 (March 1968), pp. 32, 34, 128–130.

Modern School Shop Planning. Ann Arbor, Mich.: Prakken Publications, 1971.

Olsen, Jerry C. "Forces and Factors Producing Change in Industrial-Education Facilities," *School Shop*, vol. XXVIII, no. 8 (April 1969), pp. 86–90.

Planning Industrial Arts Facilities, Eighth Yearbook, American Council on Industrial Arts Teacher Education. Bloomington, Ill.: McKnight and McKnight Publishing Company, 1959.

Schmitt, Marshall L. and James L. Taylor. *Planning and Designing Functional Facilities for Industrial Arts Education.* U.S. Department of Health, Education and Welfare, Office of Education OE-51015. Washington, D.C.: U.S. Government Printing Office, 1968.

———. "Programming the I-A Plant for New Instructional Thrusts," *School Shop*, vol. XXVIII, no. 8 (April 1969), pp. 94–97, 128–129.

School Shop Development—Research and Planning. Pittsburgh, Pa.: Rockwell Manufacturing Company, 1966.

Sterner, Herman. "Moon Rockets Boost Space-Age High School," *Industrial Arts and Vocational Education*, vol. 60, no. 3 (March 1971), pp. 24–27.

Svendsen, Clarence R. "Planned Storage of Materials," *Industrial Arts and Vocational Education*, vol. 59, no. 3 (March 1970), pp. 47–49.

Underwood, Roy T. "New $12 million-plus Huron High School Features Flexible Labs," *Industrial Arts and Vocational Education*, vol. 59, no. 3 (March 1970), pp. 59–62, 65.

Wenig, Robert E. "A Synthesis of Industrial Arts Laboratory Facility Planning," *Industrial Arts and Vocational Education,* vol. 58, no. 3 (March 1969), pp. 55–57.

Whitney, Gregory. "Plan Ahead for Wood Shop Dust Control," *Industrial Arts and Vocational Education,* vol. 57, no. 3 (March 1968), pp. 36–40.

Chapter **20**

The Instructional and Planning Centers

The titles of these two topics, the instructional center and the planning center, were cited in the previous chapter because they merited inclusion under the heading, "Factors to be Included in Planning." It was not considered feasible at that point, however, to provide detailed consideration since it is the purpose of this chapter to do so. These topics are considered together not only because of their vital importance to pupil achievement of the goals of industrial arts education, but also because of their close relationship to each other.

THE INSTRUCTIONAL CENTER

It is clearly evident that the major objectives of industrial arts cannot be achieved without considerable emphasis on the informative aspects of the learning activities. For this reason, as has been stressed throughout the text (see especially Chapter 7), the informative aspects are fully as important, if not more so, than the actual tool manipulations in industrial arts. For laboratory planning this means simply that a designated area must be equipped (1) for class lessons and discussions and (2) for individual learning activities of an informative nature.

THE PLANNING CENTER

Two major points on pupil planning have been made in previous chapters: (1) it is more important from an educational point of view that a pupil be able to plan his project well than be able to execute its construction skillfully (see Chapter 2) and (2) planning is probably the most important phase of the problem-solving method of teaching because it causes students to think (Chapter 8). It will be recalled that the development of the ability to think in students is the central purpose of industrial arts education. Since planning is so vital to the attainment of the objectives of industrial arts, space must be designated in the laboratory to carry on this activity. Experience has shown that better results are obtained from planning and that it is easier to encourage students to plan when a definite area in the laboratory is set aside for this purpose.

LOCATION OF THE CENTERS

In laboratory planning the instructional center can be separate from the planning center or they can be combined into one learning center. In either case the learning activities can be conducted in an auxiliary room or within the open-shop area. Each location has its advantages and disadvantages.

1. Use of an Auxiliary Room

In some schools it is possible to utilize an auxiliary room of the laboratory or a room separate from the main laboratory which is devoted exclusively to instruction and/or planning. The advantages of such an arrangement are freedom from distraction, and the quietness which can prevail. There are also several disadvantages. Chief among these is that when the teacher is in the laboratory, a separate planning room is not under his direct observation and supervision; this could lead to problems of class control. Also, leaving a laboratory unattended while pupils work with tools and equipment for any length of time is a questionable practice. Then too, the instructor may weary of the constant shuffling back and forth between the two rooms.

2. Use of an Open-Shop Area

Many new laboratory designs incorporate a combined instructional center and planning center as an integral part of the main shop area. Since great emphasis is being placed on instruction and planning, the center where this work is conducted should be a place of prominence. Just as the engineering department often becomes the center around which an industry is built, so the instructional-planning center becomes the heart of the industrial arts program. Some seclusion can be secured for this area by the strategic placement of equipment while at the same time the area will be "open" enough to permit supervision and assistance by the teacher.

EQUIPPING THE INSTRUCTIONAL CENTER

The primary purpose of the instructional center is to serve as an optimal site for group instruction and for self-paced individual instruction. In planning the instructional area, consideration should be given to the following:

1. Seating Facilities

Some means should be planned for students to be seated during classwork and for small group discussions as well as for individual study. While tablet arm chairs are satisfactory for most classwork, tables and chairs arranged in seminar-style lend themselves better for group discussions. Carrels are best for individual study. Figure 20-1 shows a row of carrels for individual instruction; these may be purchased or constructed in the laboratory.

2. Demonstration Bench

Some kind of demonstration bench should be provided. It may be desirable, depending on the learning activities involved, to equip the bench with both wood and metal vises. An electrical outlet is desirable for plugging in portable tools and machines.

3. Instructional Media

The instructional area should be *permanently* equipped with such instructional media as the following: tape recorder and audio cassettes; 8-mm and 16-mm movie projectors; film loop projectors; slide, overhead, filmstrip, and opaque projectors; and screens. Closed-circuit television, videotape recording equipment, and a telephone with speakers for telelectures are desirable. Consideration should be given to arranging this equipment for sequential or simultaneous use in an educational media center. This center should be the focus of attention in the instructional area. The actual arrangement of the equipment is dependent on the space available, the needs of the teaching situation, and the media available. Many plans feature a master console from which the teacher presents his instruction and controls the use of the

Fig. 20-1. Individual study carrels are ideal for problem-solving and self-paced instruction. (Courtesy, Brodhead-Garrett Company)

equipment. The console should include light-control switches for darkening the instructional area when certain kinds of sensory aids are used.

4. Miscellany

Chalkboards, both fixed and portable types, as well as bulletin boards should be provided. Storage facilities will be needed for programed lessons and other instruction sheets, film loops, audio tapes, tests, and so on. These may be stored in files, lockers, cabinets or on open shelves as appropriate. The instructional center may well be a desirable site for the teacher's "headquarters." In this case a desk, chairs, file, storage cabinet, duplicating machine, and other equipment are desirable.

EQUIPPING THE PLANNING CENTER

One of the most important features of the planning center is its general atmosphere. The entire area should be conducive for solving problems in a scholarly, scientific manner. Orderliness, cleanliness, and quietness should prevail. The center should be a pleasant place for pupils to work.

The planning area should be as well equipped as any other area in the laboratory. In planning this center, consideration should be given to the following:

1. Resource Materials

The planning center should be rich in resource materials which are carefully selected to stimulate the thinking of pupils as they plan with the tools, materials, and processes of industry and technology. This can be achieved in part by the use of materials displayed on bulletin boards of both the wall and mobile types. Displays of completed student work in the form of unusual solutions to problems or simply examples of fine craftsmanship are stimulating. Models, pictures, and objects as well as a host of other instructional media have their place in the planning center.

Every planning center should have its library of resource materials. Just what particular books, references, and periodicals will be selected depends on the type of laboratory under consideration, the amount of money available, and the initiative of the teacher in keeping this library up-to-date. It is recommended that the shop library contain as many as the budget will allow of the books available for the areas represented. These should be both project-oriented books as well as those of a technical nature. A planning center is incomplete without periodicals. Two types which should be included are (1) professional magazines and (2) those of the hobby-science-experimental type as represented by publications such as *Popular Science, Popular Mechanics,* and *Popular Home Craft.* In addition, certain industrial concerns publish periodicals which may be useful. These magazines not only furnish ideas for projects, but also contribute to the development of recreational and hobby interests.

Another important resource in the planning center is the file of project ideas. These materials may be in the form of drawings, blueprints, project sheets, or pictures. The items should be filed systematically and indexed for easy accessibility by the students.

All of the resource materials available in the laboratory probably should be indexed in a master card file. The maintenance of this index could well be incorporated into the duties and responsibilities of a member of the pupil personnel organization.

2. Seating Facilities

Work stations are needed where students can peruse the available resource materials and translate their ideas into plans for experiments, projects, and other manipulative experiences to be performed later in the laboratory. Study carrels seem ideal for individual work of this nature. Note in Fig. 20-1 that each carrel is well lighted and has some storage space. The carrels may be equipped with drawing boards and other planning aids. If the planning center is combined with the instructional center, the carrels may be equipped with slide projectors with rear-view screens, 8-mm film loop projectors, audio cassettes, and other equipment.

More modest seating arrangements in the planning center may involve a reading or work table where students can study the available resource material. A table shown in Fig. 20-2 may also double as a conference site for team-research efforts or pupil seminars. Drawing boards may be used on this table, or regular drafting tables may be used if space permits.

Fig. 20-2. A large table in the planning area is useful for individual reading or group conferences. (Courtesy, Franklin L. Busch, Dover High School, Dover, Pennsylvania

3. Drawing Equipment

Whether used in study carrels, at work tables, or at regular drafting tables, drawing equipment including boards, T-squares, triangles and instruments will be needed as well as supplies of drawing and tracing paper. A tracing box or light table (see Fig. 20–3) and an opaque projector are also useful in preparing drawings and sketches.

Fig. 20–3. A light table facilitates planning. (Courtesy, The Pennsylvania State University)

4. Storage Space

Shelves or book cases will be needed to house books and periodicals. Figure 20–4 shows a well-organized rack for periodicals. Filing cabinets are useful for storing project designs, blueprints, pictures, and the like. A system must be devised for checking out instructional materials for take-home use; perhaps the checking-in and checking-out of these materials can become a part of the student personnel system.

COMBINED INSTRUCTIONAL AND PLANNING CENTER

The instructional center and the planning center share several common purposes and uses, particularly as these relate to individual learning activities. It is obvious from an examination of the equipment used in each center that a considerable saving in floor space with less duplication of equipment will result if the centers are combined into one learning center. For example, study carrels can double as centers for individual problem-solving efforts (planning) as well as sites for self-paced individual instruction.

More important than space or cost considerations, however, are the

Fig. 20-4. Current and bound volumes of periodicals are made easily available for use by this attractive rack. (Courtesy, California State Department of Education)

pedagogical advantages of a combined instructional center and planning center. In the first place, the combination will make the task of teaching easier for the teacher because the "tools of his trade" will be organized in a central place ready for regular or remedial instruction when needed. This will be especially true if the idea of an educational media center is adopted. The art of teaching is both a complex and arduous process. Hence, any means to simplify, improve, and facilitate the learning process will contribute to quality education. Secondly, a combined center will make planning easier for pupils. At best the important problem of teaching students how to think, how to plan, and how to solve problems is difficult. Where planning is done at haphazard places around the laboratory, it is impossible to surround the student with resource materials and aids which will help him to think. A true planning center is itself a stimulating learning environment. Besides, planning is greatly facilitated if the equipment for planning is readily available in one place. And finally, a combined instructional-planning center will impress students, school administrators, and visitors with the importance of pupil planning and the informative learning activities of industrial arts education. It will be clearly evident that this learning

The Instructional and Planning Centers

center is the very heart of the laboratory from which emanate all learning activities of the program.

SUGGESTED FLOOR PLANS

No single floor plan can be expected to meet all requirements or to fit into all laboratory organizations. General features can, however, be indicated and ideas from several plans can be reorganized and combined to meet new situations. The following illustrations show how design problems for a planning center and for a combined instructional-planning center have been successfully met under varying conditions.

Figure 20-5 indicates how an auxiliary room may be utilized for planning. Much can be done by way of decoration and use of color in a small room of this type. Note the windows on the inside wall which aid the instructor in the supervision of the room while he is in the main shop area.

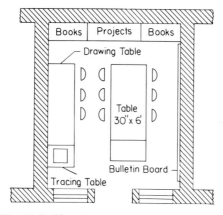

Fig. 20-5. Planning center in a separate room.

The planning center shown in Fig. 20-6 is intended for the small laboratory having a limited floor area. It is compact, yet effective. Most of the essential features are represented and it will accomodate six or seven students. This is probably the maximum number that would be expected to use the planning center at any given time in a typical shop class.

In Fig. 20-7 the planning center is designed for a long, narrow room. It takes little room from the width of the laboratory, yet is compact enough to be treated as a unit. Additional work stations in the form of chairs can be provided, if necessary.

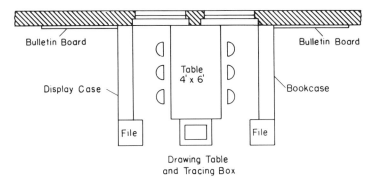

Fig. 20-6. Planning center enclosed with cases.

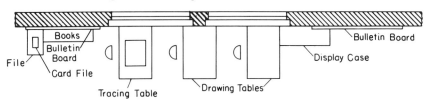

Fig. 20-7. Planning center along a wall.

Figure 20-8 shows a combined instructional center and planning center. The table arrangement is suitable for group learning activities, especially lectures and class discussions. It can be used also by individuals pursuing independent studies or in project planning. Or, individual study carrels can be added along the main laboratory wall near the book shelves. When the center is used for instructional purposes on a class basis, the teacher may prefer to use a movable chalkboard. In this center instructional media equipment is carried to the site when needed.

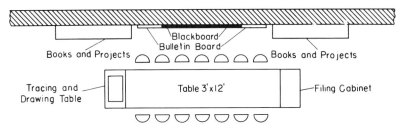

Fig. 20-8. A combined instructional center and planning center.

Figure 20-9 shows another method for combining the planning and the instructional centers effectively. All the materials and equipment needed for both instruction and planning on a group and individual

The Instructional and Planning Centers

basis are easily available. Combining functions in this way provides a workable solution to the space problem in the laboratory and at the same time promotes educational efficiency. A prime advantage of the audio-visual features supporting the instructional center is that standard machines and equipment are used in a "set-on" arrangement. Custom, built-in features are eliminated, thus reducing installation costs and facilitating easy operation and maintenance. Also, the teacher may begin to operate the center with whatever audio-visual equipment may be available, and add new and different media as funds permit. Note in the figure that the A/V table is 8 inches lower than the main table. This is to improve general visibility throughout the viewing area. The same is true of the videotape recorder which is mounted so that its top is level with that of the demonstration bench.

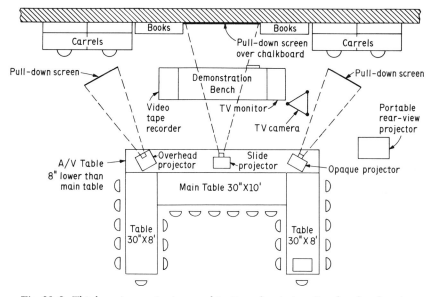

Fig. 20-9. This learning center is a combination of an instructional and a planning center.

SUMMARY

This chapter reemphasized the tremendous importance of the informative aspects and pupil planning in the attainment of the goals of industrial arts education. The sites where these learning activities can best take place are referred to as the instructional center and the planning center respectively. These centers are so important that they merit prominent, well-equipped spaces in the laboratory which are specially equipped for these purposes. It is recommended that the instructional center and the planning center be combined physically into a learning center. This area then becomes the heart of the industrial arts program from which all learning activities emanate. Factors to be considered in planning and equipping the instructional-planning center were presented and several floor plans were suggested.

DISCUSSION TOPICS AND ASSIGNMENTS

1. List the important items to be included in a fully developed instructional-planning center.
2. Compare the advantages and disadvantages of the "separate room" type and the "within-the-shop" type of learning centers.
3. Select any shop plan that does not have a learning center and redesign it to include one.
4. List all the books you feel should be in a planning center for a junior high school laboratory.
5. Discuss and give examples of how color may be used in a learning center.
6. Design an ideal layout for a combined instructional-planning center.

SELECTED REFERENCES

Ames, Jr., William E. "Make Your Own Study Carrels," *School Shop*, vol. XXIX, no. 6 (February 1970), pp. 46–47.

Finsterbach, F. C. "Industrial Arts Planning Center," *Industrial Arts and Vocational Education*, vol. 35, no. 7 (September 1946), pp. 290–292.

Knoll, Carl J. "The Recitation-Planning Center," *Industrial Arts and Vocational Education*, vol. 51, no. 3 (March 1962), pp. 48–49.

Planning Industrial Arts Facilities. Eighth Yearbook, American Council on Industrial Arts Teacher Education. Bloomington, Ill.: McKnight and McKnight Publishing Company, 1959.

Reed, Howard O. "The General Library and the School Shop," *Industrial Arts and Vocational Education*, vol. 37, no. 1 (January 1948), pp. 9–10.

Riva, David F. "An Audio Visual Learning Center," *Education Bulletin*, AFROTCRP 50–1, No. 5, May 1969.

Rudisill, Alvin E. "Packaging the Media into a Console-Controlled Classroom," *School Shop*, vol. XXVII, no. 8 (April 1968), pp. 74–77.

Chapter **21**

Equipping the Laboratory

The amount and type of equipment available for an industrial arts laboratory will be a determining factor in the program that can be developed. If the equipment is adequate and well suited to the requirements of the teacher and the needs of the students, meeting the objectives of the course will be greatly facilitated. On the other hand, if the equipment is inadequate or of a type not suited to the learning activities, the achievement of desired purposes becomes difficult if not impossible.

FACTORS AFFECTING EQUIPMENT SELECTION

The choice of equipment for the industrial arts laboratory will depend on a number of important factors which must be seriously considered. These include: (1) objectives of the program, (2) type of laboratory organization, (3) the sizes of anticipated classes, (4) the size of the laboratory to be used, and (5) the amount of money available for equipment purchases. If the laboratory is already in operation, the adequacy of its equipment may be checked against these factors. If, however, the shop is being newly equipped, each item will need to be carefully considered in light of these criteria prior to the placement of purchasing bids.

1. Objectives of the Program

Before decisions on equipment can be made, it is first necessary to consider the aims of the program. It is unwise to develop a program on the basis of equipment which may be on hand or available. Careful study of the objectives to be achieved, together with the learning activities to be utilized for their attainment, will serve as one sound criterion for the selection of equipment. For example, the type of equipment required for a course having prevocational objectives will differ from one where the avocational aim is stressed.

2. Type of Laboratory Organization

The type of organization of the laboratory will have an important influence on the selection of equipment. A general shop organization will need a wide variety of equipment, but a limited quantity of any one type; a unit shop will require more duplication of items and, perhaps, more specialized equipment in one particular area. The view of the unit machine shop depicted in Fig. 21–1 shows five metalworking lathes. One, or possibly two, of these lathes probably would be adequate for a general shop. Obviously, it is important to make a decision concerning the exact type of shop organization to be employed before considering equipment.

Fig. 21–1. Unit shop organization requires much duplicate equipment which is an important consideration in laboratory planning. (Courtesy, Clausing Corporation)

3. Class Size

The maximum number of students expected in any one class is also a determining factor in planning equipment. As noted previously, sufficient work stations must be provided to accommodate the largest class that will be using the facilities. The larger the class, the greater the number of benches and machines that will be needed. For example, if a class of twenty-five is anticipated in a general shop, it will probably be necessary to plan at least thirty-five possible work stations because it must be assumed that not all stations will be used at any one time.

4. Size of Laboratory

The total floor area available for the placement of equipment has a decided influence on what is planned or purchased. If the laboratory room is small, the purchase of large equipment is obviously out of the question. Both the types and the numbers of machines, benches, and similar equipment are affected. The problem of how closely various machines and other equipment may be crowded together without endangering the safety of the students becomes a matter of extreme importance.

5. Budgetary Considerations

The availability of funds for school shop equipment is an important consideration. Where funds are lacking to buy everything which appears to be essential, it is recommended that a full equipment list be prepared and then divided according to a priority system so that the items of most immediate importance may be purchased the first year, those having the next highest rating the next year, and so on until the full list is accounted for. In this way the program can be started on a limited basis and enriched as more equipment becomes available.

MACHINE TOOL SELECTION

Weaver defines the term *machines* as "stationary or movable devices usually power driven for cutting, shaping, and processing various materials used in the trade."[1]

---

[1] Gilbert G. Weaver, *Shop Organization and Management* (New York, N.Y.: Pitman Publishing Corporation, 1955), p. 49.

Machine tools should be purchased on the basis of the purpose for which they are to be used. For example, a unit shop may require heavy, production-type machines, whereas a general shop will probably require only a machine of medium or even light weight. Many of the mediumweight and lightweight machines are now capable of standing up under the strain of classroom use, and in most instances, they adequately meet the needs of the school shop.

1. Factors in Machine Tool Selection

Some of the important factors to be considered in the choice of machine tool equipment are: (1) maximum pupil participation, (2) safety, and (3) obsolescence. These are described in detail by Bollinger as follows:

> Each tool, each machine, each bench, and each piece of apparatus must provide for a maximum of pupil participation in its use. In other words, mere pressing of a button to perform an operation defeats learning. In addition to this, each item of equipment must represent a basic industrial process. It is only through an understanding of these processes and their effects that the pupil is enabled to interpret the infinite number of applications so characteristic of American life. Learning through experience for purposes of orientation in an industrial society requires a variety of equipment rather than a duplication of specialized items.
>
> Safety is secondary only to the educational criteria. This means that all equipment must be designated and built, and in turn selected and used, with specific reference to the size, height, strength, mental development, and experience of the individuals who are to use it, Thus, the capacity, weight, power, speed, and size of machines for industrial arts classes should be determined by the nature of the pupils who use them.
>
> Obsolescence is also a factor which industrial arts programs and school officials must learn to recognize and face. American industry "retools" periodically for a world where change is the only certainty. The school is faced with the same certainty. The equipment manufacturer is now producing less costly equipment of smarter design to encourage periodic replacement.[2]

In addition to these general factors, Bollinger also has identified 39 principles for selecting machine tools. These are classified under the headings: (1) functional features, (2) safety features, and (3) design features, as follows:

---

[2]Elroy W. Bollinger, "Principles of Industrial Arts Equipment Selection," *Industrial Arts and Vocational Education*, vol. 31, no. 3 (March 1942), p. 128.

Equipping the Laboratory

2. Functional Features

1. Machines should be of the unit type in order to provide maximum efficiency, safety, and flexibility of arrangement. Combining a circular saw, mortiser, and jointer on a single standard reduces the usefulness and efficiency of each machine besides resulting in unnecessary interference and hazards.

2. Machines should be designed and used for only one type of work. A drill press, for example, which is in such constant demand, should not be expected to serve as a router, a shaper, a sander, and a hollow-chisel mortiser in addition to its principle function of drilling.

3. Automatic feed and control devices obscuring the principles of a machine should be avoided for industrial arts classes except for reasons of safety. Automatic feeds on mortising machines and drill presses, quick gear changes on lathes, and automatic press feeders are cases in point.

4. Machines should be mounted on individual bases, preferably enclosed on all sides of the floor to facilitate cleaning both machine and floor. A machine mounted on a bench destroys in part the usefulness of both bench and machine.

5. The equipment provided in a laboratory should be of a size or capacity which will take care of the bulk, but not necessarily all, of the work which anyone would like to do. For example, there may be a distinct educational value in having a pupil himself take some work directly to an industrial shop or plant, where he will make new contacts, observe industrial methods, evaluate production costs, note merchandising procedures and experience being handled as a customer.

3. Safety Features

6. Essential safety features, such as, circular guards, jointer guards, pulley and belt guards, should be designed and supplied as an integral part of the equipment.

7. All moving parts of power-driven machines must be guarded or enclosed except those used directly in the operation involved.

8. All hand-operated machines that present hazards, such as, squaring shears, punch presses, paper cutters, etc., should be provided with effective guards.

9. All moving parts of motor-driven equipment, whether guarded or not, should be free of projections, such as, setscrews, knobs, keys, etc.

10. Guards, when used, must be simple in design and positive in action, and must interfere as little as possible with the operation of the machine. Guards that adjust themselves automatically to the work being done are to be preferred over manual types.

11. All grinding equipment should be provided with shields of plastic or laminated safety glass.

12. All motors should be equipped with overload protective devices of the thermal-relay or circuit-breaker type. These devices should be incorporated in the case with the motor switch control.

13. The size, capacity, and power of any machine should be determined with reference to the age, strength, height, and mentality of the pupils who are to operate it. It is questionable, for example, if any machine in a junior high school laboratory need be driven with a motor exceeding one horsepower.

4. Design Features

14. The average elbow height of individuals who are to use a bench or machine should be the reference point in specifying the operation level of said bench or machine.

15. All machine tools should have individual motor drives, controls, and stands.

16. Power for any machine should be adequate to operate the equipment under its full rated capacity without unreasonable overloading of the motor.

17. V-type belts are usually to be preferred to flat belts from the standpoint of power transmission and general efficiency. Flat belts, if used, should be of the endless type and may be preferred as a protective measure because of slippage on excessive overloads. This is particularly true where machines may be locked by jamming as in the case of an engine lathe or milling machine.

18. All reciprocating or revolving machine parts that work at high speeds should be balanced and counterbalanced to reduce vibrations to a minimum.

19. Machine standards should be sturdy and rigid in order to provide a solid base free from weaving and twisting for the machine it supports.

20. All handles, wheels, and mechanical controls should be of easy access to the operator, be arranged not to interfere with each other, and be electroplated with an anti-corrosive metal.

21. Speed controls should be convenient, safe, positive, and of a range sufficient for the work for which the machine was designed and the experience level of the operators; e.g., beginning printing press operators require an unusually slow press speed.

22. The use of detached knobs, wrenches, etc., for adjusting and operating a machine should be avoided as far as possible.

23. Machine parts, such as saw blades, drill spindles, mortiser bits, etc., should be easily and quickly adjusted and interchanged without damage to the parts.

24. Machines or cabinet bases should not interfere with the movements or comfort of the operator; e.g., toe room should always be considered in the design.

25. Machines should be designed to allow the maximum amount of working space around the point of operation.

26. Power machines should be provided with switches placed within the operator's natural reach and vision while the machine is in operation but so located that accidental switching is avoided.

27. Motor-driven machines of one horsepower or less should be equipped with a toggle or pushbutton type switch operating in a vertical position and placed within natural reach of the operator.

28. The quality and kind of materials used in the construction of machines for school use should be comparable to that used in machines for industry.

29. Sealed roller or ball bearing should usually be considered preferable to other types of bearings.

30. Collectors for shavings, dust, etc., should be an integral part of the machine. The machine should, however, lend itself to installation of a central dust-collecting system.

31. Where possible, motors should be housed within the machine, but made easily accessible for maintenance.

32. Flexible molded rubber power cords should be supplied and used in connection with all portable and semi-portable equipment.

33. The need for periodic lubrication should be reduced to a minimum through such means as sealed bearings packed in grease or oil. Parts needing periodic lubrication should be fitted with snap-cover oil cups or alemite zerk fittings located for ease of identification and servicing.

34. Machines should be designed to operate with a minimum noise factor. At no time should the noise of any machine exceed 70 decibels.

35. Machines should be cushioned, preferably with rubber mountings furnished as an integral part of the machine.

36. Machines embodying sheet-metal construction should be treated with a noise-absorbing or dampening medium glue or sprayed on inside surfaces.

37. Machines should be painted a distinctive color sufficiently light to have a light-reflection factor of at least 40, be without glare, easy to clean, and of neat appearance.

38. Local lighting should be incorporated as an integral part of machines wherever possible.

39. Simplicity of construction and design should be considered desirable in all equipment.[3]

---

[3] *Ibid.*, pp. 128–130.

WRITING SPECIFICATIONS

In many cases it is both necessary and desirable to write specifications covering equipment to be purchased for the industrial arts shop. Wherever bids are required, specifications must be prepared.

A specification is nothing more than a detailed description of a hand tool, machine, or piece of equipment. Depending on the nature of the item to be purchased, a specification may include reference to one or more of the following: general description including details of construction and minimum specific requirements, details on motor and electrical controls, itemization of accessories, safety standards and guards, delivery conditions including date and whether prepaid or f.o.b.

Weaver states that the purposes of a specification are to:

Eliminate any item that does not meet the requirements.
Include all possible items that do meet the requirements.
Assure reliable inspection by qualified impartial inspectors.
Prevent shipment of items claimed "just as good."[4]

Usually the teacher knows approximately what he desires in a machine, but finds it difficult to put his ideas in the form required for competitive bids. The following suggestions may be helpful.

1. Decide upon the most important features desired in any given piece of equipment.

2. Find the machine (or other equipment) which comes the closest to meeting all requirements.

3. Write specifications around the selected item in such a manner that all desired general features are included, but not so specifically that no other make or model can meet the requirements.

4. Sometimes the name, model, and catalog number of an item are specified with the phrase, *or equal*, following. This means that the bidder must offer equipment which has the same size, capacity, general construction, safety features, etc., as the item specified. In such cases, care must be exercised to make sure that such bids are actually equal to the item desired.

If the phrase "no substitute" follows the name, model, and catalog number of the item, it means that only the specified item will be acceptable. This eliminates bids from other manufacturers on that item.

Typical specifications for a 24-in. scroll saw are offered to illustrate some of the points covered in this section:

Scroll saw with arm-to-saw capacity of 24-in.; thickness of cut 1¾-in.; tilting table, 45° right and front, 15° left; removable overarm

[4] Weaver, *op. cit.*, p. 54.

for unlimited cutting capacity with saber blade; variable speed from 650 to 1,700 cutting strokes per minute. Machine to be mounted on heavy-gauge steel stand. Motor to be ⅓ hp 208–220/440-volt, a-c, 50–60 cycle three-phase, 1,425–1,725 rpm. Starter to be three-phase across the line magnetic starter with no-voltage, low-voltage, and overload protection; stop, start, and reset button to be mounted in cover. Machine to be complete with motor base and bracket, belt, pulley guard, lamp attachment, 1 saber blade, and 3 scroll-saw blades. This item to be Delta No. 40–305 or equal.

If desirable, the following may be either a part of an individual specification or part of a covering sheet called "General Conditions for Bids on Machine Tools."

Delivery date to be: August 1, 19___; postpaid to *(specify school and address)*. Delivery at school site shall include uncrating and complete assembly ready for electrical installation.

SELECTION OF HAND TOOLS

A hand tool is defined by Weaver as:

The small tools of the occupation to process materials or service work by hand manipulations.

Examples: chisels, hammers, brushes, scissors, nippers.[5]

There are several factors to be considered in specifying and choosing hand tools. Among the more important of these are:

1. Buy standard-make tools made by a known and reputable manufacturer. The slightly higher prices which may be paid for such tools will be repaid many times in their additional life expectancy.

2. Buy tools that fit the students. This does not mean to purchase toys. However, somewhat smaller saws and lighter hammers are essential for very young students.

3. Buy only what is needed. There is a tendency to buy more of some tools than may be necessary. For example, it is doubtful if any laboratory will need more than two rip saws or one 1-in. auger bit. Try to determine the largest number of students that will ordinarily be using a given tool at any one time.

4. Buy diversified types and sizes. Particularly in a general shop, it is essential to have a wide variety of kinds and sizes of tools to meet all possible requirements. It is much better to have a set of chisels from ⅛ to 2-in. than to have three ¼-in. chisels and three ¾-in. chisels.

[5] *Ibid.*, p. 49.

TOOL STORAGE

After the hand tools have been selected and specifications prepared for their purchase, it is necessary to give some consideration to how they will be stored. Tool storage is an important part of laboratory planning and layout.

In early times toolrooms were considered an essential part of industrial arts shops. Today, however, they are regarded as vestigial remains of a bygone era and their place has been taken these many years by a central tool panel either of the open or the closed type. The central tool panel itself may someday be relegated to the past if the multitool panel system continues to gain prominence. A trend in this direction was noted some years ago:

> Within recent years, however, there has been an evident trend toward still further decentralization and the use of specialized or area tool panels. This has been particularly true in the many variations of the general shop. For example, one now finds the woodworking tools concentrated on a panel near the woods area, the metalworking tools near the metals area, and the craft tools near the place where they will be used.
>
> In many instances, this trend toward greater decentralization of tool storage has been carried even further. Tools and equipment which are to be used with a given machine or bench are located adjacent and convenient to it. Thus, all of the chucks, toolholders, wrenches, etc., to be used on a lathe may either be mounted on a panel attached to the frame of the lathe itself or on a cabinet or stand nearby.[6]

Figure 21-2 depicts a typical tool panel in an industrial arts laboratory, in this case a woods technology laboratory. This is an example of an area tool panel cited above. Figure 21-3 shows two tools panels tailored to fit industrial arts activities at the elementary school level. Note that the panels are mobile and hold the tools at convenient height for elementary school pupils.

For many years tool panels, tool mountings, and tool arrangements have been neglected aspects of industrial arts laboratories. Examination of the tool panel in the average industrial arts shop will reveal that, in most cases, major improvements could be made. In too many laboratories the hand tools are hung up haphazardly or arranged with little evidence of careful planning even from the standpoint of student safety. Often there is no visible system of tool accountability.

[6]Gordon O. Wilber, "Educational Implications of Storage Methods," *Modern School Shop Planning* (Ann Arbor, Mich.: Prakken Publications, 1957), p. 37.

Equipping the Laboratory

Fig. 21-2. A typical tool panel in an industrial arts laboratory. (Courtesy, California State Department of Education)

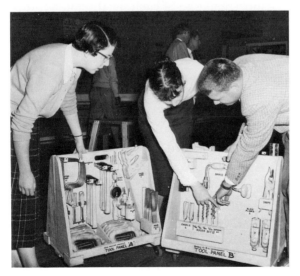

Fig. 21-3. These mobile tool panels are ideal for elementary school pupils. (Courtesy, California State Department of Education)

A research study was initiated to identify principles for arranging and mounting hand tools in industrial arts laboratories.[7] The principles listed below were validated by a 50-man jury of national experts in the field. After being subjected to statistical analysis, these principles were found to meet the test for reliability at the one percent level of confidence.

1. Principles Related to Tool Panels

1. A tool panel should be adequately lighted, preferably by both natural and artificial light.

2. A tool panel should be pleasing to the eye and give a feeling of orderliness.

3. A tool panel and its toolholders should be of simple, clean-cut design and construction to require a minimum of maintenance and cleaning.

4. A tool panel or cabinet should be economical of wall or floor space, yet large enough so that (a) several students can use the panel simultaneously and (b) it will hold all present shop tools with sufficient space for future expansion and acquisition of additional tools.

5. A tool panel should be of such height and depth and so located that the average student will be able to remove and replace any tool.

6. A tool panel should be located in the shop so that (a) it will be convenient (require short travel distance) to work stations where the tools will be used (see Fig. 21–4), (b) it will have sufficient space on front and sides to prevent student congestion, and (c) it is on or near natural traffic routes.

7. A tool panel should be designed so that each tool has a definite place.

8. A tool panel should be designed to facilitate tool accountability.

9. A tool panel should have smooth, unbroken surfaces to permit the mounting of tools in any location on these surfaces.

10. A tool panel should utilize a simple method of tool identification to differentiate among (a) tools in different areas of the same shop and (b) tools in different shops of the same school.

11. A tool panel should be designed and constructed so that it can be locked.

[7]Norman C. Pendered, "Principles for Arranging, Mounting, and Storing Hand Tools in Industrial Arts Shops." A research study sponsored by the Central Fund for Research, The Pennsylvania State University, University Park, Pa., 1962.

Equipping the Laboratory

Fig. 21-4. Note the short travel distance to this woodworking tool panel. (Courtesy, John Mitchell, Gorham State College of the University of Maine, Gorham, Maine)

2. Principles Related to Tool Arrangements

1. Tools should be arranged on the panel so that their withdrawal or replacement will not interfere with the other tools.

2. Large, heavy tools should be arranged near the bottom of the tool panel.

3. Small, light tools should be arranged higher on the panel or on doors (if doors are utilized), but in such a manner that they will not cause interference or annoyance in opening or closing the doors.

4. Tools should be arranged on the panel in logical, organized groupings based on a balanced judgment involving the following: (a) group together all tools of one area or activity (i.e., all woodworking tools together, all metal working tools together), and (b) group together all tools of one area that are used together (i.e., brace and bits, rules and scribers). In Fig. 21-5 note the layout tools are grouped together as are the cutting tools, the smoothing tools, and other related types.

5. Tools used most frequently should be located in the most accessible positions.

6. Tools should be arranged so that students are not subject to safety hazards in removing or replacing tools.

Fig. 21–5. A well-balanced, organized tool arrangement. (Courtesy, Elmer Hemberger, Downington Joint School District, Downington, Pennsylvania. Photo by William J. Wilkinson)

7. Tools should be arranged to conserve space on the panel without overcrowding.

3. Principles Related to Toolholders

1. Toolholders should be sturdily constructed of durable material to withstand heavy, frequent use.
2. A toolholder should be designed so that only the proper tool can be returned to the empty holder.
3. A toolholder should protect the tool from damage or undue wear.
4. A toolholder should be designed so that the tool can be replaced in one position only.
5. A toolholder should facilitate the swift, easy removal and replacement of the tool. (a) the simplest means of tool removal is a straight off-the-holder motion; replacement is a straight on-the-holder motion. (b) The next simplest means of tool removal is an up-and-off the holder motion; replacement is a down-and-on the holder motion.
6. A toolholder should support the tool so that its handle or other natural grasping point may be utilized in tool removal and replacement.

7. A toolholder should support the tool firmly so it cannot be jarred from its proper place by shop vibration, or by opening or closing of doors.

8. A toolholder should permit quick and easy identification and selection of sized tools.

9. A toolholder should require simple installation (preferably from the front) to facilitate easy replacement or relocation.

10. A toolholder should be designed to facilitate a brief visual inspection for missing parts of the tool.

11. A toolholder should offer the maximum safety protection to those who use the panel.

12. A toolholder which supports one of several tools of the same size and type should be constructed so that any one of the designated tools will fit into it.

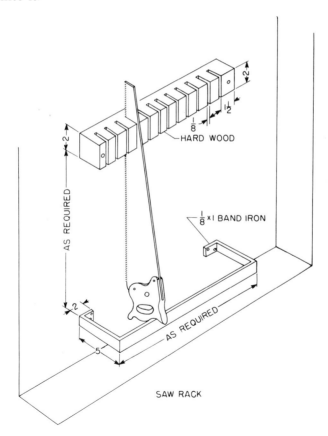

Fig. 21-6. An example of a poorly designed toolholder.

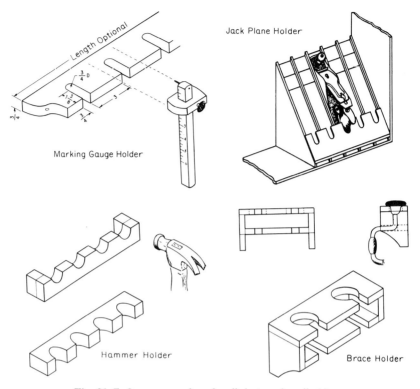

Fig. 21-7. Some examples of well designed toolholders.

13. A toolholder which supports more than one tool of the same size and type should have a means of indicating how many tools belong on this toolholder.

14. A toolholder should support the tool so that its weight causes the tool to hang straight and in proper alignment.

Figure 21-6 illustrates a saw rack which does not satisfy the criteria for a good toolholder. Figure 21-7 shows several examples of toolholders which seem to embody the principles previously outlined.

SUMMARY

Industrial arts equipment should be purchased on the basis of the requirements of the program, the type of shop organization, the size of classes, the room size, and the amount of funds available. Three factors for selecting machine tools include (1) maximum pupil participation, (2) safety, and (3) obsolescence. Thirty-nine principles for selecting machine tools were presented under these headings: (1) functional features, (2) safety features, and (3) design features. Whenever bids are required, written specifications must be prepared. Included in the chapter were suggestions for writing specifications together with a sample specification.

Important considerations in selecting hand tools are to buy standard-make tools to fit the students and to buy diversified types and sizes of tools, but only what is needed. Tool panels have long ago taken the place of toolrooms in industrial arts education. Decentralization of tool panels is common practice today. Thirty-two principles related to tool panels, tool arrangements, and toolholders were identified in this chapter.

DISCUSSION TOPICS AND ASSIGNMENTS

1. Assume that you have been employed to teach general shop activities in a junior high school. You wish to add an area called visual communications. Prepare an equipment list for this new area.

2. Prepare specifications for a major item of machine tool equipment of your choice.

3. Visit an industrial arts laboratory. Make a critical analysis of the equipment from the standpoint of (1) fitness to purpose, (2) placement, and (3) safety features.

4. Prepare a list of hand tools for teaching construction activities to a class of 15 junior high school students.

5. In accordance with the principles outlined in this chapter, design an ideal holder for a tool of your choice.

6. Prepare pros and cons for this statement: A tool panel should be designed so that it can be locked.

SELECTED REFERENCES

Bollinger, E. W. "Principles of Industrial Arts Equipment Selection," *Industrial Arts and Vocational Education*, vol. 31, no. 3 (March 1942), pp. 128–130.

Coverdill, Ernest J. "Specification Writing for Equipment Procurement," *Industrial Arts and Vocational Education*, vol. 54, no. 3 (March 1965), pp. 70–72.

Dean, C. Thomas. "When You Select and Specify Equipment," *Industrial Arts and Vocational Education*, vol. 59, no. 3 (March 1970), pp. 112–114.

Fitz, C. F. "Protective Toolholders," *School Shop*, vol. XXX, no. 6 (April 1968), p. 74.

Fox, Carl C. and Lester A. Wright. "Individual Tool Panels for the Auto Shop," *School Shop*, vol. XXVIII, no. 9 (May 1969), p. 52.

Modern School Shop Planning. Ann Arbor, Mich.: Prakken Publications, Inc. 1971.

Pendered, Norman C. "Principles for Arranging, Mounting, and Storing Hand Tools in Industrial Arts Shops," *Contributions to Industrial Arts Education*, vol. 1, no. 1. Jermyn, Pa.: The Tarleton Press, 1963.

———. "Toolholder Evaluation Check List," *School Shop*, vol. XXIII, no. 1 (September 1963), p. 15.

Planning Industrial Arts Facilities. Eighth Yearbook, American Council on Industrial Arts Teacher Education. Bloomington, Ill.: McKnight and McKnight Publishing Company, 1959.

"Toolholder of the Month," *School Shop*, vol. XXIII, no. 2 (October 1963), p. 80; vol. XXII, no. 3 (November 1963), p. 44; vol. XXIII, no. 4 (December 1963), p. 34; vol. XXIII, no. 5 (January 1964), p. 36; vol. XXIII, no. 6 (February 1964), p. 57; vol. XXIII, no. 7 (March 1964), p. 46; XXIV, no. 1 (September 1964), p. 54.

Weaver, Gilbert G. *Shop Organization and Management*. New York, N. Y.: Pitman Publishing Corporation, 1955.

Wilber, Gordon O. "Educational Implications of Storage Methods," *Modern School Shop Planning*. Ann Arbor, Mich.: Prakken Publications, Inc. 1957.

Chapter **22**

Completing and Evaluating the Layout

There is probably no such thing as a finished layout. The process of laboratory planning is a continuous cycle of planning and revision and, like curriculum construction, the cycle is a never-ending one. Even after a new laboratory is constructed or an old one remodeled and the program is well in operation, continual improvements in the layout doubtless will be made. It must be remembered that all things in life are changing; the only thing constant in life itself is change. This applies equally well to laboratory planning.

In this chapter several important topics will be considered to help bring the plan to tentative completion. These include (1) use of color in the environment, (2) safety features to be considered in planning, and (3) evaluation of physical facilities.

COLOR IN THE ENVIRONMENT

Within recent years much as been printed concerning the use of color both in industry and in the school laboratory. Insofar as industry is concerned, Lytle notes:

> Industry . . . credits improvement of employee morale, elimination of work accidents, and increased productivity to color-engineered plants. Why? Because visibility means safety. Safety means increased production. Increased production means more money. And, so it goes.[1]

Contending that color should be used in the school environment, this same author states:

[1] Robert B. Lytle, Jr. "Color: Its Effect on Teaching and Learning," *School Shop*, vol. XXII, no. 8 (April 1963), p. 43.

For a learning situation, the utilization of color principles seems a necessary function. . . . The consideration devoted to the controlling of color in the shop environment can result in more and better learning by students if even only through the notable Hawthorne effect.[2]

It is well established that color can create certain physiological and psychological effects. From a physiological point of view, industry has discovered that color is related to fatigue, tension, and perhaps other physical ailments. A prominent paint manufacturer reports the following:

When the color of the material being fabricated is too similar to the color of the working parts of his machine, the operator is often subjected to eyestrain. His eyes become fatigued. This tiredness usually is not felt within the eye itself. It is communicated to muscles and nerves in other parts of the body. A person suffering from eye fatigue feels tired all over. His alertness is dulled, his mental attitude suffers and often he may fall victim to headache, nervousness, digestive upsets, and other physical and mental disturbances.[3]

In addition to eyestrain and tiredness brought on by long exposure to similar colors, research has also shown that colored lights can produce muscular tension. For example:

Investigation showed that muscular tension rose slightly under blue light. Green light increased it a little more. Yellow light raised it to thirty units from a normal twenty-three. When a person is subjected to a given color, his psychological and physical condition may both be affected.[4]

Some of the psychological effects of color are well known. Color can be used effectively to confuse the eye and to fool the intellect of man. From the beginning of time nature has used color to camouflage and thus protect wildlife. Man himself has also used color in warfare to camouflage both soldiers and equipment.

An advancing color such as red or orange makes an object seem to be closer than it really is. Likewise, a receding color like blue causes an object to appear to be farther away than it actually is. For this reason blue should never be used in hazardous areas of the shop or applied to moving parts of a machine. The American homemaker knows well that the use of advancing or receding colors can give the illusion of affecting

---

[2] *Ibid.*, p. 91.
[3] *Energy in Color* (Pittsburgh, Pa.: Pittsburgh Plate Glass Company, no date), p. 4.
[4] *Color Dynamics* (Office buildings, auditoriums, apartments, churches, hotels) (Pittsburgh, Pa.: Pittsburgh Plate Glass Company, no date), p. 5.

room size. These same color principles can be applied to shop planning. For example, a square room can be made to appear more nearly rectangular, a long room can be made to appear wider, or the room height can appear to be higher or lower than it really is.

Light-colored objects seem lighter in weight compared with dark-colored objects, which seem heavier. Combinations of color can cause objects to be either conspicuous or indistinct. This fact alone makes color important in accident prevention.

Although some laboratory research on color has been conducted, field studies on color as a factor in the school shop environment seem conspicuous by their absence. Some years ago the Pittsburgh Plate Glass Company sponsored an investigation into the use of planned color in the school environment of elementary and junior high school pupils. This project was conducted by the psychological laboratory of The Johns Hopkins University Institute for Cooperative Research in selected schools of Baltimore, Maryland.

Some of the results of this study are reported by Rice who states:

> The two-year study in Baltimore's public schools produced substantial evidence that color environment, if correctly planned, has a favorable effect on the behavior and performance traits of children in the elementary grades. Observations conducted in three elementary schools showed that color has the greatest beneficial results on children in kindergartens, that boys show greater response to color than girls and that improvement in *scholastic* achievement is even more noticeable than improvement in *behavior* traits.[5]

In three junior high schools where a similar study was conducted there were too many uncontrolled variables so that no significant conclusions could be drawn. In commenting on the inconclusive results at the junior high school level, Rice states the following:

> We discovered that this seemingly contradictory evidence from the junior high schools was in reality a confirmation of what other research has encountered.
>
> We learned that, repeatedly, experiments in Wisconsin, Texas, and other areas in the study of color and other environmental factors in the junior high age range showed contradictory trends. Said the conductor of some of these other experiments: "During the junior high school age, there is a physical growth spurt that tends to prevent the correlating of achievement in performance records with the single physical factor of the environment."[6]

---

[5] Arthur H. Rice, "What Research Knows About Color in the Classroom," *The Nation's Schools*, vol. 52, no. 5 (November 1953), p. III.
[6] *Ibid.*, p. VIII.

In the *Accident Prevention Manual For Shop Teachers*, a tentative conclusion on color in the school environment is drawn as follows:

> Experiments in color research in the classroom indicate that color does exert favorable benefits on certain levels of school children. While data are inconclusive, the weight of evidence at the moment seems to favor the use of planned color environments in certain schools. The benefits accruing from colored environments seem obvious to many teachers of shop subjects, for they take great pride in keeping their shops, tools, and equipment neatly painted in attractive colors. The potential of a colored environment for attitude development at the secondary level especially in the case of school shops is unknown, but perhaps should be given the benefit of any doubt at this time.[7]

Often the industrial arts laboratories are one of the showplaces of the school and a place to which administrators bring school visitors. An attractively colored learning environment is impressive to both administrators and visitors alike and probably will leave them with a favorable impression of the entire program.

SELECTION OF SHOP COLORS

The application of color to the school shop should be based on sound principles of color harmony tempered by a standard color code for marking safety hazards. Only flat paints should be used to avoid the possibility of glare and unwanted reflection from glossy surfaces. Important recommendations for the use of color in the industrial arts laboratory are:

1. Ceilings

Light colors, such as, white, off-white, ivory, or cream should be used on ceilings because these colors reflect light well and ensure an even distribution with a minimum of shadows. Where the maximum reflection of light is needed, the ceiling should be painted white. This will also reduce the brightness contrast with ceiling lighting. Beams, ducts, and overhead projections or obstructions should be painted the same color as the ceiling. If it is desired to minimize visual clutter overhead, pipes, wires, and other lines may be painted the ceiling color. This will not only obscure these objects, but will also make the shop ceiling seem higher.

---

[7]Norman C. Pendered, "Developing Safety Consciousness," in William A. Williams (ed.), *Accident Prevention Manual For Shop Teachers*, National Association of Industrial Teacher Educators (Chicago, Ill.: American Technical Society, 1963), p. 73.

2. Walls

If the room has good natural lighting, the walls should be painted a receding color, such as light green. If the room is darker, the walls should be painted a lighter color—yellow or aquamarine.

Posts and columns should be regarded as part of the shop walls and painted the same color as the walls.

3. Floors

Usually it is unnecessary to paint wood, linoleum, or asphalt tile floors. Concrete floors may be painted a light gray color.

4. Machines

All machines should be painted a color that will harmonize with the walls. For example, if the walls are green, the machines may be a darker shade of green. Additional colors may be applied to key areas of machines as follows:

Handwheels (except rims) and control levers should be painted beige or buff. The knobs of control levers should be painted orange.

Working surfaces which are not machine-finished should be painted a light color to provide visual contrast with the work. For example, the cross-slide of a lathe may be painted cream or ivory so that the work will show up easily.

Guards which protect cutting edges should be painted orange. Examples include the guards on jointers and circular saws. The interior surfaces of movable guards protecting gears and pulleys should be painted orange too. Orange may be applied to such other key parts of machines as the lathe tool rest and the miter gauge on both circular saw and band saw.

The electrical switch box should be painted orange, but this does not apply to the front of the switch, that is, to the cover plate or the pushbutton controls. If the switch box is built into the machine so that the cover plate and buttons are relatively flush with the exterior surface, then a band of orange should be painted around the cover plate.

Stop buttons on electrical switches should be painted red; starting buttons should be painted orange.

Moving parts of machines such as the ram of a shaper or other parts that present a bump or collision hazard should be painted yellow with black stripes.

5. Equipment

Shop equipment is best painted the same color as the machines. Undersurfaces should be painted white to reflect light into dark areas. The vertical edges of table tops and bench tops should be painted yellow. Safety cans for flammable liquids should be painted red with the name of the contents painted yellow.

STANDARD SAFETY COLOR CODE

In selecting and applying colors to the industrial arts environment, the planner must not let his enthusiasm run wild. It must be kept in mind that too many color identifications will cause confusion and eye fatigue, thus defeating the primary purpose for which color is used. In other words, safety color markings should be kept at a minimum to focus attention on those markings which are in use.

The American Standards Association has adopted six standard colors, plus black and white, for marking physical hazards and for identifying certain equipment. The use of this standard safety color code is recommended for industrial arts laboratories. The essential elements of the ASA Safety Color Code are as follows:[8]

RED is the basic color for the identification of (1) fire protection equipment, (2) danger, and (3) emergency stops on machines.

Suggested applications for *red* include fire-protection equipment including buckets, exit signs, extinguishers, and hose locations; safety cans for flammable liquids; stop bars on hazardous machines and stop buttons for electrical switches for emergency stopping of machinery.

ORANGE is the basic color for designating dangerous parts of machines or energized equipment which may cut, crush, shock, or otherwise injure. Orange also emphasizes hazards when enclosure doors are open or when gear, belt, or other guards around moving equipment are open or removed, exposing unguarded hazards.

Suggested applications for *orange* include safety starting buttons, inside of movable guards for gears, pulleys, and chains; exposed edges of pulleys, gears, rollers, cutting devices, power jaws, and the like.

YELLOW is the basic color for designating caution and for marking physical hazards, such as striking against, stumbling, falling, tripping, and "caught in between." Solid yellow, yellow-and-black stripes, or yellow-and-black squares may be used interchangeably, in a combina-

---

[8] *Safety Color Code for Marking Physical Hazards and the Identification of Certain Equipment* (Z53.1–1953), American Standards Association, sponsored by the National Safety Council (New York, N.Y.: American Standards Association, Inc., 1953).

Completing and Evaluating the Layout

tion that will create the most attention in the particular environment.

Suggested applications for *yellow* include: handrails, guardrails, or top and bottom treads of stairways; pillars, posts, or columns which might be struck; low overhead obstructions such as beams or pipes; exposed or unguarded edges of platforms or pits; and projections or doorway markers. Waste containers for explosive or highly combustible materials should have a yellow band around the container. The contents should be indicated on the yellow band in red letters.

GREEN is the color for designating safety and the location of first aid equipment.

Suggested applications for *green* include safety bulletin boards, first-aid kits, stretchers, and the dispensary. A white cross edged with green is used to identify a cabinet containing first-aid materials.

BLUE is the color for designating caution, limited to warning against the starting, use of, or movement of equipment under repair or being worked upon.

Suggested applications for *blue* include warnings, such as painted barriers or flags. These should be located at the starting point or power source of machinery and displayed conspicuously on ovens, tanks, electrical controls, valves, and ladders. Many industrial arts laboratories use large blue disks with "Out of Order" painted in white letters for marking equipment that is down for repairs.

PURPLE is the basic color for designating radiation hazards. Since it is highly unlikely that industrial arts laboratories will handle radioactive material, suggested applications for purple will not be given here.

BLACK and WHITE. Black and white, or a combination of these are the basic colors for the designation of traffic and housekeeping markings. Solid white, solid black, single color striping, alternate stripes of black and white or black-and-white checkers should be used in accordance with local conditions.

Suggested applications for *black and white* include (1) traffic areas such as dead ends of aisles or passageways; location and width of aisleways; stairways (risers, direction and border limit lines) and directional signs; and (2) housekeeping such as location of refuse cans, white corners for rooms or passageways; drinking fountains and food-dispensing-equipment locations; clear floor areas around first-aid, fire-fighting, and other emergency equipment.

SAFETY FEATURES IN SCHOOL SHOP LAYOUTS

It goes without saying that the industrial arts laboratory should be a *safe* place to work with the tools, materials, and processes of industry.

The designing phase is a critical one in planning safe physical conditions because many safety features can be incorporated while the layout is still on the drawing board.

Throughout these several chapters on school shop planning, numerous recommendations have been advanced that will help make the shop a safe place for students to learn. Safety should not be marked by casual planning; instead, each shop plan should be examined critically in terms of both general and specific safety features. The layout should undergo continual evaluation with a view to accident prevention. Campbell has identified three accident-prevention goals in the following order of preference:

> Elimination of the hazard from the machine, method, material, or shop structure.
>
> Guarding or otherwise minimizing the hazard at its source if the hazard cannot be eliminated.
>
> Guarding the person of the operator through the use of personal protective equipment and devices if the hazard cannot be eliminated or guarded at its source.[9]

As the layout begins to unfold on the drawing board it should be subjected to critical analysis for the purpose of identifying safety hazards which may be inherent in the design. If possible, these hazards should be eliminated at once. Sometimes this can be accomplished by revising the space arrangements of machines and equipment. Or, if the hazard cannot be eliminated, then it should be minimized at its source by guards, railings, or aisles. Of course, these should be properly color-coded. And finally, attention should be given to providing personal protective equipment for the operator. Although this latter item may be more properly in the domain of methods or shop management, nevertheless, protective gear may require space consideration, hence, it becomes a matter of shop layout too.

In a research study in safety education, Williams identified 181 shop safety practices and categorized them under nine headings. Some of the safety practices which are especially applicable to shop planning are reproduced here. It is advisable to check each shop layout against these safety features.

> Make an analysis of all hazards in the shop involving machines, hand tools, and general environment.
>
> Provide 4-ft. minimum aisles throughout the shop for general travel.
>
> Have all toolrooms, storage rooms, and shop doors swing outward.

---

[9]Edward A. Campbell, "Built-in Safety for I-E Facilities," *Industrial Arts and Vocational Education*, vol. 50, no. 3 (March 1961), p. 47.

Completing and Evaluating the Layout

Avoid overhead storage of supplies, equipment, or shop projects. If balcony must be used, provide a toe-board and railing around it.

Provide fluorescent lights for the general lighting of the shop.

Provide submaster switches at convenient locations throughout the shop to disconnect power-driven machines.

Provide a ground on all motors, fuse boxes, switch boxes, and other electrical equipment.

Provide overload protection on all motors.

Provide individual cutoff switches for each machine—separate from operator-control switch.

Provide individual lights on each machine.

Fasten all machines securely to the floor.

Provide a nonskid floor area for students to stand on when operating hazardous machines.

Enclose all gears, moving belts, and other power transmission devices with permanent guards or barricades.

Provide and require the use of point-of-operation guards for operation and involving machine cutting, drilling, shaping, and forming.

Make all equipment control switches easily accessible to the operator.

Provide storage for accessories to machines in specially designed racks conveniently located for the operators.

Use color-coding on hazardous machines to emphasize danger areas.

Use alternate yellow-and-black stripes on protruding parts, low beams, and tripping hazards.

Use painted lines on the floor around each hazardous item of equipment.

Paint all corners of the shop with white paint to expose any dirt that might accumulate.

Provide properly marked boxes or bins for various kinds of scrap stock.

Fasten all benches securely to the floor.

Provide fire extinguishers in the shop area.

Mark the location of fire fighting equipment in the shop with a large "bright red" square, arrow, or bar, high enough to be seen all over the shop.

Store flammable liquids in approved safety containers.

Provide for the bulk storage of flammable materials (gasoline, paint thinner, etc.) in an area away from the main school building.

Provide brushes for the cleaning of equipment after each use.

Provide a safety suggestion box for student use.

Provide a first-aid kit in the shop.[10]

---

[10] William A. Williams, "Present Status and Preferred Practices in Safety Education in Pennsylvania Vocational Industrial School Shops" (unpublished doctoral dissertation, University of Pittsburgh, 1959).

EVALUATION OF PHYSICAL FACILITIES

The physical environment of a school shop can be evaluated by means of several rating devices, score cards, and check lists which are presently available in the field. In one research study a valid and reliable instrument was developed to measure the quality of an industrial arts program. This instrument appraised four major factors, one of which was the physical conditions of the shop. The entire rating device consisted of 77 objective-type items which were classified under these headings or major factors: Instructional Program (*what* is taught), Physical Conditions (*where* it is taught), Methods and Management (*how* it is taught), and The Teacher (*who* does the teaching). The validity of these items was established by three separate national juries of recognized leaders and prominent personalities in industrial arts throughout the United States. The reliability of this descriptive rating scale was tested experimentally in selected school systems in two states and was found to be .967. Major Factor 1 of this evaluative instrument consists of 29 items and is reproduced in the following section:

Major Factor 1—Physical Conditions[11]
(items 1 through 29—Value: 100 points)

Directions: The two types of items in this rating scale are called "check 1" items and "check each" items. The type of item is indicated in the parenthesis following the item heading. The following examples illustrate how to mark each type of item.

How to mark "check 1" items:
Check or encircle the response which *best* describes the condition or situation in the shop. Then, transfer the number preceding that response to the circle provided at each item. This number represents the weighted point value of the item.

EXAMPLE: 1. FLOOR LEVEL OF SHOP: (check 1)
  0 ___ total basement without areaways
  .6 ___ third floor or above
  1.2 ___ total basement with areaways
  2.4 ___ partial basement
  3 ___ first floor or street level

(3)

How to mark "check each" items:
Check each response under the item *if it is present in the shop or if the teacher or pupils do the thing asked for in the item.* Then count the number of check marks and place that number on the line provided under the last response. Complete multiplication as indicated and transfer product to the usual circle provided at the item.

---

[11]Norman C. Pendered, *Upgrading the Industrial Arts Program* (State College, Pennsylvania: College Science Publishers, 1959), pp. 3–9.

Completing and Evaluating the Layout

EXAMPLE: 18. FOLLOWING PRESENT IN SHOP? (check each)
- ___ drinking fountain
- ___ washing facilities
- ___ first aid facilities
- ✓ chalkboard
- ✓ fire extinguisher
- ___ safety-type waste cans
- ✓ scrap or waste boxes
- ___ shop project display (may be in hall, etc.)
- ✓ master switch for power machines

(2) 5 × .4 equals 2.0

1. FLOOR LEVEL OF SHOP: (check 1)
   - 0 ___ total basement without areaways
   - .6 ___ third floor or above
   - 1.2 ___ total basement with areaways
   - 1.8 ___ second floor
   - 2.4 ___ partial basement
   - 3 ___ first floor or street level

2. GENERAL SHAPE OF SHOP ROOM: (check 1)
   - 0 ___ U-shaped
   - 1 ___ L-shaped
   - 2 ___ square
   - 3 ___ rectangular

3. OUTSIDE EXIT: (check 1)
   - 0 ___ no exit to outside
   - 1.5 ___ single door as outside exit
   - 3 ___ double door as outside exit

4. SHOP ROOMS TAUGHT IN SIMULTANEOUSLY BY ONE TEACHER: (check 1)
   - 0 ___ two or more rooms not adjacent
   - 1 ___ adjacent rooms, poor visibility between rooms
   - 2 ___ adjacent rooms, good visibility between rooms
   - 3 ___ single room, fair visibility, some obstructions to full view
   - 4 ___ single room, good visibility, no obstructions to full view

5. FLOOR OF SHOP: (check 1)
   - 0 ___ not appropriate to activities
   - 3 ___ appropriate to activities taught

6. RANGE OF TEMPERATURE IN SHOP: (check 1)
   - 0 ___ little or no control of heat
   - 1 ___ varies widely, below 65° to over 70°
   - 2 ___ usually between 60° and 70°
   - 3 ___ usually between 65° and 68°

7. VENTILATION SYSTEM: (check 1)
    0 ____ no ventilation system in shop
    1.5 ____ system unsatisfactory, or out of order, does not force sufficient clean air into shop
    3 ____ system satisfactory, sufficient clean air forced into shop
8. LIGHTING: (check 1)
    Using only artificial illumination, check below average reading in foot-candles for ten different work stations at work level:
    0 ____ 4–7 ft-c
    1 ____ 8–11 ft-c
    2 ____ 12–15 ft-c
    3 ____ 16–19 ft-c
    4 ____ 20-up ft-c
9. WALLS OF SHOP: (check 1)
    0 ____ soiled and dark in color
    1 ____ soiled but light in color
    2 ____ clean and light in color
10. CEILING OF SHOP: (check 1)
    0 ____ soiled and dark in color
    1 ____ soiled but light in color
    2 ____ clean and light in color
11. SHOP LIBRARY. (check 1)
    0 ____ no shop library
    2 ____ meager, inadequate number of books and periodicals
    4 ____ variety of books and periodicals in sufficient amounts
12. PROJECT PLANNING SPACE: (check 1)
    0 ____ no area or space specifically provided
    3 ____ special area devoted to project planning
13. SHOP SPACE WHERE ALL PUPILS CONCERNED MAY GATHER FOR INFORMATIVE LESSON: (check 1)
    0 ____ no space available in shop
    2 ____ regular shop benches used
    4 ____ special space and seating provided
14. WHEN USING AUDIO-VISUAL AIDS, CAN SHOP BE DARKENED SUCCESSFULLY? (check 1)
    0 ____ no
    2 ____ yes, or special room available
15. INDIVIDUALLY ASSIGNED WORK STATIONS PROVIDED FOR LARGEST CLASS? (check 1)
    0 ____ no
    3 ____ yes

Completing and Evaluating the Layout

16. STORAGE SPACE: (check each)
    ____ raw materials and supplies
    ____ special tools and machine accessories
    ____ projects under construction
    ____ finished projects
    ____ × 1 equals ____

17. SQUARE FOOTAGE OF FLOOR AREA PER PUPIL IN AVERAGE SIZE CLASS: (check 1)
    0   ____ 0–49 sq ft per class pupil
    .8   ____ 50–59 sq ft per class pupil
    1.6 ____ 60–79 sq ft per class pupil
    2.3 ____ 80–89 sq ft per class pupil
    3   ____ 90–100 sq ft per class pupil

18. FOLLOWING PRESENT IN SHOP? (check each)
    ____ drinking fountain
    ____ washing facilities
    ____ first aid facilities
    ____ chalkboard
    ____ first extinguisher
    ____ safety-type waste cans
    ____ scrap or waste boxes
    ____ shop project display (may be in hall, etc.)
    ____ master switch for power machines
    ____ × .4 equals ____

19. QUANTITY OF HAND TOOLS: (check 1)
    0   ____ not sufficient in any activity, pupils forced to idleness
    1.7 ____ not sufficient in most activities; pupils must take turns
    3.4 ____ sufficient in most activities; pupils seldom delayed
    5   ____ sufficient in each activity; pupils not delayed

20. CONDITION OF HAND TOOLS: (check 1)
    0 ____ worn out with replacements needed
    2 ____ some sharp, many dull
    4 ____ all in good sharp condition

21. CONDITION OF HAND TOOLS WITH CUTTING EDGES: (check 1)
    0 ____ some broken, many nicked, damaged
    2 ____ some sharp, many dull
    4 ____ all in good sharp condition

22. CONDITION OF POWER MACHINES: (check 1)
    0   ____ old, obsolete, replacements needed
    1.3 ____ worn, defaced, damaged, repairs needed
    2.6 ____ worn, but still serviceable condition
    4   ____ new or nearly equivalent condition

Completing and Evaluating the Layout

23. PERCENTAGE OF POWER MACHINES IN SATISFACTORY OPERATING CONDITION: (check 1)
    - 0 ___ 0
    - 1 ___ 25
    - 2 ___ 50
    - 3 ___ 75
    - 4 ___ 100

24. PERCENTAGE OF POWER MACHINES EQUIPPED WITH EFFECTIVE SAFETY GUARDS: (check 1)
    - 0 ___ 0
    - 1 ___ 25
    - 2 ___ 50
    - 3 ___ 75
    - 4 ___ 100

25. OPERATOR OF ONE MACHINE WORKS IN DANGER ZONE OF ANOTHER MACHINE? (check 1)
    - 0 ___ yes
    - 3 ___ no

26. OPERATOR PROTECTED BY NON-SKID AND SHOCK-PROOF FLOOR COVERING: (check 1)
    - 0 ___ at 25% of power machines
    - 1 ___ at 50% of power machines
    - 2 ___ at 75% of power machines
    - 3 ___ at 100% of power machines

27. CONDITION OF EQUIPMENT: (check 1)
    - 0 ___ worn out, replacements needed
    - 1.3 ___ defaced, damaged, repairs needed
    - 2.3 ___ worn, but still serviceable
    - 4 ___ new or nearly equivalent condition

28. EQUIPMENT IN SHOP ADEQUATE FOR: (check 1)
    - 0 ___ none of the activities
    - 2 ___ some of the activities
    - 4 ___ all or each activity

29. SHOP SUPPLIES: (check 1)
    - 0 ___ meager, inadequate for each activity
    - 2 ___ insufficient for most activities
    - 4 ___ sufficient for all activities

TOTAL POINTS SCORED IN THIS MAJOR FACTOR. TRANSFER THIS TOTAL TO "SUMMARY OF POINTS"

Summary of Points

MAJOR FACTOR 1 — *Physical Conditions*

(a) ___ points scored (b) Possible score: 100 points

Efficiency is $\frac{(a)}{(b)}$ or $\frac{}{100} \times 100$ ___ %

Obviously it is not necessary to wait until the laboratory is constructed before evaluating the physical conditions. If the layout is as detailed as it should be, then it may be evaluated with the rating scale just described. In fact, it is highly desirable to make such an evaluation, and especially before final drawings are submitted to the school architect.

SUMMARY

Color is known to create certain physiological and psychological effects on individuals. Industry is said to use color to improve employee morale, to eliminate work accidents, and to increase production. One study of color in the elementary school showed that (1) kindergarten children benefited the most, (2) boys showed greater response to color than girls, and (3) improvement in achievement was greater than in behavior traits. Investigations at the secondary school level yielded inconclusive results. The Safety Color Code of the American Standards Association was presented as well as recommendations for the selection of shop colors.

Plans for industrial arts laboratories should be analyzed critically in terms of general and specific safety features to prevent accidents. Twenty-nine safety practices applicable to laboratory planning were reported.

The physical conditions of the school shop layout can be evaluated objectively while still on the drawing board by the measuring device presented in the chapter.

DISCUSSION TOPICS AND ASSIGNMENTS

1. Select a machine of your choice and describe in detail how it should be painted in conformity with the suggestions made in this chapter including the ASA Safety Color Code.
2. Make a standard check list of safety features which should be incorporated in all school shop plans.
3. Select a plan for an industrial arts facility from a periodical in the field and evaluate it with the rating device given in this chapter.

SELECTED REFERENCES

Birren, Faber. "The Psychology of Color for the Classroom," *The Nation's Schools*, vol. 57, no. 4 (April 1956), pp. 92–94.

Campbell, Edward A. "Built-in Safety for I-E Facilities," *Industrial Arts and Vocational Education*, vol. 50, no. 3 (March 1961), pp. 47, 52.

———. "How to Use Color in the Industrial Laboratory," *Industrial Arts and Vocational Education*, vol. 51, no. 10 (December 1962), pp. 24–25.

Color Dynamics (grade schools, high schools, colleges) Pittsburgh, Pa.: Pittsburgh Plate Glass Company, no date.

Color Dynamics (office buildings, auditoriums, apartments, churches, hotels) Pittsburgh, Pa.: Pittsburgh Plate Glass Company, no date.

Energy in Color. Pittsburgh, Pa.: Pittsburgh Plate Glass Company, no date.

Lytle, Jr., Robert B. "Color: Its Effect on Teaching and Learning," *School Shop*, vol. XXVII, no 8 (April 1963), pp. 43–44, 88, 90–91.

Modern School Shop Planning. Ann Arbor, Mich.: Prakken Publications, 1967.

Pendered, Norman C. "Developing Safety Consciousness," *Accident Prevention Manual for Shop Teachers.* William A. Williams, ed., National Association of Industrial Teacher Educators. Chicago, Ill.: American Technical Society, 1963.

———. *Upgrading the Industrial Arts Program.* State College, Pa.: College Science Publishers, 1959.

Planning Industrial Arts Facilities, Eighth Yearbook, American Council on Industrial Arts Teacher Education. Bloomington, Ill.: McKnight and McKnight Publishing Co., 1959.

Rice, Arthur H. "What Research Knows About Color in the Classroom," *The Nation's Schools*, vol. 52, no. 5 (November 1953), pp. I–VIII.

Safety Color Code for Marking Physical Hazards and the Identification of Certain Equipment, Z53.1–1953, American Standards Association, sponsored by the National Safety Council. New York, N.Y.: American Standards Association, Inc., 1953.

Williams, William A. "Present Status and Preferred Practices in Safety Education in Pennsylvania Vocational Industrial School Shops" (unpublished doctoral dissertation, University of Pittsburgh, Pittsburgh, 1959).

Appendix I

Sample Unit from NDEA Institute

Gorham State College, Gorham, Maine
July 1 to August 9, 1968*

Title of Unit: Introduction to Technology and Industrial Arts

Teaching Team: Eugene S. Chaplin, Richard O. Gilpatrick, John Kelley, Dennis D. Maust, Frank W. Reed.

Introduction: Technological development in the twentieth century is proceeding at a rate of speed unheard of in the history of man. In America today, we depend upon technology to supply us with not only our basic needs, but all the luxuries we can demand of it. Although necessary to our way of life, this rapid development must be controlled, and to control it, we must understand it. We can do this only by studying the environment within which it was created—industry.

A historical study of American industry would reveal the need for technological advancement. An awareness of industry's role in creating and advancing technology must be an integral part of the educational development of our youth. Industry's influence is felt by everyone in our society, and industrial arts, as a part of general education, has an obligation to provide a basic knowledge of industry to our youth. This knowledge must be imparted to pupils in order to better prepare them for the complexities of life.

Scope: This unit will provide an introduction to industrial arts, indus-

---

*Units for the Laboratories of Industries. Gorham State College, Maine. Office of Education (DHEW), Washington, D.C. EDRS Microfiche ED-031-554, John Mitchell, Institute Director, August 1968, 486 pp.

try and technology for junior high school pupils who have had no previous experience in this type of program.

This unit should be completed in one week of five single class periods or two double class periods. The unit may be adapted to all ability levels in the three grades of the junior high school.

Objective 1: To develop an understanding of the meaning and purpose of industrial arts and its relationship to industry.

Expected Behavioral Changes	*Pupil Activities*	*Teacher Lessons*
The pupils will be able to:	Have pupils:	
1. Recognize industrial arts as a part of general education	1. a. Discuss correlation of industrial arts with other subject matter	Industrial Arts and General Education
2. Recognize the need for industry in our society	2. a. List local industries b. List articles of above c. Examine products and materials of local industries	The Role of Industry in our Society
3. Compare industrial arts with industry and technology	3. a. Make paper product b. Visit small industry c. Tour shop	The Relationship between Industry, Industrial Arts and Technology

Objective 2: To develop desirable attitudes toward the organization and operation of the industrial arts program.

Expected Behavioral Changes	*Pupil Activities*	*Teacher Lessons*
The pupils will be able to:	Have pupils:	
1. Recognize the need for proper conduct and safe practices in the industrial arts laboratory	1. a. Discuss rules for general safety in the laboratory b. View film	Safe Practices in the Industrial Arts Laboratory
2. Demonstrate a knowledge of the administrative practices of the industrial arts laboratory	2. a. Complete student contract form b. Discuss laboratory practices	Administration of the Industrial Arts Laboratory

Appendix I 517

| 3. Analyze the physical arrangement of an industrial arts laboratory | 3. a. Participate in laboratory tour

 b. Discuss floor plan | Physical Arrangement of the Industrial Arts Laboratory |

Approach:

1. Divide the pupils into groups and manufacture a simple product from paper. Using only a plain sheet of paper, they will be able to produce geometric designs—animals, birds, airplanes, and others. This experience should assist in their gaining an understanding of "industry."

2. Arrange a display of materials and products of local industries.

3. Arrange a display of old and modern tools and use the chart, "Tools that Created Civilization."

4. Exhibit pictures of industries and related material on the bulletin board.

Resource Materials:
- A. Reference and research materials:

    Bethel, Lawrence, *Industrial Organization and Management,* New York: McGraw-Hill Publishing Co., 1962.

    Gerbracht, Carl and Frank E. Robinson, *Understanding America's Industries,* Bloomington, Ill.: McKnight and McKnight, 1958.

    Maine State Department of Education, *Industrial Arts Technology,* Augusta, Maine: The Department of Education, 1965.

    Silvius and Curry, *Teaching Successfully in Industrial Education,* Bloomington, Ill.: McKnight and McKnight, 1967.

    United States Department of Labor, *Occupational Outlook Handbook,* Washington, D.C.: U.S. Government Printing Office, 1967.

- B. Teaching aids or devices:

    1. Chart: *Tools That Created Civilization,* Wilkie Foundation, Des Plaines, Illinois

    2. Strip film, *Basic Shop Safety,* Jam Handy Corporation, Detroit, Michigan, $5.00

    3. Film, *The Factory,* University of Michigan, B&W, 13 min., rental, Audio Visual Education Center, 416 4th Street, Ann Arbor, Michigan, $4.50

    4. Manufactured articles from local industries

    5. Information sheets: (a) lab floor plan, (b) "Industrious Art" work sheet

6. Job application form
7. Time card
8. Floor plan transparency

Tools and Equipment: (1) tools found in a typical industrial arts laboratory, (2) overhead projector, (3) viewing screen, (4) movie projector, (5) film strip projector.

Materials and Supplies: This being an introductory unit, no special materials and supplies are needed for its implementation.

Lessons To Be Taught:
1. Industrial arts and general education
2. The role of industry in our society
3. The relationship between industry, industrial arts and technology
4. Administration of the industrial arts laboratory
5. Safe practices in the industrial arts laboratory
6. Physical arrangement of an industrial arts laboratory

LESSON TITLE: INDUSTRIAL ARTS AND GENERAL EDUCATION

Presentation:
  I. Introduce team to class
     A. Write names of team on chalkboard
     B. Have each team member stand as name is called
  II. Read names of class members
     A. Have students stand and tell where they live
     B. Discussion
  III. Ask what students expected when they signed up for course
     A. Call on individual students for response
     B. Discussion
  IV. Show where industrial arts fits into broad field of general education
     A. Major subject areas
       1. Math
         a. Measurement
         b. Formulas
         c. Shapes
       2. Science
         a. Experiments
         b. Practical application
         c. Chemicals

3. English
    a. Reading and writing
    b. Job application
    c. Communication
4. Social studies
    a. Development of tools and machines
    b. Industrial Revolution
B. Have students participate in discussion

LESSON TITLE: THE ROLE OF INDUSTRY IN OUR SOCIETY

Presentation:
I. Define industry
    A. Industry is organization of society
    B. Purpose is to produce goods and services
II. Common types
    A. Manufacturing
    B. Construction
    C. Service
III. Needs of industry
    A. Raw materials
    B. Machines and tools
    C. Money
    D. People to work
    E. Factories
IV. Need for industries
    A. Products and services
    B. Jobs
    C. Money

Reference:
Carl Gerbracht and Frank E. Robinson, *Understanding America's Industries*, pp. 2–5.

LESSON TITLE: THE RELATIONSHIP BETWEEN INDUSTRY, INDUSTRIAL ARTS AND TECHNOLOGY

Presentation:
I. Define industry, industrial arts and technology
    A. Produces goods and services
    B. Study of industry
    C. Technique or systematic method of production

II. Industrial arts and industry
  A. Place of industry in our society
  B. Types of industry
  C. Organization
  D. Materials used
  E. Products
  F. Processes
III. Technology and industry
  A. Early history of industry
  B. Mass production and automation of modern industries

Reference:

Lavon B. Smith and Marion E. Maddox, *Elements of American Industry,* pp. 256–259.

LESSON TITLE: ADMINISTRATION OF THE INDUSTRIAL ARTS LABORATORY

Presentation:
I. Student contract form
  A. Explain job application
    1. Personal data
    2. Student competencies
  B. Relate to industry
    1. How to fill out application
    2. Interview
II. Time card
  A. Attendance record
  B. Relate to industry
III. Identification card
  A. Communication in laboratory
  B. Department assignment
    1. Engineering
    2. Production
    3. Marketing
  C. Relate to industry
    1. Entrance to plant
    2. Department identification
IV. Assign student number and locker
  A. Number
    1. Instruction seat

Appendix I

   2. Locker compartment
   3. Tool checks (if necessary)
  B. Locker
   1. Apron
   2. Plans
   3. Materials

Reference:
 G. Harold Silvius and Estell H. Curry, *Teaching Successfully in Industrial Education*, pp. 220–251.

LESSON TITLE: SAFE PRACTICES IN THE LABORATORY

Presentation:
 I. Why is safety important in our lives?
  A. Safety on the highway
  B. Safety in our homes
 II. Safety in the industrial arts laboratory
  A. Safety is really common sense
  B. Personal habits
   1. Attitude about safety
   2. Clothing
    a. Remove sweaters
    b. Wear a shop apron
   3. Keep hands clean
   4. Always walk in the laboratory
   5. Horseplay not tolerated, no fooling around
   6. Report any injuries immediately
  C. General safety rules
   1. Avoid putting nails, etc., in the mouth
   2. Leave vises in a closed position
   3. Store products in a safe place
   4. Keep oily rags in a covered metal container
   5. Keep floors clean
  D. Tool safety
   1. Carry tools with the points down
   2. Never put tools in your pocket
   3. Hand tools to others by the handle first
   4. Report broken or damaged tools to instructor

                5. Wear safety glasses when working on dangerous tools or machines
                6. Respect power machines and persons operating them

Reference:
G. Harold Silvius and Estell H. Curry, *Teaching Successfully in Industrial Education,* pp. 397–484.

LESSON TITLE: PHYSICAL ARRANGEMENT OF AN INDUSTRIAL ARTS LABORATORY

Presentation:
   I. Compare activities of classroom with those of laboratory
        A. Classroom
            1. Organize ourselves
            2. Decide on products
            3. Plan construction
            4. Terms: planning and designing
        B. Laboratory
            1. Make products
            2. Terms: manufacture
  II. Point out location of materials and supplies
        A. Wood, metal, plastics and leather
        B. Nuts, bolts, wood screws, washers, etc.
 III. Illustrate factors affecting laboratory arrangement
        A. Convenience
        B. Safety
  IV. Point out section of laboratory where pupils will carry on most of their activities
        A. General purpose benches
        B. Tool panel, hand tools
   V. Describe laboratory as a place of vast activities
        A. Many materials used
        B. Many operations carried on

Reference:
Maine State Department of Education, *Industrial Arts Technology,* pp. 10–17.

Appendix 2

Programed Instruction Sheet on the History of Mass Production
Gerald E. Brown, Malvern, Pennsylvania

INSTRUCTIONAL OBJECTIVES:

Upon completion of this programed unit, the student should be able to:
1. Name five prominent men who influenced the mass production process and tell in one sentence what each man's major contribution was.
2. List the four stages through which mass production has evolved.
3. Cite two changes in our society which came about because of mass production.

Answer	Information	Response
	1. The division of labor was the first stage in the development of mass production process.	1. Mass production was first started with a d_____of l_____.
1. Division of labor	2. The spinning jenny which spun many strands of yarn instead of one, was the first machine of major importance in mass production.	2. A machine which spun many strands of yarn and started mechanical mass production was the s_____ j_____.
2. Spinning jenny	3. The spinning jenny was developed by an Englishman, James Hargreaves in the year 1770.	3. The man who invented the machine that spun strands of yarn at one time was J_____ H_____.

Appendix II

3. James Hargreaves

4. Richard Arkwright

5. Water power

6. Steam Power

7. James Watt

8. Industrial Revolution

9. Evans' Flour Mill

10. Oliver Evans

11. Eli Whitney

12. Eli Whitney

13. Standardized-interchangeable parts

14. Way of life

4. A new system of manufacture, the factory system, was devised by an Englishman, Richard Arkwright.

5. At this stage of machine production, water power was the only way to power machines.

6. The second stage came with the development of steam power.

7. James Watt refined existing means of steam power and produced the steam engine.

8. The Industrial Revolution came into being when machines were combined with the steam engine.

9. Evans' Flour Mill was the first totally automatic factory. This mill was established in America in 1787.

10. The automatic flour mill was invented and operated by Oliver Evans

11. Eli Whitney, when making cotton gins, was the first to use guides and holding devices for tools and workpieces.

12. Eli Whitney was also the first to use standardized-interchangeable parts in the production of muskets in 1798

13. The idea of standardized-interchangeable parts was the third important stage of mass production.

14. Standardized-interchangeable parts in mass production brought about a new social and economic way of life.

15. The fourth important stage of mass production was scientific management, devised by Frederick W. Taylor in 1890.

4. The factory system of manufacture was first designed by R_____ A_____.

5. One stage in the history of mass production was the stage when w_____ p_____ was used to drive machinery.

6. S_____ P_____ was the next important stage in mass production.

7. The steam engine was refined by J_____ W_____.

8. The combination of steam power and machines brought about the I_____ R_____.

9. The first completely automatic factory was E_____ F_____ M_____ established in 1787.

10. The automatic flour factory was invented in 1787 by O_____ E_____.

11. Devices which guide and hold tools and workpieces were first used by E_____ W_____ in the production of cotton gins.

12. The idea of standardized-interchangeable parts was used first in the mass production of muskets by E_____ W_____.

13. Another important stage in the history of mass production came with the idea of S_____-I_____ P_____.

14. Standardized-interchangeable parts made possible a new _____ of _____.

15. Scientific management was started by _____ _____.

Appendix II

15. Frederick W. Taylor	16. By this time, the craftsmen were gone, men had specific jobs rather than a craft.	16. The working man in industry no longer had a craft he had a s_____ j_____.
16. Specific job	17. The assembly line, where parts were added by different men as the product moved by was developed by Henry Ford I.	17. The assembly line technique was refined by H_____ _____.
17. Henry Ford I	18. Sub-contracting is buying portions of the whole product already made by another industry to save time and money.	18. Sometimes an industry can save money by buying parts of their product from another company. This is called _____-_____.
18. Sub-contracting	19. Henry Ford I was the first to use subcontracting to buy parts or sub-assemblies for his automobiles.	19. Sub-contracting was started by _____ _____ _____.
19. Henry Ford I	20. At first a man's job was to run the machines, now his job has changed to that of keeping the machines running.	20. Man's job is to k_____ the _____ _____.
20. Keep the machines running	21. With mass production has come shorter working hours giving man more leisure time to develop new social activities.	21. Man can develop more and new social activities because of the increased amount of _____ _____.
21. Leisure time	22. Recently some men have feared that the machine will take over and leave man an outcast in his own society.	22. With the advent of automatic machines some men fear that the_____ _____ _____ _____.
22. Machine will take over.	23. Have no fear. Refined mechanical operations and complex equations have not successfully excluded one intricate factor—MAN.	23. The factor which is still controlling all of the automatic operations performed during mass production is_____.
23. Man.		

CRITERION TEST

Directions: After you have completed the programed instruction sheet, answer the following questions in the space provided.

1. Write the names of five men who influenced the mass production process and write in one sentence what each man's contribution was.

Men	*Major Contribution*
1._____	_____
2._____	_____

3._____   _____
4._____   _____
5._____   _____

2. List four stages through which mass production has evolved.
    1._____ 3._____
    2._____ 4._____
3. What two changes in our society occurred because of mass production in industry?
    1._____
    2._____

Index

Accident reporting, 296–297. *See also* Records, laboratory
Administrative records, 291–298. *See also* Records, laboratory
Adult classes, 49–51, 73, 435–437, 454, 455
Aesthetic appreciations, 88–89
Aids, visual. *See* Instructional media
Allen, Charles R., 192–193
Arts and Crafts Movement, 162
Assemblies, school, 432
Assembly area, 451, 454
Assignment sheets, 105, 198–199, 369
Attendance records, 291–292
Audio casettes, 320–322, 471, 472
Audio recordings, 320–322
Audio-visual aids. *See* Instructional media
Automation, 19–21
Avocation, 49–51, 87–88, 108–111

Babcock, Robert J., 39, 40, 41
Behavior changes
  evaluation of, 397, 398–402
  examples of, 84–96
  in learning activities, 102–103, 132–133
  as outcomes, 80–83
  in programed learning, 210–212
  progress chart for, 300, 302
  typical, 83–84
Bensen, M. James, 206, 215, 216–218, 403
Blackboards. *See* Chalkboards; Instructional media
Bollinger, Elroy W., 193, 484–487
Bonser, Frederick G., 14, 78
Brown, Alan D., 174–175
Bulletin boards, 332–333, 334

Campbell, Edward A., 506
Cardinal principles of education, 3–4

Career education, 32, 43, 45, 61, 406. *See also* Vocational guidance
Carlyle, Thomas, 17
Central purpose of education, 7–8, 26–30, 52, 73, 97–98
Chalkboards, 333–334
Class demonstration, 217, 218, 224–255, 384. *See also* Demonstration
Closed-circuit television, 253, 342–344. *See also* Instructional media
Cognitive domain, 402–409
College-level industrial arts, 51–52
Color code, safety, 504–505
Color in the laboratory, 499–505
Colors, shop, 502–504. *See also* Laboratory planning
Community resources
  definition of, 377
  how to locate, 388–390
  importance of, 379–382
  roadblocks to using, 378–379
  service opportunities, 391–392, 438–439
  types of, 382–388
Community service, 391–392, 438–439
Completing the layout. *See* Laboratory planning; Equipping the laboratory
Comprehensive general shop, 57–59, 263–266
Computers, 19, 21, 204–205
Computer-assisted instruction, 204–205
Computers, 19, 21, 204–205
Conant, James B., 15
Concept films, 352–356. *See also* Film loops, single concept; Instructional media; Research findings
Construction, 45, 47, 48, 55, 60
Consumer knowledge, 90, 113–116, 406
Course of study, construction of

527

Course of study *(continued)*
 definition of, 155
 importance of, 155
 steps in, 155–156
Craftsmanship, 77, 95–96, 128–130, 236, 245–250, 315, 355–356, 357. *See also* Skill, development of; Research findings
Creative expression, 91–92
Cremin, Lawrence A., 22
Critical thinking, 7, 26–30, 47–48, 52, 67, 73–74, 91–92, 97–98, 119–123, 159, 161, 179, 180, 396, 406–407
Culture
 improving the, 7–8, 26–30, 73–74
 nature of American, 17–22
Curriculum construction. *See* Course of study; Lesson planning

Dello-Russo, Robert, 175–177
Democracy
 basic precepts of, 1–3
 industrial, 17
Democratic society, 6, 68
Demonstration
 advantages of, 228–229
 area in shop, 240
 as an audio-visual aid, 319, 320
 bench, 240, 471
 class, 218, 224–255, 384
 definition of, 224
 follow-up on, 240
 group, 227
 individual, 227
 lesson plan for, 150, 153–154
 making effective, 229–230
 multiple, 264, 265
 negative instruction in, 250–251
 new developments in, 242–252
 preparing for, 230–234
 presenting the, 234–238
 reasons for giving, 225–226
 records, 299
 research findings, 218, 245–252
 single, 265–266
 starting the general shop with, 263–266
 starting the unit shop with, 258–259
 television in, 241–243
 terminating the, 239
 types of, 227–228
 uses of, 226
 using instruction sheets, 194

Demonstration *(continued)*
 using overview films, 251–252
 using teaching machines, 246–250
 videotaping, 243–245
DeVore, Paul W., 21
Drawing
 area, 53
 new terminology, 55
 in planning, 185–186
 standardized test, 412
 See also Planning; Visual communications

Education
 adult, 49–51, 73, 435–437, 454, 455
 cardinal principles of, 3–4
 career, 45, 48, 91, 116–119, 385
 central purpose of, 7–8, 26–30, 52, 73–74. *See also* Critical thinking; Educational Policies Commission
 in a free society, 15
 general, 3
 goals of quality, 4
 purposes of, 3–11, 81
 relationship of industrial arts, 14–35
 Report of Harvard Committee, 15
Educational Policies Commission
 central purpose of education, 7–8, 26–30, 52, 73–74. *See also* Critical thinking
 critical thinking, 7–8, 26–30, 52, 73–74. *See also* Critical thinking
 needs of youth, 10–11
 purposes of education, 4, 81
 social responsibility in pupils, 2
Educational specifications, 444
Educational television, 340–342. *See also* Television; Videotape recording
Electricity
 area, 54
 new terminology, 55
 standardized test, 412
 vocabulary test, 408–409
Electronics, 54, 55, 412
Elementary industrial arts, 39–43, 173
Equipment, laboratory. *See* Equipping the laboratory
Equipping the laboratory
 completing the layout, 443–468
 factors affecting equipment selection, 481–483
 hand tool selection, 489
 machine tool selection, 483–487

Index

Equipping the laboratory *(continued)*
   tool storage, 461, 490–496
   See also Laboratory planning
Etsweiler, W. H., Jr., 292, 294–295
Evaluation instruments
   behavioral change evaluation, 398–402
   critical thinking, 396
   examples of test items, 402–409
   physical facilities, 508–513
   rating manipulative skills, 409–411
   standardized tests, 411–412
   student achievement and progress, 413–414
   use and care of tools, 410–411
   vocabulary test, 408–409
Evaluation, laboratory, 508–513
Evaluation of program, 256–257, 508–513
Evaluation, pupil
   achievement, 413–414
   behavioral change evaluation, 397–402
   in cognitive domain, 402–409
   critical thinking, 396
   examples of test items, 402–409
   final grades, 303
   of manipulative skills, 409–411
   measurement vs. evaluation, 395
   in psychomotor domain, 409–411
   records, 290, 299–303
   standardized testing, 411–412
   technique of, 397–402
   in terms of objectives, 396
   traditional practice, 395–396
   vocabulary, 408–409
Evaluation, project, 187–188, 409–411
Exhaust systems, 457
Exhibits, 425–428
Exploration of industry, 71, 84–87, 104–108, 272–273, 314

Fairs, project, 425–428
Field trips, 316–319, 382
Film loops, single concept, 351–356
Films
   how to use effectively, 349–351
   8mm, 351–356
Filmstrips, 356–358
Financial records, 303, 304
Flannelboards, 334–335
Floor plans
   instructional/planning centers, 477–479
   laboratory, 463–465
   See also Laboratory planning

Friese, John F., 155
Funderburk, Earl C., 15

General education
   behavioral goals of, 81
   cardinal principles of, 3–4
   central purpose of, 7–8, 26–30, 52, 56, 73, 97–98, 396. *See also* Critical thinking
   definition of, 3
   in a free society, 15
   goals of quality education, 4
   purposes of, 3–11, 81
   relationship of industrial arts to, 14–35, 67–68, 133
   Report of Harvard Committee, 15
General shop
   characteristics of, 59
   comprehensive, 57–59
   general unit, 59–60
   standardized test, 412
   starting the, 263–266
   typical layouts, 447, 449, 464, 465
General unit shop, 59–60
Gerbracht, Carl, 39, 40, 41
Gilchrist, Robert S., 413
Grading. *See* Evaluation, pupil; Evaluation instruments; Records, laboratory
Graphic arts, 54, 321
Greer, Arthur G., 272–273
Group demonstration, 218, 224–255, 258, 262, 264–266, 384. *See also* Demonstration
Group rotation, pupil, 267

Hackett, Donald F., 429
Hand tools
   definition of, 489
   factors in selecting, 489
   storage of, 490–496
Haney, George M., 146–147, 174–175
Harvard Committee, Report of, 15
Herbert, Harry A., 342
High school, industrial arts, 45–49, 175
Hofer, Armand G., 245–246
Hoots, William R., Jr., 39, 40

Imperative needs of youth, 10–11
Improving the culture, 7–8, 26–30, 73–74. *See also* Critical thinking
Independent studies, 384
Individual demonstration, 227. *See also* Demonstration

Individual rotation, pupil, 267–278
Individual worth, 2
Industrial
　crafts, 54–55
　democracy, 17
　development, 17–22, 69–72
　exploration, 71, 84–87, 104–108, 272–273, 314, 403–405
　production, 17–22
　society, 69–72
Industrial arts
　in adult programs, 49–51, 435–437
　for adults, 49–51, 435–437
　aesthetic appreciations, 88–89
　avocational, 72, 87–88, 108–111
　basis for selecting learning activities, 101–102
　central purpose of, 26–30, 97–98, 396. *See also* Critical thinking
　at college level, 51–52
　comprehensive general shop, 57–59
　creative expression, 77, 91–93, 119–123
　critical thinking, 7, 26–30, 47–48, 52, 56, 67, 73–74, 91–92, 97–98, 119–123, 396, 406–407
　as a curriculum area, 48–49, 51–52
　definition of, 16, 39
　demonstration, 150, 153–154, 218, 224–255, 258–259, 264–266, 299, 319, 320, 384, 471
　elementary school, 37, 39–43
　in elementary school, 39–43, 173
　enrollment, 38
　exhibits, 425–428
　exploration of industry, 84–87, 104–108, 272–273, 314
　fairs, 425–428
　as a free elective, 48–49, 51–52
　general unit shop, 59–60
　historical phases of, 14, 162–163, 192–193
　instructional centers, 469–479
　instructional program, 37
　in junior high school, 45
　laboratory planning, 443–468, 469–480, 481–498, 499–514
　learning activities, 101–131, 132–158
　in middle school, 43–45
　nationwide picture of, 36–38
　objectives, 64–79, 80–100, 481–482. *See also* Critical thinking
　organization, 56–60, 482

Industrial arts *(continued)*
　in our schools, 36–63
　planning centers, 469–479
　pre-professional, 46–48
　pre-technical, 48
　project, 136, 159–191, 409–411
　public relations, 418–442
　records, 288–310
　relationship to general education, 14–35, 67–68, 133
　safety education, 72, 82, 94–95, 125–128
　in senior high school, 45–49, 175
　service courses, 51
　sizes and types of schools, 37
　skill, development of, 77, 95–96, 409–411
　social relationships, 93–94
　subject areas, 52–56
　teacher education, 51
　teachers, 38
　terminology, 55–56
　unit shop, 56–57
　vocational guidance, 45, 48, 91, 116–119, 385
Industrial crafts, 54–55
Industrial visits, 316–319, 382
Instructional area, 453, 469–480. *See also* Laboratory planning; Instructional/planning centers
Instructional materials, written
　assignment sheets, 198–199
　information sheets, 199–201
　job sheets, 197
　limitations of, 195–196
　operation sheets, 193, 198, 216–218
　origin of, 192–193
　project sheets, 197
　programed instruction sheets, 201–220, 523–526
　purposes of, 194–195
　student planning sheets, 181–186, 198
　types of, 196–201
Instructional media
　audio casettes, 320–322, 471–472
　bulletin boards, 332–333
　center, 471
　chalkboards, 333–334, 472
　chalkboard, substitute, 364
　closed-circuit television, 342–344
　demonstrations, 319–320
　educational television, 340–342
　film loops, 8mm, 351–356, 472

Index

Instructional media *(continued)*
 films, 347–356, 471–472
 filmstrips, 356–358, 471–472
 flannelboards, 334–335
 free, inexpensive materials, 335–336
 industrial visits, 316–319, 382–383
 in laboratory planning, 471, 474
 mockups, 323, 330–331
 models, 323–329
 new thrusts in, 311–314
 objects, 322–323
 opaque projection, 370, 371, 471–472
 overhead projections, 361–368
 photography for the teacher, 371–372
 polysensory learning, 311–314
 process charts, 332
 project charts, 332
 research findings, 216–217, 218–220, 246–252, 317–318, 342, 349–351, 355–356, 357
 screens, 348, 471–472
 single concept film loops, 351–356
 slides, 358–360
 specimens, 322–323
 still pictures, 368–370
 study prints, 368–369
 tape recorders, 320–322, 471–472
 telespeaker, 386
 videotape recording, 345–347, 386
Instructional/planning centers
 combined, 475–477
 demonstration bench, 471
 definition of, 469
 equipping instructional center, 471–472
 equipping planning center, 472–475
 floor plans, 477–479
 instructional center, 454, 469
 instructional media, 471–472
 location of, 470–471
 planning center, 454, 469–470
Instructional records, 290, 299–303
Instruction sheets. *See* Instructional materials, written
Intelligence of common man, 2
Inventory records, 289, 292–295

Job analysis, 192–193
Job sheets, 197
Junior high school industrial arts, 45, 174

Kauffman, Henry J., 166
Kruppa, J. Russell, 251–252, 356

Laboratory equipment. *See* Equipping the laboratory
Laboratory management
 importance of, 256
 personnel organization, 271–287
 pupil rotation, 267–268
 record keeping, 288–310
 shop tour, 260–261
 starting the class, 257–258
 starting the general shop, 263–266
 starting the unit shop, 258–263
 See also Evaluation, laboratory
Laboratory planning
 assembly areas, 451, 454
 auxiliary rooms, 453, 470
 color in the environment, 499, 505
 determining shops needed, 448–450
 display areas, 433–434, 459
 educational specifications, 444
 equipping the laboratory, 481–498
 equipping the planning/instructional center, 472–476
 evaluation of physical facilities, 508–513
 exhaust systems, 457
 factors to be considered, 452–459, 481–483
 floor plans, 447, 449, 464–465, 477–479
 instructional area, 453, 469–480
 instructional media, 471–472
 layout pointers, 459, 462
 lighting, 456, 458
 locating facilities, 448
 machine tool selection, 483–487
 minimum space requirements, 450–452
 planning center, 454, 469–480
 planning for the future, 462–463
 role of industrial arts teacher, 445
 safety color code, 504–505
 safety features in layout, 505–507
 selection of hand tools, 489
 selection of shop colors, 502–504
 shape of shop, 452
 soundproofing, 457, 459
 specifications, 488
 steps in, 445–448
 storage areas, 453, 455, 472, 475
 teacher's desk area, 454
 tool storage, 461, 490–496
 typical layouts, 447, 449, 464–465, 477–479
 utilities, 456–457
 wiring, 457–458

Laboratory planning *(continued)*
  work stations, 453
  See also Equipping the laboratory; Evaluation, laboratory; Instructional/planning center
LaRue, James P., 250
Learning activities
  basis for selecting, 101–102
  center, 469–480
  examples of, 102–130
  for industrial arts, 102–130
  organizing, 132–158
  student and teacher, 104–130, 142
Learning center, 475–479
Lecture-discussion method, 218–219
Lesson planning
  demonstration plan, 153–154
  examples of, 151–152, 153–154
  headings, 149–150
  informative plan, 151–152
  using community resources, 377–394
  See also Course of study; Unit teaching
Library, shop, 473, 476. *See also* Instructional/planning center
Lighting, laboratory, 456–457, 463
Line production, 172. *See also* Production

Machine tools, 483–487
Mager, Robert F., 82, 210–211
Maine State Plan, 82, 145–146, 515–522
Maley, Donald, 146
Management, laboratory. *See* Laboratory management
Manual arts, 14, 162
Manual training, 14
Manufacturing
  in America, 17–22, 69–72
  industrial arts area, 55, 60
  school plan, 447, 449, 464
Maryland Plan, 146–147
Mass production
  history of, 523–526
  industrial, 17–22
  programed sheet, 523–526
  in school, 55, 60, 172–177
Metalworking
  description, 53–54
  new terminology, 55
  research, 218–219, 245–250
  standardized test, 412
Methods. *See* Demonstration; Project method; Lecture-discussion method;

Methods *(continued)*
  Unit method
Micheels, William J., 412
Middle school, 43
Mitchell, John, 135–136, 147, 515
Mockups, 323, 330–331, 332
Models, 323–329
Moegenburg, Louis A., 220
Mossman, Lois C., 16, 78

National Safety Council, 296–297
NDEA Institute, 147, 515–522
Needs of pupils
  biological, 76–77
  contributions of industrial arts to, 74–77
  imperative needs of youth, 10–11
  occupational, 32–33
  personal, 30–32
  psychological, 74–75
  sociological, 75–76
Newspapers, 430–432

Objectives
  aesthetic appreciations, 77, 88–89
  analysis of, 80–100
  avocational, 77, 87–88, 108–111
  cardinal aim, 97–98
  career education, 71, 77, 116–119
  consumer knowledge, 72, 77, 90, 113–116
  craftsmanship, 72, 77, 111–113, 405–406
  creative expression, 77, 91–93
  critical thinking, 77, 91–93, 97–98, 119–123
  definition of, 82
  in the demonstration, 230
  derivation of, 64
  evaluation of, 395–417
  exploration of industry, 71, 77, 84–87, 104–108, 272–273, 314, 403, 405
  importance of, 65–67
  industrial arts, 77
  in laboratory planning, 481
  of most worth, 97–98
  recreational, 72, 77, 108–111
  safety education, 72, 77, 94–95, 102–104, 125–128, 398–402, 407
  skill, 77, 95–96, 128–130, 407
  social relationships, 77, 93–94, 123–125
  use of, 65
  use of personnel organization in, 276
  vocational guidance, 77, 91, 116–119,

Index

Objectives *(continued)*
   142, 385–387, 406–407
   See also Critical thinking; Education; Educational Policies Commission
Objects, 322–323
Opaque projection, 370–371, 471–472
Open House, 429–430
Operation sheets, 193, 198, 216–218
Organization
   affecting equipment selection, 482
   comprehensive general, 57–59
   general unit, 59–60
   of learning activities, 132–158
   personnel, 271–287
   record keeping, 288–310
   of subject matter, 132–158
   types of laboratory, 56–60
   See also Personnel organization
Osburn, Burl N., 23
Overhead projection
   adding color to, 364–365
   advantages of, 361–363
   animated effects, 367
   chalkboard substitute, 364
   overlays, 365–366
   pointing, 363–364
   revelation, 364
   simulated motion, 367–368
   techniques for using, 361–368
   transparent objects, 366

Pace, C. Robert, 82
Pendered, Norman C., 191, 256–257, 492, 502, 508–513
Personnel organization
   definition, 271
   devices for depicting, 284–285
   examples of, 279–284
   features of, 276–279
   how to, 274–276
   purposes of, 271–274
Photography for the teacher, 371–372
Planning
   centers, 469–480
   course of study, 155–156
   demonstration, 153–154
   instructional center, 469–480
   lesson, 147–151
   project, 27, 179–180, 185–186, 261–262
   student plan sheets, 181–186, 198, 303
   See also Demonstration; Drawing; Laboratory planning

Planning sheets, student, 181–186, 198, 303
Polysensory learning, 311–314. *See also* Instructional media
Power technology, 55
Prakken, Lawrence W., 419, 420
Principles of education. *See* Cardinal principles of education; General education; Educational Policies Commission
Problem solving, 28–30. *See also* Critical thinking; Project
Process charts, 332
Production
   history of, 523–526
   industrial, 17–22, 69–72
   industrial arts area, 55
   line, 172–173
   mass, in school, 172–173
   programed sheet, 523–526
Programed instruction
   background, 201–202
   branching programing, 207, 210
   computer-assisted, 204–205
   criterion test, 212
   cueing, 213, 214
   examples of, 206–207, 208–209, 214–215, 523–526
   fading, 213
   features of, 202–203
   field testing, 215
   format, 205–210
   linear programing, 205–207
   research findings, 216–221, 245–250
   review frames, 213
   revision of, 214–216
   sheets, 201, 204–205, 523–526
   steps in developing, 210–216
   teaching machines, 203
   teaching skills with, 245–250
   terminal behavior, 210–212
   testing, 214–216
   textbooks, 203
   types of, 203–205
   use of illustrations, 214–216
Programed textbooks, 203. *See also* Programed instruction
Progress chart, 299–300, 301, 302
Project
   assigned, 169
   charts, 332
   choice within groups, 169–170
   class, 171

Project *(continued)*
  criteria for selecting, 166–169
  definition of, 159
  evaluation of, 187–188, 409–411
  exhibits, 425–428
  fairs, 425–428
  free selection of, 170
  functions of, 163–164
  historical aspects of, 162–163
  individual, 171
  methods for using, 169–171
  misuses of, 187
  planning, 179–180, 185–186
  production, 172–177
  records, 290, 299–303
  sources of, 177–178
  steps in project method, 164–166
  types of, 171–177
  See also Learning activities; Project method
Project charts, 332
Project method
  definition, 160
  misuses of, 187
  steps in, 164–166
  See also Learning activities; Project
Project sheets, 197
Psychomotor domain, 409–411
Public relations
  adult classes, 435–437
  building industrial arts image, 439–440
  community service, 438
  definition, 418
  developing a program of, 423–438
  display cases, 433–434
  exhibits and fairs, 425–428
  importance of, 419–423
  keeping gap closed, 438
  newspapers, 430–432
  Open House, 429–430
  radio and television programs, 435
  school assemblies, 432
  school publications, 430–433
  talks to organizations, 428–429
  types of, 425
  versus publicity, 418–419
Publicity. *See* Public relations
Pupil personnel organization, 271–287

Radio and television
  in public relations, 435
  See Television; Videotape recording

Record keeping
  importance of, 289–291
  scope of problem, 288–289
  using personnel organization, 307–308
  See Records, laboratory
Records, laboratory
  administrative, 291–298
  attendance, 291–292
  closed versus open, 305
  demonstration, 299
  designing a system of, 304–307
  final grades, 303
  financial, 303–304
  format, 306
  importance of, 289–291
  informative lesson, 299
  instructional, 299–303
  inventory, 292–295
  keeping, 307–308
  loan, 297–298
  longevity of, 305
  materials ticket, 303–304
  mode of response, 305–306
  preparation of, 307
  progress chart, 299–300
  project, 300, 303
  requisitions, 295–296
  safety, 291, 296–297
  storage of, 306–307
  test, 299
  See also Record keeping
Recreation, 72, 77, 87–88, 108–111
Related information. *See* Learning activities
Repp, Victor E., 218
Requisitions, 295–296. *See also* Records, laboratory
Research findings
  class demonstration and performance guide, 218
  closed-circuit television, 342
  color in shop environment, 501
  critical thinking test, 396
  field trips, 317–318
  field trips, vicarious, 317–318
  films in class demonstrations, 251–252
  filmstrips in teaching a skill, 357
  film loops in teaching a skill, 355–356
  film research summary, 349–351
  instruction sheets, 218
  inventory system, 292–295
  laboratory management, 256–257

Index 535

Research findings *(continued)*
　lesson planning and teacher behavior, 148
　overview films, demonstrations, concept films, 251–252, 356
　positive versus negative instruction, 250–251
　practices in safety education, 506–507
　principles for arranging, mounting, storing hand tools, 492–496
　programed learning and teacher demonstrations, 245–246
　programed learning versus lecture-discussion, 218–219
　programed learning versus videotape, 219–220
　programed operation sheets, 216–217
　scale to evaluate physical facilities, 508–513
　school news in local papers, 430–431
　teacher demonstration versus illustrated performance guides, 218
　teaching machines in teaching a skill, 246–250
Resources, community. *See* Community resources
Richards, Charles R., 14, 133
Rotation, pupil
　group, 267
　individual, 267–268
Russell, James E., 14
Russian system, 162

Safety
　color code, 504–505
　education, 72, 77, 94–95, 102–104, 125–128, 398–402, 407
　features in layouts, 505–507
　features in machine selection, 485–486
　records, 296–297
Schmitt, Marshall L., 36
School assemblies, 432
School publications, 433
Selvidge, Robert W., 193
Sheets, student planning
　advantage of, 180–181
　characteristics of, 181, 184–185
　example of, 182–183
　methods for using, 185–186
Shemick, John M., 152–153, 154, 236, 246–250, 355–356
Sherk, Dennis H., 355–356

Single-concept film loops, 351–356, 472.
　See also Instructional media; Research findings
Shop layout. *See* Laboratory planning
Shop tour, 260
Skill, development of, 77, 95–96, 128–130, 236, 245–250, 315, 355–356, 357. *See also* Craftsmanship; Research findings; Skills, teaching
Skills, teaching
　using instructional media, 315
　See also Demonstration; Research findings; Skill, development of
Slides, 358–360
Social–Industrial Theory, 14
Social relationships, 77, 93–94, 123–125.
　See also Personnel organization; Social responsibility
Social responsibility, 2
Sommer, Seymour A., 356
Soundproofing, 457–459
Specifications
　educational, 444
　writing, 488–489
Specimens, 322–323
Standardized tests, 411–412
Still pictures, 368–370
Storage, tool, 490–496
Struck, F. Theodore, 147, 148
Student needs. *See* Needs of pupils; Youth
Student personnel organization, 271–287
Student planning sheets, 181–186, 198
Study prints, 368–369
Subject matter. *See* Learning activities; Organizing learning activities
Swedish sloyd, 162

Tape recorders, 320–322, 471–472
Teaching machines
　computer-assisted, 204–205
　in the demonstration, 246–250
　with lecture-discussion, 218–219
　See also Programed instruction; Research findings
Teaching unit. *See* Unit method; Unit teaching
Telespeaker, 386
Television
　closed-circuit, 241–243, 342–344
　in the demonstration, 241–243
　educational, 340–342
　monitor, 24

Television *(continued)*
  in public relations, 435
  See also Videotape recording
Testing. *See* Evaluation, pupil
Tests
  records, 299
  standardized, 411–412
  See also Evaluation, pupil
Thieme, Eberhard, 41
Thinking. *See* Critical thinking; Education, central purpose of; Educational Policies Commission; Problem solving
Toolholders, 494–496
Tool panels, 492–497
Tools
  hand tool selection, 489
  machine tool selection, 483–487
  storage, 461, 490–497
Tool storage, 490–497
Transmitting a way of life, 4–6, 22–26, 66–72
Traum, Emil F., 219–220
Trips, field/industrial, 316–319, 382–383.
  See Shop tour

Unit method
  advantages of, 136
  class project, 172–177
  definition, 134–136
  how to plan, 140–145
  phases of, 138–139
  starting the general shop with, 263–264
  See also Unit teaching
Unit shop
  characteristics, 56–57
  definition of, 56
  equipping, 482, 484
  starting the, 258–263
Unit teaching
  activity phase, 138
  advantages of, 136
  bibliographies, 144
  class project, 172–177
  culminating activities, 144
  culminating phase, 139–140
  definition, 134–135
  evaluation of, 144
  example of, 145–147, 515–522
  in the field, 145–147
  how to plan, 140–145

Unit teaching *(continued)*
  instructional aids, 145
  length of, 137–138
  lessons to be taught, 144, 147–154
  Maine State Plan, 145–146
  Maryland Plan, 146–147
  NDEA Institute, 147, 515–522
  objectives of, 142
  orientation phase, 138
  outline of content, 142
  phases of, 138–140
  pre-planning phase, 138
  student and teacher activities, 142
  title of unit, 141
  types of, 137
  See also Unit method

Videotape recording
  in community resources, 386
  in the demonstration, 243–245
  as instructional media, 345–347
  research findings, 219–220
  See also Television
Visual aids. *See* Instructional media
Visual communications, 55
Vocabulary test, 408–409. *See also* Evaluation instruments; Evaluation, pupil
Vocational guidance, 77, 91, 116–119, 142, 385–387, 406–407

Warner, Richard A., 218–219
Weaver, Gilbert G., 483, 488–489
Weber, Earl M., 49
Wilber, Gordon O., 490
Williams, William A., 507
Woodworking
  area, 53
  new terminology, 55
  test, 412
Work stations, 453
Wrinkle, William L., 413
Written instructional materials. *See* Instructional materials, written

Youth
  biological needs, 76–77
  imperative needs of, 10–11
  occupational needs, 32–33
  personal needs, 30–32
  psychological needs, 74–75
  sociological needs, 75–76